VOLUME FOUR HUNDRED AND TWENTY-NINE

METHODS IN
ENZYMOLOGY

Translation Initiation:
Extract Systems and
Molecular Genetics

METHODS IN ENZYMOLOGY

Editors-in-Chief

JOHN N. ABELSON AND MELVIN I. SIMON

Division of Biology
California Institute of Technology
Pasadena, California

Founding Editors

SIDNEY P. COLOWICK AND NATHAN O. KAPLAN

VOLUME FOUR HUNDRED AND TWENTY-NINE

METHODS IN ENZYMOLOGY

Translation Initiation: Extract Systems and Molecular Genetics

EDITED BY

JON LORSCH
Johns Hopkins University School of Medicine
Department of Biophysics and Biophysical Chemistry
Baltimore, Maryland

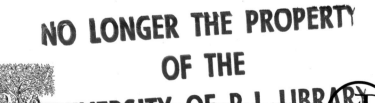

AMSTERDAM • BOSTON • HEIDELBERG • LONDON
NEW YORK • OXFORD • PARIS • SAN DIEGO
SAN FRANCISCO • SINGAPORE • SYDNEY • TOKYO
Academic Press is an imprint of Elsevier

ELSEVIER

Contents

15. Analysis of Ribosomal Shunting During Translation Initiation in Eukaryotic mRNAs   323

Vincent P. Mauro, Stephen A. Chappell, and John Dresios

Contributors

Nadia Amrani
Department of Molecular Genetics and Microbiology, University of Massachusetts Medical School, Worcester, Massachusetts

Katsura Asano
Molecular, Cellular, and Developmental Biology Program, Division of Biology, Kansas State University, Manhattan, Kansas

Diane Barth-Baus
Department of Biochemistry, School of Medicine, Case Western Reserve University, Cleveland, Ohio

Stephen A. Chappell
Department of Neurobiology, The Scripps Research Institute, and The Skaggs Institute for Chemical Biology, La Jolla, California

Jeff Coller
Center for RNA Molecular Biology, Case Western Reserve University, Cleveland, Ohio

Thomas E. Dever
Laboratory of Gene Regulation and Development, National Institute of Child Health and Human Development, National Institutes of Health, Bethesda, Maryland

Tzvetanka D. Dinkova
Departamento de Bioquimica L-103, Facultad de Quimica Conjunto "E," Paseo de la Inv. Científica, Universidad Nacional Autonoma de Mexico, Mexico D.F.

John Dresios
Department of Neurobiology, The Scripps Research Institute, and The Skaggs Institute for Chemical Biology, La Jolla, California; Science Applications International Corporation, San Diego, California

Daniel R. Gallie
Department of Biochemistry, University of California, Riverside, California

Fátima Gebauer
Centre de Regulació Genómica (CRG-UPF), Barcelona, Spain

Matthias W. Hentze
Gene Expression Unit, European Molecular Biology Laboratory, Heidelberg, Germany

Alan G. Hinnebusch
National Institute of Child Health and Human Development, National Institutes of Health, Bethesda, Maryland

Allan Jacobson
Department of Molecular Genetics and Microbiology, University of Massachusetts Medical School, Worcester, Massachusetts

Rosemary Jagus
Center of Marine Biotechnology, University of Maryland Biotechnology Institute, Baltimore, Maryland

Paul Lasko
Department of Biology and DBRI, McGill University, Montreal, Quebec, Canada

Bumjun Lee
Molecular, Cellular, and Developmental Biology Program, Division of Biology, Kansas State University, Manhattan, Kansas

Nicolas Locker
MRC Laboratory of Molecular Biology, Cambridge, United Kingdom

Peter J. Lukavsky
MRC Laboratory of Molecular Biology, Cambridge, United Kingdom

Michael B. Mathews
Department of Biochemistry and Molecular Biology, University of Medicine and Dentistry of New Jersey, New Jersey Medical School, Newark, New Jersey

Vincent P. Mauro
Department of Neurobiology, The Scripps Research Institute, and The Skaggs Institute for Chemical Biology, La Jolla, California

William C. Merrick
Department of Biochemistry, School of Medicine, Case Western Reserve University, Cleveland, Ohio

Klaus H. Nielsen
Department of Molecular Biology, University of Århus, Århus C, Denmark

Andrew M. Parrott
Department of Biochemistry and Molecular Biology, University of Medicine and Dentistry of New Jersey, New Jersey Medical School, Newark, New Jersey

Chingakham Ranjit Singh
Molecular, Cellular, and Developmental Biology Program, Division of Biology, Kansas State University, Manhattan, Kansas

Robert E. Rhoads
Department of Biochemistry and Molecular Biology, Louisiana State University Health Sciences Center, Shreveport, Louisiana

Matthew S. Sachs
Department of Environmental and Biomolecular Systems, OGI School of Science and Engineering, and Department of Molecular Microbiology and Immunology, School of Medicine, Oregon Health and Science University, Portland, Oregon

Byung-Sik Shin
Laboratory of Gene Regulation and Development, National Institute of Child Health and Human Development, National Institutes of Health, Bethesda, Maryland

Nahum Sonenberg
Department of Biochemistry and McGill Cancer Center, McGill University, Montreal, Quebec, Canada

Yuri V. Svitkin
Department of Biochemistry, McGill University, Montreal, Quebec, Canada

Bela Szamecz
Institute of Microbiology, AS CR, Prague, Czech Republic

Gritta Tettweiler
Department of Biology and DBRI, McGill University, Montreal, Quebec, Canada

Tsuyoshi Udagawa
Molecular, Cellular, and Developmental Biology Program, Division of Biology, Kansas State University, Manhattan, Kansas

Leos Valášek
Institute of Microbiology, AS CR, Prague, Czech Republic

Melissa R. Walsh
Department of Biochemistry and Molecular Biology, University of Medicine and Dentistry of New Jersey, New Jersey Medical School, Newark, New Jersey

Marv Wickens
Department of Biochemistry, University of Wisconsin, Madison, Wisconsin

Cheng Wu
Department of Environmental and Biomolecular Systems, OGI School of Science and Engineering, Oregon Health and Science University, Beaverton, Oregon

Preface

Over the past 15 years, it has become clear that translation initiation is a key regulatory point in the control of gene expression. Loss-of-control of protein synthesis has been implicated in a variety of diseases ranging from cancer to viral infection, and there is increasing interest in the development of new drugs that target translation initiation. Despite the profound biological and medical importance of this key step in gene expression, we are only beginning to understand the molecular mechanics that underlie translation initiation and its control, and much work remains to be done.

These MIE volumes (429, 430, and 431) are a compilation of current approaches used to dissect the basic mechanisms by which bacterial, archaeal, and eukaryotic cells assemble, and control the assembly of, ribosomal complexes at the initiation codon. A wide range of methods is presented from cell biology to biophysics to chemical biology. It is clear that no one approach can answer all of the important questions about translation initiation, and that major advances will require collaborative efforts that bring together various disciplines. I hope that these volumes will facilitate cross-disciplinary thinking and enable researchers from a wide variety of fields to explore aspects of translation initiation throughout biology.

Initially, we had planned to publish a single volume on this subject. However, the remarkable response to my requests for chapters allowed us to scale up to three volumes. I would like to express my sincerest appreciation and admiration for the contributors to this endeavor. I am impressed with the outstanding quality of the work produced by the authors, all of whom are leaders in the field. I am especially grateful to John Abelson for giving me the opportunity to edit this publication and for his support and advice throughout the project. Finally, I am indebted to Cindy Minor and the staff at Elsevier for their help and wisdom along the way.

Jon Lorsch

METHODS IN ENZYMOLOGY

VOLUME 206. Cytochrome P450
Edited by MICHAEL R. WATERMAN AND ERIC F. JOHNSON

VOLUME 207. Ion Channels
Edited by BERNARDO RUDY AND LINDA E. IVERSON

VOLUME 208. Protein–DNA Interactions
Edited by ROBERT T. SAUER

VOLUME 209. Phospholipid Biosynthesis
Edited by EDWARD A. DENNIS AND DENNIS E. VANCE

VOLUME 210. Numerical Computer Methods
Edited by LUDWIG BRAND AND MICHAEL L. JOHNSON

VOLUME 211. DNA Structures (Part A: Synthesis and Physical Analysis of DNA)
Edited by DAVID M. J. LILLEY AND JAMES E. DAHLBERG

VOLUME 212. DNA Structures (Part B: Chemical and Electrophoretic
Analysis of DNA)
Edited by DAVID M. J. LILLEY AND JAMES E. DAHLBERG

VOLUME 213. Carotenoids (Part A: Chemistry, Separation, Quantitation,
and Antioxidation)
Edited by LESTER PACKER

VOLUME 214. Carotenoids (Part B: Metabolism, Genetics, and Biosynthesis)
Edited by LESTER PACKER

VOLUME 215. Platelets: Receptors, Adhesion, Secretion (Part B)
Edited by JACEK J. HAWIGER

VOLUME 216. Recombinant DNA (Part G)
Edited by RAY WU

VOLUME 217. Recombinant DNA (Part H)
Edited by RAY WU

VOLUME 218. Recombinant DNA (Part I)
Edited by RAY WU

VOLUME 219. Reconstitution of Intracellular Transport
Edited by JAMES E. ROTHMAN

VOLUME 220. Membrane Fusion Techniques (Part A)
Edited by NEJAT DÜZGÜNEŞ

VOLUME 221. Membrane Fusion Techniques (Part B)
Edited by NEJAT DÜZGÜNEŞ

VOLUME 222. Proteolytic Enzymes in Coagulation, Fibrinolysis, and Complement
Activation (Part A: Mammalian Blood Coagulation Factors and Inhibitors)
Edited by LASZLO LORAND AND KENNETH G. MANN

USE OF RETICULOCYTE LYSATES FOR MECHANISTIC STUDIES OF EUKARYOTIC TRANSLATION INITIATION

William C. Merrick *and* Diane Barth-Baus

Contents

Abstract

This chapter describes how commercially available, nuclease-treated rabbit reticulocyte lysates can be used to study different types of translation initiation (cap-dependent initiation, reinitiation, internal ribosome entry site-mediated initiation) and the influence of different initiation factors on these translation mechanisms. Additionally, with the use of sucrose gradients, it is possible to use

Department of Biochemistry, School of Medicine, Case Western Reserve University, Cleveland, Ohio

Methods in Enzymology, Volume 429
ISSN 0076-6879, DOI: 10.1016/S0076-6879(07)29001-9

nuclease-treated reticulocyte lysates to monitor the formation of ribosomal complexes for their content of mRNA, initiator met-tRNA$_i$, and initiation factors. The advantage of using nuclease-treated lysates rather than purified initiation factors is that reactions occur at or near the *in vivo* rate in contrast to rates observed in reactions with purified components, which are generally 10- to 1000-fold lower. The disadvantage is not being able to accurately control the amount of individual initiation factors, although the use of either factor additions or specific inhibitors can be helpful in assessing the role of specific individual initiation factors.

1. Introduction

The original development of the rabbit reticulocyte system took advantage of two discrete characteristics. First, by the use of phenylhydrazine, rabbits could be treated to the point at which their blood contained approximately 95% reticulocytes and, second, as reticulocytes, the predominant protein being made was 85 to 90% hemoglobin (Borsook *et al.*, 1952; Kruh and Borsook, 1956). In early studies, one of the concerns was whether the amino acids were accurately incorporated into hemoglobin chains. Given that both the sequence of the chains and the procedures for resolving the tryptic peptides of hemoglobin were well established, it was easily determined that a cell-free system using reticulocyte lysates did indeed perform accurate protein synthesis. In addition, the rate of protein synthesis was nearly the same as the *in vivo* rate. With the clever use of the Ca^{2+}-dependent micrococcal nuclease, Jackson and Hunt (1983) made the reticulocyte lysate a system dependent on exogenous mRNA for protein synthesis. This translation system is now available commercially from five different sources (GE Healthcare Life Sciences, Promega, Ambion, Alator Biosciences, and Pel-Freeze; note that the names of these companies have been subject to change and are the ones presently listing reticulocyte lysate) and the direct testing of several of these systems indicates that they can yield reliable results with the appropriate controls (Kozak, 1990).

Much of the use of reticulocyte lysates has focused on studies that have examined regulation via eIF2α phosphorylation or the characteristics of mRNAs that either enhance or inhibit their translation. Although many studies have examined mRNAs for factor requirements, the need for exogenous proteins, or for regulation via RNA-binding proteins, these studies have primarily used fractionated systems and either mRNA binding to ribosomes or toe printing as the readout for analysis. The limitation here is the loss of kinetics as the reactions generally proceed rather slowly and not with good molar yield. As a consequence, it is often useful to check the individual systems in reticulocyte lysates to determine if consistent results are obtained when assayed at the *in vivo* rate.

This chapter will cover the general conditions for the use of nuclease-treated rabbit reticulocyte lysates, optimization, types of reporters, basic experiments and then more refined experiments that vary the standard conditions for mRNA utilization, and the influence of changing the effective activity for various translation initiation factors (either by the addition of exogenous initiation factors or by the direct inhibition of endogenous initiation factors).

2. MATERIALS

2.1. Nuclease-treated reticulocyte lysates

This lysate can be obtained from a number of commercial vendors (GE Healthcare Life Sciences, Promega, Ambion, Alator Biosciences, and Pel-Freeze). However, we have used the lysate provided by Promega and, thus, all of the comments below will relate to this product. In general, this should also apply to the lysates provided by the other vendors. The lysate comes in several kit forms depending on your choice of reporter and whether you want to use radioactivity, fluorescence, or light emission as the readout for protein synthesis. The technical manual from Promega titled "Rabbit Reticulocyte Lysate System: Instructions for Use of Products L4960 and L4151" is very complete in describing the reagents required for these various readouts.

2.2. tRNA

The reticulocyte lysates are optimized to synthesize the α and β chains of hemoglobin, which have an unusual amino acid composition relative to most proteins. Therefore, to ensure that your mRNA of choice is not restricted by codon usage, often researchers add tRNA from a general tissue to balance the tRNA isoacceptors. Standard preparations of tRNA from rabbit or beef liver or from yeast are commercially available. We have found that these tRNAs are more active if first extracted with phenol and then purified on a Sephadex G-100 column that removes any large RNAs (either intact or fragments of rRNA or mRNA).

2.3. mRNA

For most researchers, mRNAs are generated by the use of T7 RNA polymerase to synthesize a particular mRNA or dicistronic mRNA. Several commercial vendors (Ambion, Promega, Invitrogen) have kits available that require a plasmid containing the T7 promoter $5'$ of the desired RNA sequence. Alternatively, Promega has a plasmid available that contains two different luciferase reporters in which the nucleic acid sequence between the reporters may be varied to assess possible internal ribosome entry site (IRES) activity.

3. Methods

3.1. Preparation of tRNA

If the original source is a tissue, the tissue is homogenized at 4° (all steps are performed at 4° unless noted otherwise) in a buffer containing 20 mM Tris·HCl, pH 7.5, 1 mM dithiothreitol, 100 mM KCl (other standard homogenizing buffers are also acceptable). The homogenate is centrifuged at 10,000×g for 30 min to pellet cellular debris. The supernatant is then shaken vigorously with an equal volume of water-saturated phenol for 10 min and the phases are separated by centrifugation at 10,000×g for 20 min. The aqueous phase (the top phase) is removed and one-tenth volume of 20% potassium acetate, pH 4.5, and two volumes of chilled 95% ethanol are added. This solution is allowed to sit overnight at −20° and the precipitated RNA is collected by centrifugation at 10,000×g for 20 min. The supernatant is discarded and the pellet suspended in a minimal volume of homogenizing buffer. To eliminate any contaminating phenol, the solution can either be extracted with ether or a second ethanol precipitation can be performed. The purified RNA is then applied to a Sephadex G-100 column equilibrated with 20 mM Tris·HCl, pH 7.5, 1 mM dithiothreitol, and 1 mM MgCl. The tRNA will elute in the back half of the column and can be monitored by any aminoacylation reaction (Merrick, 1979a) or by gel electrophoresis. rRNA and mRNA will elute either in the void volume or near the void volume. The tRNA is concentrated by ethanol precipitation (as above), collected by centrifugation, and suspended in the column-equilibrating buffer used above. For convenience, the tRNA is usually taken up at a concentration of about 50 to 200 A_{260} units/ml. Good quality tRNA should have an absorption profile where $A_{260} = A_{220} = 2 \times A_{230}$ or $2 \times A_{280}$.

If the starting material was a crude tRNA (or soluble RNA) purchased commercially, the tRNA is taken up in the previously described column equilibration buffer, extracted with an equal volume of water-saturated phenol, and then precipitated with one-tenth volume of potassium acetate, pH 4.5, and two volumes of 95% ethanol as described. This tRNA is then subjected to gel filtration on Sephadex G-100 and concentrated as described.

It is important to note that phenol is quite caustic and extreme care should be taken to avoid contact with the skin or eyes. In the event the skin is contacted with phenol, rinse immediately with 95% ethanol, as phenol is infinitely soluble in ethanol. Failure to do this quickly (within a minute) will result in scarring of the skin. If phenol gets into the eye, flush the eye extensively with water and then seek immediate medical attention.

3.2. Preparation of mRNA

The preparation of mRNA from a plasmid is the most common source of mRNAs now used, although mRNAs prepared by the use of oligo(dT) selection from natural sources is fine as well. Since the various buffers and enzymes are proprietary in nature, the researcher should follow the instructions provided by the manufacturer for generating transcript mRNAs. For our studies, we have used the enzymes and reagents from Ambion and found them to be very good. The key variable is the substrate used to generate a capped mRNA. The older compound is m^7GpppG. This reagent has the advantage of being less expensive, but can also be misincorporated with the m^7G being the first nucleotide in the RNA chain and a G as the "cap nucleotide" (Stepinski et al., 2001). The amount of RNA obtained and the degree of capping are inversely related, with an optimal degree of cap addition achieved with a cap analog-to-GTP ratio of about 1 to 8. To avoid the difficulty of a misincorporated cap analog, we use the "antireverse cap analog" (ARCA; Ambion), which is similar to m^7GpppG except that the m^7G portion has a 3' OCH_3 group. This blocking of the 3' hydroxyl group means that this analog can be incorporated only in the correct orientation (m^7G as the cap and G as the first nucleotide in the RNA). Following transcription, the mRNA is purified according to the manufacturer's directions, which usually include a phenol/chloroform extraction followed by ethanol precipitation. As with tRNA above, ether extraction or a second ethanol precipitation is required to remove any traces of phenol. It is convenient to have the final mRNA at a concentration of about 2 to 10 A_{260}/ml, which corresponds to about 80 to 400 μg/ml.

4. TRANSLATION OF AN mRNA TO YIELD A RADIOACTIVE PRODUCT

The standard reaction mixture for translation would contain the following:

1. Rabbit reticulocyte lysate[1]	35 μl
2. Amino acid mixture minus methionine[1]	1 μl
3. [35S]Methionine at 10 mCi/ml[2]	2 μl
4. RNasin (ribonuclease inhibitor)[1]	1 μl
5. mRNA	2 μl
6. Nuclease-free water[1]	9 μl
Total	50 μl

---

[1] Materials are supplied in the reticulocyte lysate kit or are available from Promega.
[2] Given the estimate that the methionine concentration in the lysate is about 5 μM, this results in the specific activity of the methionine being about 100 mCi/mmol.

If the reporter peptide to be made is from a tissue other than reticulocytes, it would be advisable to add about 0.2 A_{260} per reaction of tRNA from liver or yeast (previously described) if not included in the kit lysate. Also, for most purposes, the researcher can use a 25 μl reaction volume as the high specific activity of the [^{35}S]methionine allows for quite sensitive detection. Additionally, we have found that best results are obtained with [^{35}S]methionine that is less than 6 weeks old.

5. QUANTITATION OF REACTION PRODUCTS

A simple mechanism to quantitate protein synthesis is to determine the amount of hot trichloroacetic acid (TCA)-precipitable radioactivity. This method uses the strength of a 10% TCA solution at high temperature to hydrolyze the aminoacyl linkage between methionine and the tRNA and at the same time precipitates the protein. In this case, it is best to perform the reaction in 13 × 100-mm test tubes. At the end of the reaction, to each tube is added 2 ml of cold 10% TCA and the tubes are mixed. Next, each tube is heated to 90° for 10 min. The tubes are then placed on ice for 5 min and, finally, the precipitated protein is collected by vacuum filtration using a fine filter membrane (Millipore filter, type HAWP). After the sample has been applied, it is washed twice with 2 ml of cold 10% TCA and then finally with 2 ml of cold 95% ethanol. The ethanol wash removes the last traces of TCA; failure to do so may result in some quenching when the samples are subjected to liquid scintillation spectrometry. The filters are then dried for 10 min under a heat lamp, placed in scintillation vials, and scintillation cocktail is added. Radioactivity is then determined using scintillation spectroscopy. The advantage of this procedure is that it is relatively rapid and very quantitative. For some applications (use of ^{3}H-labeled amino acids or unusual proteins), a slightly different protocol may be preferred (see the Promega Reticulocyte Lysate manual). For example, the hemoglobin in the lysate tends to quench a low-energy emitter such as ^{3}H or some proteins, like collagen, are hydrolyzed in 10% TCA at 90°.

The most common alternative is to subject the sample to analysis by sodium dodecyl sulfate (SDS) gel electrophoresis. To this end, after the reaction time has been completed, the tubes are placed on ice and usually 2 to 4 μl of the reaction mixture is mixed with 20 μl of SDS sample buffer (50 mM Tris·HCl, pH 8.0, 2% SDS, 0.1% bromophenol blue, 10% glycerol, 10 mM dithiothreitol; note: add the dithiothreitol from a frozen 1 M stock just before the buffer is to be used), heated to 90° for 10 min, and then the sample is ready to be applied to an SDS gel (tube or slab; see formulations following). It is not possible to apply more sample due to the very high protein content of the lysate. The gel is run until the tracking dye has nearly reached the bottom of the gel. At the end of the run, the gel is placed in stain (0.01% Coomassie blue, 40% methanol, 7% acetic acid: 40 min for

0.75-mm-thick gels and 60 min for 1.0-mm-thick gels). After this, the gel is briefly rinsed with distilled H_2O, placed in destain (7% acetic acid, 5% methanol), and destained overnight with several changes of destain. Just prior to drying, the gel is soaked in destain made 1% in glycerol (to prevent cracking of the gel as it dries). Using one of several methods, the gel is dried. (Technical note: the gel is well dried when there is no smell of acetic acid.) The radioactivity in the gel is visualized by the use of X-ray film or the use of a PhosphorImager. The dried gel is exposed for 6 to 20 h to obtain a reasonable level of signal. As the energy of emission for either ^{35}S or ^{14}C is rather low, the dried gel must be directly in contact with the measuring device (X-ray film or PhosphorImager plate). Paper or Saran wrap will block detection. The following indicates that the experiment was successful: there is no protein band in the absence of added mRNA, the protein band is present when mRNA is added, and there is only a single protein band (in particular, no protein bands that are of lower molecular weight).

Note: if the researcher has used ^{14}C-labeled amino acids to label the protein, it may require longer exposure times to obtain a good signal given the reduced specific activity compared to [^{35}S]methionine. If ^{3}H is used, the gel may need to be soaked in a fluorographic solution (i.e., Amplify, GEHealthcare Inc.) to permit detection of the very low-energy tritium emissions. In this case, treatment of the gel and subsequent drying of the gel should follow the manufacturer's protocol.

Gel formulations (12.5% SDS gels): note that the volume required will depend on the gel electrophoresis system used.

Separating gel (the lower gel):

1. 40% acrylamide (w/v)[3] (ratio 29:1 acrylamide:bisacrylamide)	5 ml
2. 4× separating buffer (1.5 M Tris·HCl, pH 8.8, 0.4% SDS)	4 ml
3. Double deionized water	6.86 ml
4. 10% ammonium persulfate (w/v)[4]	0.13 ml
5. TEMED	0.013 ml
Total	16 ml

Stacking gel (the upper gel):

1. 40% acrylamide (w/v)[3] (ratio 29:1 acrylamide: bisacrylamide)	0.33 ml
2. 4× stacking buffer (0.5 M Tris·HCl, pH 6.8, 0.4% SDS)	1.0 ml
3. Double deionized water	2.63 ml
4. 10% ammonium persulfate[4]	0.04 ml
5. TEMED	0.004 ml
Total	4.0 ml

[3] Acrylamide and bisacrylamide are neurotoxins with cumulative effects. These solutions are absorbed directly through the skin. Therefore, gloves should be worn at all times when handling these solutions. In addition, even the polymerized gels should be handled with gloves as they may still contain some unpolymerized material.

[4] Add the ammonium persulfate solution last as this initiates the polymerization reaction. The stock solution can be stored at 4° for 2 weeks and for 4 months at −20°.

Running buffer:

1. Tris base   30.28 g
2. Glycine   144.13 g
3. SDS   10 g
4. Double deionized water   About 9.9 liters
Total   10 liters

The advantage of using gel electrophoresis is that the protein band in question is readily seen and possible evaluation of "side products" is easily achieved (something not observable with hot TCA-precipitable radioactivity). The disadvantage is the lack of absolute, quantitative control. This disadvantage can be corrected for by the use of a reliable mRNA included in the series as an internal control for the gels to be run.

6. Optimization of Translations

Although the use of a "kit" gives the impression that all is controlled for and that it is just necessary to add the mRNA and go, nothing could be further from the truth. In reality, no two mRNAs appear to be the same. Therefore, a standard series of optimizations needs to be done. For the usual mRNA, the first optimization is for the amount of mRNA to be added to the translation mixture. For a variety of reasons, it is best to determine the extent of translation using SDS gel electrophoresis as this will also make it possible to ascertain if any aberrant products are being made. Ideally, the researcher should be adding a level of mRNA that is about one-third to one-half of saturation, which for many mRNAs is about 0.2 to 0.5 μg per reaction. However, as noted in Fig. 1.1A, the exact amount will vary depending on the mRNA [in part as the 5' and 3' untranslated regions (UTRs) and in part the coding region]. We have no theoretical explanation for the differences, but the differences are quite real and very reproducible. We usually use a titration range that goes from 0.05 to 1.0 μg per reaction. Figure 1.1A and B shows the best, worst, and average mRNAs from a much larger number of mRNAs that we have studied. The second variable to optimize is time. Depending on the mRNA in use, we have found that the optimal time varies from 20 to 60 min. The key feature is to find a time at which there continues to be a linear increase in product made. This is important as the use of an extended time (such as 60 or 90 min) may allow for slow and less favorable translations to occur and this will contribute to the total amount of product significantly if the expression of the

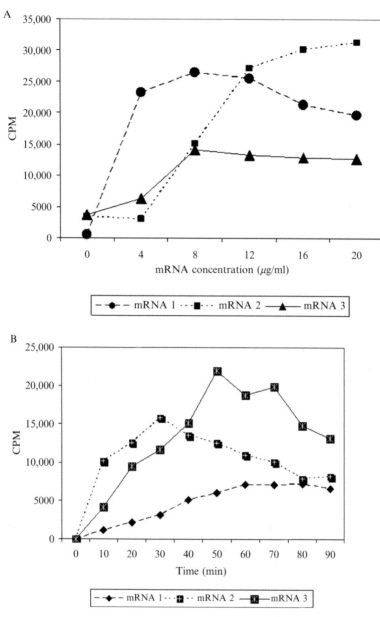

Figure 1.1 Optimization of the translation of different monocistronic mRNAs. In (A), a titration of three different monocistronic mRNAs is performed, each having a differ-ent 5′ and 3′ and coding region. Protein synthesis was monitored by hot trichloroacetic acid (TCA) precipitation of [^{35}S]methionine. (B) Time course of incorporation of [^{35}S] methionine into hot TCA-precipitated protein. Different mRNAs were used in (A) and (B). These patterns represent the best, average, and worst translation expression patterns that we have observed in our studies with a much larger number of mRNAs and should serve as a guide for what individual researchers might observe.

desired protein was only linear for the first 30 min. The third variable is added exogenous tRNA, especially for mRNAs that encode proteins that use a much different codon bias [i.e., especially any of the proteins from bacteria such as thymidine kinase (TK), chloramphenicol acetyltransferase (CAT), or β-galactosidase (βgal), or proteins with an unusual codon usage]. The titration range here would be from 0.05 to 0.5 A_{260} units per reaction.

7. Reporter Proteins for Translation

A number of reporter proteins have been used to monitor protein synthesis in reticulocyte lysates; the most common are luciferase (firefly and *Renilla*), TK, CAT, and βgal. The advantage of these reporters is that they are often used by others (so there is a basis for comparison) and they can all be used in standard *in vivo* eukaryotic cell systems with essentially no background (only the luciferase proteins are from a eukaryotic source). That said, the luciferase proteins are the most enzymatically active and can be readily quantitated in the cell-free translation system where usually only 5 to 10 pmol (or less) of product might be made (roughly 0.2 to 0.5 μg of a protein with a molecular weight of 50,000 Da). Although any of these proteins may be used to monitor the incorporation of a radioactive amino acid, the extreme length of βgal makes it less suitable (approximately 1000 amino acids in length). The real utility of the luciferase proteins is that their synthesis can be independently monitored in the presence of other mRNAs. This is either useful or necessary when examining nonnuclease-treated extracts or when measuring luciferase production in the presence of an mRNA preparation that encodes proteins of a similar molecular weight. Essentially any other protein could also be used for monitoring expression. The key concern would be whether there is any contribution of the coding region to the efficiency of translation of the mRNA (that is, a sequence or structure within the coding region that might interact with either the 5′ or 3′ UTRs to influence translation). Although this is not expected, some mRNAs do contain an IRES element within their coding sequence that can lead to the synthesis of multiple protein products (Komar *et al.*, 2003). Others may contain weak initiation start sites and yield more than one protein product (either with overlapping reading frames or different reading frames). As should be obvious, these proteins/mRNAs may be the direct target of studies to determine the differential utilization of the various start sites or elements that might biologically regulate their expression.

8. EXPERIMENTAL USE OF NUCLEASE-TREATED LYSATES

8.1. Initiation mechanisms

There are a variety of initiation mechanisms used by various eukaryotic systems. The most predominant by far is "cap-dependent" initiation (Johannes *et al.*, 1999); the general scheme for this pathway has been presented in a number of reviews (for example, see Hershey and Merrick, 2000). For this pathway, the mRNA is recognized by its 5′ m^7G cap structure and then bound to the 43S subunit complex. Subsequent scanning of the mRNA in a 3′ direction leads to the identification of the initiating AUG codon through base pairing with the met-tRNA$_i$ present in the ternary complex (eIF2·GTP·met-tRNA$_i$). Curiously, mRNAs that have been shown to use this pathway are reasonably translated in reticulocyte lysates even if the mRNA lacks an m^7G cap. It is anticipated that this promiscuity is the result of the mRNA not having to compete for translation and that should the level of a capped mRNA added to the reaction be saturating, then the uncapped mRNA would be poorly translated.

There are two general strategies when examining the cap dependence of the translation of an mRNA. The first is to synthesize an mRNA that has a structure similar to the m^7G cap, but lacks the methyl group (either ApppG or GpppG is used in the synthetic synthesis in place of m^7GpppG). These mRNAs behave as if they are uncapped for translation, but the presence of the nonmethylated nucleotide protects against RNA degradation from the 5′ end. While this is a real concern when using extracts from tissues or cells for *in vitro* translations, we have not found that the mRNA is degraded in reticulocyte lysates during a routine 30- to 40-min incubation, so that this may be an unnecessary concern. The second method to examine the cap-dependent translation of an mRNA is to monitor the reduction in translation of the mRNA when a cap analog is added to the reaction mixture (100 μM m^7GTP). Generally, a 60% reduction in translation is observed, which compares poorly with the reduced translation of an uncapped mRNA that is 10 to 20% the level of the capped mRNA. In part, this may reflect the presence of bits of the 5′ end of the globin mRNA resulting from the initial nuclease treatment (globin mRNA is about 0.1 μM in untreated lysate) such that there is some inhibition at the beginning.

Reinitiation has been well characterized for only a few mRNAs (for a review, see Geballe and Sachs, 2000; Hinnebusch, 2000). In general, most of these mRNAs can be translated in the reticulocyte lysate system. However, to observe the biological regulation associated with the mRNA, there may be an additional requirement. In the case of using the GCN4 $5'$ UTR, up-regulation of expression is achieved by reducing the level of the ternary complex. The simplest way to do this is to take advantage of the high levels of the interferon-induced eIF2α protein kinase, PKR. This kinase requires double-stranded RNA for activation; a commercially available double-stranded RNA that is often used is poly(I)·poly(C). To obtain the optimal level of up-regulation of reinitiation, the poly(I)·poly(C) should be titrated into the lysate, although the optimal level is usually around 5 to 10 ng/ml. Other mRNAs that utilize reinitiation are influenced by polyamines or arginine (Geballe and Sachs, 2000), and these components can also be titrated into the lysate to evaluate their influence. In all of these, controls with normal cap-dependent translation should be performed in parallel to determine the effect of these additions on normal translation.

IRES-mediated translation has been well characterized for viral IRES elements, but less well for cellular IRES elements. To monitor the expression of mRNAs containing an IRES element, the original mRNAs are usually made in both a capped form and an uncapped form to show that the presence of the m^7G cap does not enhance translation. As a control, IRES-mediated translation is usually compared with an internal control, a cap-dependent mRNA. This can be done in two ways. First, the two mRNAs can be mixed together so that each is translated as a monocistronic mRNA (note that this needs to be done at nonsaturating levels of the mRNA mixture). The second way is to generate an mRNA where the cap-dependent coding region follows the $5'$ UTR and the IRES and its associated coding region are $3'$ of the first reading frame. This classic bicistronic mRNA makes it possible to use a single mRNA and evaluate both cap-dependent and IRES-mediated expression at the same time, in the same reaction (Pelletier and Sonenberg, 1988). The only limitation here is that most commonly, the level of expression from the IRES element is roughly three to eight times lower than what is observed if the IRES element is in a monocistronic mRNA. The reason for this reduction is not clear, but may reflect steric hindrance for accessing the IRES element when there are translating ribosomes in the vicinity (i.e., on the same mRNA). Standard manipulations of the translation of these mRNAs include the effect of adding m^7GTP, dsRNA, omission of the m^7G cap, omission of the poly(A) tail, insertion of elements in either the $5'$ or $3'$ UTR suspected/known to alter the translation of other mRNAs, and the influence of ionic strength.

8.2. Competition between mRNAs

A common concern about different mRNAs is how efficiently they are translated. This cannot be determined from simply observing how much radioactivity is incorporated per microgram of RNA added to the translation mixture. It can be determined by comparing two or more different mRNAs in the same translation reaction mixture (Brendler *et al.*, 1981;

Godefroy and Thach, 1981). Starting with a mixture that is roughly one to one on a molar basis for each mRNA, the mixture is titrated into the lysate until well past saturation (Fig. 1.2A). What is observed in this process is the following. At limiting mRNA concentrations, the proteins expressed are roughly in proportion to the amount of mRNA present (as noted above, in the bicistronic mRNA, IRES-mediated translation is artificially suppressed). However, as saturation is approached, the more competitive mRNA begins to dominate at the expense of the less competitive mRNA and the ratio of protein products changes reflecting this competition (Fig. 1.2B). From the competition shown in Fig. 1.2B, it is evident in the bicistronic mRNA that the cap-dependently expressed protein represents the more competitive mRNA (i.e., in comparing cap-dependent translation to IRES-mediated translation; Anthony and Merrick, 1991). This general observation on competition is more thoroughly reflected in the mathematical treatment of translation published by Godefroy and Thach (1981). Unfortunately, this direct competition experiment gives only a relative readout, such that multiple comparisons need to be made if the investigator wishes to rank order a number of different mRNAs.

8.3. Synthesis of proteins of high specific radioactivity

For some specific uses, access to a highly radioactive protein can be of value. This can be achieved with the correct mRNA template and the use of amino acids of high specific activity (the most commonly used is the ^{35}S mixture of methionine and cysteine, although some may choose to use [^3H] leucine, which is also available at high specific activity). The key here is to have an affinity purification system available, as there is very little protein made and the lysate is generally several hundred milligrams of protein per milliliter. Alternatively, if the protein of interest will react with ligand, substrate, or binding protein to yield a complex detectable by gel shift, pull down, or resolution by some column matrix (most commonly gel filtration), then it may not be necessary to purify the labeled protein from the lysate. Under the standard conditions as defined above (considering the content of cold amino acid in the lysate), a protein of 50,000 molecular weight should have a specific activity of approximately 1000 cpm/pmol or 20,000 cpm/μg (assuming methionine is 1% of the incorporated amino acid). This value can be increased by using a ^{35}S mixture of cysteine and methionine or by increasing the number of millicuries of [^{35}S]methionine added to the reaction mixture.

8.4. Influence of variations of factor activity

For those who have access to purified factors (Benne *et al.*, 1979; Merrick, 1979b; Staehelin *et al.*, 1979), one can ask whether changing the concentrations of initiation factors influences start site selection (when initiation is

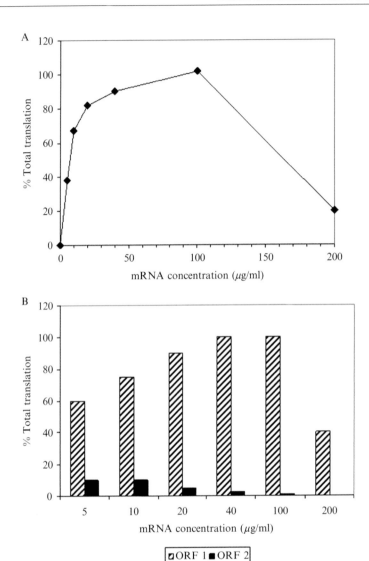

Figure 1.2 Competition between cap-dependent and internal ribosome entry site (IRES)-mediated translation. A T7 transcript of the bicistronic mRNA TK/P2CAT, kindly provided by Dr. Nahum Sonenberg, was titrated into reticulocyte lysates. (A) Hot trichloroacetic acid (TCA) precipitated [^{35}S]methionine representing the sum of thymidine kinase (TK) and chloramphenicol acetyltransferase (CAT) protein synthesis following a 60-min incubation at 30°. (B) Relative synthesis of each of the polypeptides (TK and CAT) as determined from sodium dodecyl sulfate (SDS) gel electrophoresis and scanning laser densitometry. % Translational Efficiency represents the relative amount of protein made in comparison to the maximal total synthesis observed at 100 μg/ml RNA (with correction for the methionine content in TK and CAT). See Anthony and Merrick (1991) for a more complete description of this experiment.

occurring at two different start codons). Although we have yet to complete the studies examining a larger number of mRNAs and different factors, it does appear that increased concentrations of either eIF5 or eIF5B can influence start site selection with increased activity favoring the 5′ start site, although with a noted decrease in overall expression (unpublished observation). This is similar to what Donahue and coworkers have observed in yeast using a hyperactive eIF5 (Huang et al., 1997). Addition of eIF4F (and to some extent eIF4B or eIF4A) to a translation reaction with a bicistronic mRNA enhances expression more from the IRES element than the 5′ cap-dependent coding region (Anthony and Merrick, 1991). These same factors also dramatically stimulate the translation of uncapped mRNAs. Similarly, the addition of eIF4A is required if the amount of secondary structure in the 5′ UTR is systematically increased (Svitkin et al., 2001). Others have used dominant negative mutants of the translation factors, such as eIF4A (Pause et al., 1994). This mutant effectively inhibited both cap-dependent and IRES-mediated expression.

However, most laboratories do not have access to all the translation factors and may have access only to those factors that can be expressed in *Escherichia coli* as single polypeptide chains. As an alternative, it is possible to gain insights into the translation mechanisms outlined above by the use of inhibitors. Two inhibitors are readily available, m^7GTP and poly(I)·poly (C). Although m^7GTP would be anticipated to inhibit cap-dependent translation (as it does), it also tends to stimulate expression from IRES elements in a bi-cistronic mRNA (Anthony and Merrick, 1991). This stimulation may reflect that more eIF4F is available since it cannot bind to the m^7G cap of the mRNA or it may indicate that eIF4F is more active when bound to m^7GTP. The addition of poly(I)·poly(C) leads to the activation of PKR, the phosphorylation of eIF2, and the subsequent reduction of the levels of ternary complex available to initiate protein synthesis. In general terms, this would be expected to inhibit cap-dependent and IRES-mediated translation while stimulating the translation of mRNAs containing upstream (and regulatory) open reading frames (ORFs) as in the case of GCN4. Our experience is that this reduction in ternary complexes does not reduce the level of expression from IRES-mediated translation (Hui et al., 2003). Thus, there would appear to be either an alternate pathway for IRES-mediated translation or an alternate rate-limiting step, as this observation is not consistent with the use of the standard, ordered 80S initiation pathway where the ternary complex binds to the 40S subunit prior to the binding of mRNA.

Recent studies from the Pelletier laboratory have indicated that a number of small molecules can be used as inhibitors of translation initiation or elongation (Bordeleau et al., 2005, 2006; Chan et al., 2004; Kumar et al., 2004; Malina et al., 2005; Novac et al., 2004; Robert et al., 2006a,b). Although the general mechanism of inhibition has been established for most of these inhibitors, their differential effect on various aspects of translation initiation has not been examined and could be useful in extended studies of reinitiation, IRES-mediated initiation, cap-dependent initiation, or alternate start site selection. Use of one of these inhibitors confirmed the observation that IRES-mediated translation appears to be refractory to reduced levels of ternary complexes (Robert *et al.*, 2006a). Although some of these small molecule inhibitors have restricted availability, it is hoped that in the near future all will become accessible.

More sophisticated options are to use protein inhibitors of translation. The three best characterized proteins are 4E-BP (which binds eIF4E and blocks cap-dependent translation), Pdcd4 (isolated as a tumor suppressor), which binds to eIF4A, and the p56 family of interferon-induced proteins that bind to different subunits of eIF3 and, depending on the protein, appear to inhibit either eIF2 or eIF4F activity. As each of these proteins is a single polypeptide chain, each can be (and has been) expressed in *E. coli* from appropriate plasmids.

In summary, a large number of inhibitors of translation have been identified and most are rather readily available. These inhibitors have already been useful in probing initiation reactions and show great promise for continued use to more accurately define the steps (or different steps) used in the various initiation schemes previously described. Given the uncertainty of most initiation schemes for everything except normal cap-dependent translation, these inhibitors provide an excellent alternative mechanism to examine the less utilized initiation pathways.

8.5. Sucrose gradients

The use of sucrose gradients in the study of protein synthesis has been extensive and has been the foundation for many of the initiation schemes proposed to date. This methodology allows for the separation of both free met-tRNA$_i$ and mRNA from higher molecular weight complexes and, as such, makes it possible to determine the components associated with complexes based upon their resolution according to sedimentation rate (roughly, molecular weight). For most studies, the sedimentation rate of any of the ribosomal complexes is 40S or greater, whereas tRNA and mRNA have sedimentation rates of 4S and 10 to 30S, respectively. While it is possible to follow the presence of met-tRNA$_i$ as [^{35}S]methionine and the mRNA as a ^{32}P label (body labeled or end labeled), the availability of antibodies to each of the translation initiation factors (Santa Cruz Biotechnologies) makes it possible to monitor the presence of the factors as a function of the complex they are in or as a function of the mRNA under study (i.e., perhaps an unusual mRNA such as a GCN4-type mRNA that uses reinitiation). For those with access to individually purified factors, these can be radiolabeled by reductive methylation using [^{14}C]formaldehyde so that the proteins can be monitored by radioactivity (either as direct scintillation counting or by the use of SDS gel electrophoresis) (Benne et al., 1979; Peterson et al., 1979).

For standard sucrose gradients that will resolve 40S, 60S, and 80S complexes, we routinely use the SW28 rotor (Beckman) and buckets that hold tubes of about 16.5 ml. Gradients are made from 10 to 30% sucrose in a buffer that contains 20 mM Tris·HCl, pH 7.5 (or 20 mM HEPES, pH 7.4), 3 mM MgCl$_2$, 100 mM KCl, 2 mM dithiothreitol, and either 100 μM GTP or GDPNP (the inclusion of a guanine nucleotide in the gradient increases the yield of ternary complexes associated with 40S subunits; Peterson et al., 1979). The sample, generally 50 to 100 μl, is applied to the top of the chilled gradient and centrifugation is for 20 h at 16,000 rpm at 4°. The gradients are fractionated using upward displacement with 60% sucrose with an ISCO model 640 fractionator, which makes it possible to monitor the absorbance of the subunits at A_{254}. Various fraction sizes can be used depending on the researcher's need; however, we have found that rarely is resolution enhanced by collecting more than about 20 individual fractions. If the desired analysis is a determination of radioactivity, aliquots from each fraction can be mixed with a scintillation solution for aqueous samples and radioactivity determined by scintillation spectroscopy. If the position or quantitation of proteins bound to the subunits/ribosome is desired, aliquots can be analyzed by SDS gel electrophoresis (as previously described). As the concentration of the proteins is often dilute in the gradient, we commonly precipitate the proteins with cold 10% TCA. In a typical example, 200 μl of each fraction is mixed in a microfuge tube with 2 μl of a 1 mg/ml solution of soybean trypsin inhibitor and 20 μl of 100% cold TCA (100 g/100 ml). The mixture is held on ice for 30 min and then microfuged for 10 min at 4°. The supernatant is carefully decanted and the pellet is vigorously mixed with 200 μl acetone (this step will remove traces of TCA). The solution is again centrifuged for 10 min at 4°. The supernatant is carefully decanted and the acetone allowed to evaporate. To the pellet is added 20 μl of SDS sample buffer and the tube is heated to 90° for 10 min. These samples are now ready for SDS gel electrophoresis. The soybean trypsin inhibitor that is included acts as a carrier protein to facilitate quantitative precipitation. However, it also serves as an internal control for equivalent recovery and loading of each gradient fraction. Soybean trypsin inhibitor was chosen because its 21,000 molecular weight does not overlap with any of the translation initiation factor proteins (or their subunits) and thus will not interfere with evaluating whether a given protein (peptide) is present. Depending on whether the researcher wants to identify the protein of interest by subsequent Coomassie blue (or silver) staining or by Western blot, the following guideline may be useful in deciding which to use. A milliliter of reticulocyte lysate contains about 20 A_{260} units of RNA, mostly rRNA. This corresponds to 160 μg of ribosomes or 40 pmol in the 100 μl reaction mixture. For a protein of molecular weight 50,000, 40 pmol

Hartz, D., McPheeters, D. S., Traut, R., and Gold, L. (1988). Extension inhibition analysis of translation initiation complexes. *Methods Enzymol.* **164,** 419–425.

Hershey, J. W. B., and Merrick, W. C. (2000). The pathway and mechanism of initiation of protein synthesis. *In* "Translational Control of Gene Expression" (N. Sonenberg, J. W. B. Hershey, and M. B. Mathews, eds.), pp. 33–88. Cold Spring Harbor Laboratory Press, Cold Spring Harbor, N.Y.

Hinnebusch, A. G. (2000). Mechanism and regulation of initiator methionyl-tRNA binding to ribosomes. *In* "Translational Control of Gene Expression" (N. Sonenberg, J. W. B. Hershey, and M. B. Mathews, eds.), pp. 185–243. Cold Spring Harbor Laboratory Press, Cold Spring Harbor, N.Y.

Huang, H. K., Yoon, H., Hannig, E. M., and Donahue, T. F. (1997). GTP hydrolysis controls stringent selection of the AUG start codon during translation initiation in *Saccharomyces cerevisiae.* *Genes Dev.* **11,** 2396–2413.

Hui, D. J., Bhasker, C. R., Merrick, W. C., and Sen, G. C. (2003). Viral stress-inducible protein p56 inhibits translation by blocking the interaction of eIF3 with the ternary complex eIF2·GTP·Met-tRNA$_i$. *J. Biol. Chem.* **278,** 39477–39482.

Jackson, R. J., and Hunt, T. (1983). Preparation and use of nuclease-treated rabbit reticulocyte lysates for the translation of eukaryotic messenger RNA. *Methods Enzymol.* **96,** 50–74.

Johannes, G., Carter, M. S., Eisen, M. B., Brown, P. O., and Sarnow, P. (1999). Identification of eukaryotic mRNAs that are translated at reduced cap binding complex eIF4F concentrations using a cDNA microarray. *Proc. Natl. Acad. Sci. USA* **96,** 13118–13123.

Kolupaeva, V. G., Pestova, T. V., and Hellen, C. U. (2000). An enzymatic footprinting analysis of the interaction of the 40S ribosomal subunits with the internal ribosomal entry site of hepatitis C virus. *J. Virol.* **74,** 6242–6250.

Komar, A. A., Lesnik, T., Cullin, C., Merrick, W. C., Trachsel, H., and Altman, M. (2003). Internal initiation drives the synthesis of Ure2 protein lacking the prion domain and affects [URE3] propagation in yeast cells. *EMBO J.* **22,** 1199–1209.

Kozak, M. (1990). Evaluation of the fidelity of initiation of translation in reticulocyte lysates from commercial sources. *Nucl. Acids Res.* **18,** 2828.

Kozak, M. (1998). Primer extension analysis of eukaryotic ribosome-mRNA complexes. *Nucl. Acids Res.* **26,** 4853–4859.

Kruh, J., and Borsook, H. (1956). Hemoglobin synthesis in rabbit reticulocytes *in vitro.* *J. Biol. Chem.* **220,** 905–915.

Kumar, R., Garneau, P., Nguyen, N., William Lown, J., and Pelletier, J. (2004). Methionine substituted polyamides are RNAse mimics that inhibit translation. *J. Drug Target* **12,** 125–134.

Malina, A., Khan, S., Carlson, C. B., Svitkin, Y., Harvey, I., Sonenberg, N., Beal, P. A., and Pelletier, J. (2005). Inhibitory properties of nucleic acid-binding ligands on protein synthesis. *FEBS Lett.* **579,** 79–89.

Merrick, W. C. (1979a). Assays for eukaryotic protein synthesis. *Methods Enzymol.* **60,** 108–123.

Merrick, W. C. (1979b). Purification of protein synthesis initiation factors from rabbit reticulocytes. *Methods Enzymol.* **60,** 101–108.

Merrick, W. C. (1979c). Evidence that a single GTP is used in the formation of 80S initiation complexes. *J. Biol. Chem.* **254,** 3708–3711.

Novac, O., Guenier, A. S., and Pelletier, J. (2004). Inhibitors of protein synthesis identified by a high throughput multiplexed translation screen. *Nucl. Acids. Res.* **32,** 902–915.

Pause, A., Methot, N., Svitkin, Y., Merrick, W. C., and Sonenberg, N. (1994). Dominant negative mutants of mammalian translation initiation factor eIF4A define a critical role for eIF4F in cap-dependent and cap-independent initiation of translation. *EMBO J.* **13,** 1205–1215.

Pelletier, J., and Sonenberg, N. (1988). Internal initiation of translation directed by a sequence derived from poliovirus RNA. *Nature* **334**, 320–325.

Pestova, T. V., Hellen, C. U., and Shatsky, I. N. (1996). Canonical eukaryotic initiation factors determine initiation of translation by internal ribosomal entry. *Mol. Cell Biol.* **16**, 6859–6869.

Peterson, D. T., Merrick, W. C., and Safer, B. (1979). Binding and release of radiolabeled eukaryotic initiation factors 2 and 3 during 80S initiation complex formation. *J. Biol. Chem.* **254**, 2509–2516.

Robert, F., Kapp, R. F., Khan, S. N., Acker, M. G., Kolitz, S., Kazemi, S., Kaufman, R. J., Merrick, W. C., Koromoilas, A. E., Lorsch, J. R., and Pelletier, J. (2006a). Initiation of protein synthesis by hepatitis C virus is refractory to reduced eIF2·GTP·Met-tRNA$_i$ ternary complex availability. *Mol Biol. Cell* **17**, 4632–4644.

Robert, F., Gao, H. Q., Donia, M., Merrick, W. C., Hanamm, M. T., and Pelletier, J. (2006b). Chlorissoclimides: New inhibitors of eukaryotic protein synthesis. *RNA* **12**, 717–725.

Staehelin, T., Erni, B., and Schreier, M. H. (1979). Purification and characterization of seven initiation factors for mammalian protein synthesis. *Methods Enzymol.* **60**, 136–165.

Stepinski, J., Waddel, C., Stolarski, R., Darzynkiewicz, E., and Rhoads, R. E. (2001). Synthesis and properties of mRNAs containing the novel "anti-reverse" cap analogs 7-methyl(3'-O-methyl)GpppG and 7-methyl(3' deoxy)GpppG. *RNA* **7**, 1486–1495.

Svitkin, Y. V., Pause, A., Haghighat, A., Pyronnet, S., Witherell, G., Belsham, G. J., and Sonenberg, N. (2001). The requirement for eukaryotic initiation factor 4A (eIF4A) in translation is in direct proportion to the degree of mRNA 5' secondary structure. *RNA* **7**, 382–394.

CHAPTER TWO

STUDYING TRANSLATIONAL CONTROL IN DROSOPHILA CELL-FREE SYSTEMS

Fátima Gebauer* *and* Matthias W. Hentze[†]

Contents

Abstract

Classically, *Drosophila* cell-free translation systems have been used to study the response of the translational machinery to heat shock treatment. We and others have developed optimized *Drosophila* embryo and ovary extracts, and their use has expanded to the study of a variety of translational control events. These extracts recapitulate many of the aspects of mRNA translation observed *in vivo* and retain critical regulatory features of several translational control processes. Indeed, their use is rapidly improving our knowledge of molecular mechanisms of translational control. In this chapter we provide general guidelines and detailed protocols to obtain and use translation extracts derived from *Drosophila* embryos and ovaries.

1. INTRODUCTION

Cell-free translation systems derived from animal cells such as rabbit reticulocytes, HeLa, or other cultured cells have been instrumental in deciphering key aspects of translation, including the mechanism of translation initiation, the role of mRNA features [cap structure, poly(A) tail, structural and other regulatory elements] in translation, or the mechanisms

* Centre de Regulació Genòmica (CRG-UPF), Barcelona, Spain
† Gene Expression Unit, European Molecular Biology Laboratory, Heidelberg, Germany

Methods in Enzymology, Volume 429
ISSN 0076-6879, DOI: 10.1016/S0076-6879(07)29002-0

by which some RNA-binding proteins interfere with the translational machinery. In the 1980s, *Drosophila* extracts derived from embryos and cells in culture (SL-1, SL-2, and Kc cells) were typically used to study the profound change in protein synthesis caused by heat shock (Maroto and Sierra, 1988; Storti *et al.*, 1980; Zapata *et al.*, 1991). These extracts were responsive to the addition of exogenous mRNA because of pretreatment with micrococcal nuclease to destroy the endogenous mRNAs (Scott *et al.*, 1979). This treatment, however, was not always successful and often inactivated the embryo extract (Scott *et al.*, 1979).

More recently, *Drosophila* embryo and ovary extracts that translate exogenous mRNA with high efficiency have been obtained (Castagnetti *et al.*, 2000; Gebauer *et al.*, 1999; Lie and Macdonald, 2000). These extracts have been used to study translational control events that impinge on fly development, such as the regulation of the mRNAs encoding the antero-posterior axis determinants Oskar and Nanos, or of the mRNA encoding the dosage compensation complex component Msl-2 (Beckmann *et al.*, 2005; Chekulaeva *et al.*, 2006; Clark *et al.*, 2000). Similar lysates have been used to study sequence-specific mRNA deadenylation events (Jeske *et al.*, 2006) and the phenomenon of RNA interference (Tuschl *et al.*, 1999).

Translationally active extracts have been prepared from a range of stages of embryo development (0–18 h postfertilization). Embryo and ovary extracts recapitulate key properties of translation observed *in vivo*, such as the stimulatory role of the mRNA m^7GpppN cap structure and the poly(A) tail, as well as the synergism between the two (Castagnetti *et al.*, 2000; Gebauer *et al.*, 1999; Lie and Macdonald, 2000). While many mRNAs display a strong cap dependence in these systems, the overall effect of the poly(A) tail is more variable and depends not only on its length but also on the particular mRNA tested. In addition, the presence of a cap structure greatly improves the stability of the mRNA in both embryo and ovary extracts whereas the poly(A) tail does not appear to contribute significantly to mRNA stability in these extracts (Castagnetti *et al.*, 2000; Gebauer *et al.*, 1999; Lie and Macdonald, 2000). Thus, to preserve mRNA stability, we recommend that the exogenous transcripts to be evaluated in these systems carry either a canonical (m^7GpppN) or a noncanonical (ApppN) 5' cap structure, depending on the purpose of the experiment. Although, as stated above, the cap structure plays an important role in the translational efficiency of most mRNAs, translation driven by the IRESs of *Drosophila* reaper, hsp70, hid and grim mRNAs also occurs efficiently in embryo extracts (Hernández *et al.*, 2004; Vázquez-Pianzola *et al.*, 2006).

Ovary and embryo translation extracts differ in important practical aspects. First, we and others have attempted to establish large-scale preparations of ovary extracts without success (Lie and Macdonald, 2000; F. Gebauer, S. Castagnetti, M. W. Hentze, and A. Ephrussi, unpublished). Ovary extracts are obtained in limited amounts by manually dissecting flies,

Table 2.1 Conditions for *in vitro* translation assays

Reagent	Volume[a] (μl)	Optimal range (mM)
2 mM amino acids	0.3	NA
1 M creatine phosphate[b]	0.17	NA
10 mg/ml creatine kinase	0.08	NA
1 M HEPES pH 7.4	0.24	NA
10 mM Mg(OAc)$_2$	X[c]	$\geq$0.3
1 M KOAc	Y[c]	40–80
2.5 mM spermidine	Z[c]	$\leq$0.3
100 mM DTT	W[c]	$\leq$1.2
mRNA	M[d]	NA
Incubate for 90 min at 25°		

[a] Volumes are given for a total reaction volume of 10 μl. A master mix should be prepared for as many samples as required.

[b] Leftovers should be discarded after thawing.

[c] The optimal concentrations that we found for capped and polyadenylated firefly luciferase mRNA are 0.4 mM Mg(OAc)$_2$, 80 mM KOAc, 0.1 mM spermidine, without DTT.

[d] The mRNA should be capped. In general, translation improves with poly(A) tails longer than 31 residues. The mRNA should be used in the linear range of translation. We normally add 0.03 pmol mRNA per reaction.

while embryo extracts can be prepared in large quantities. Second, ovary extracts are exquisitely sensitive to freeze/thaw cycles, and are best used immediately after preparation. If necessary, they can be frozen only once and kept in liquid nitrogen. In contrast, embryo extracts are robust and withstand incubation on ice for up to 6 h and more than four freeze/thaw cycles without much loss of activity.

In this chapter, we provide detailed protocols to obtain translationally active ovary and embryo extracts for the translation of exogenous mRNAs. Because the translational efficiency is sensitive to small variations in the concentration of salts and other components (see Table 2.1), we prepare crude extracts without addition of salts and subsequently optimize the reaction conditions for specific mRNAs. Alternative protocols in which salts are added during extract preparation have been described by others (Lie and Macdonald, 2000; Tuschl *et al.*, 1999).

2. PREPARATION OF OVARY EXTRACTS

A schematic diagram depicting the preparation of ovary and embryo extracts is shown in Fig. 2.1. As mentioned above, ovary extracts are prepared in small scale.

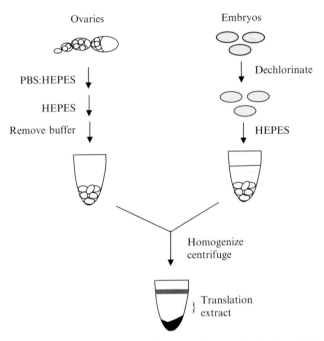

Figure 2.1 Schematic representation of the procedure to obtain *Drosophila* ovary and embryo extracts for translation. See text for details.

1. Feed adult flies on yeast for 2 to 3 days at 25°.
2. Select female flies and manually dissect them on ice-cold phosphate-buffered saline (PBS) to obtain the ovaries.
3. Place the ovaries in an Eppendorf tube containing PBS on ice. Allow the ovaries to settle by gravity and measure the volume of settled material.
4. Wash twice with 12 volumes of a (1:1) mix of cold PBS:DEI [10 mM HEPES, pH 7.4, 5 mM dithiothreitol (DTT), 1× Complete Protease inhibitors cocktail from Roche] by gently tilting the tube up and down, and allow ovaries to settle by gravity on ice. Quickly wash twice with 12 volumes of cold DEI. The idea behind this washing protocol is to gradually transfer the ovaries from an isotonic medium that is detrimental to translation (PBS) to the hypotonic translation solution (DEI). During this time, the volume of settled ovaries doubles.
5. Remove the remaining buffer and manually homogenize the ovaries using a plastic pestle that fits the Eppendorf tube or by pipetting the sample up and down.
6. Spin the homogenate for 15 min at 4° in a microcentrifuge at 12,000 rpm. Centrifugation results in the separation of the homogenate into three phases: a pellet containing debris and cell nuclei, an intermediate cytoplasmic phase, and a low-density lipidic top phase.

7. Discard the pellet and mix the two remaining phases. Add glycerol to 10% final. Optimally, the extracts should be used immediately for translation. Alternatively, they can be flash frozen and stored in liquid nitrogen.

3. Preparation of Embryo Extracts

Fertilized eggs are laid by 2- to 3-day-old adult flies on agar-apple juice plates (2.9% agar, 30% apple juice, 4.4% sugar-beet syrup, 0.25% Nipagin) spread with yeast paste (0.6% propionic acid, 0.68 g/ml dry yeast in deionized water). Embryos can be collected at any time after egg laying. Embryo collections usually consist of mixed developmental stages (e.g., an overnight collection contains 0 to 12 h embryos) unless a synchronization protocol is applied. To obtain synchronized embryos, plates are exchanged three times during the course of 3 h (1 h/exchange). Following this procedure, flies release the unsynchronized embryos they have kept inside. These plates are discarded. In the fourth exchange, flies are allowed to lay eggs for 30 min to 1 h. These embryos are expected to be synchronized with a 30 min to 1 h interval. Synchronization can be monitored with the microscope. The plate is removed from the fly chamber and embryos are allowed to develop for the desired time before collection.

A procedure for large-scale preparation of embryo extracts is provided below (see Fig. 2.1). This protocol can be scaled down.

1. Collect embryos (usually 10 to 60 ml) in a pile of sieves (Neolab): the top one with a cut-off size of an adult fly, the second with a cut-off size for fly appendages (leg, antenna, etc.), and the third with a cut-off size of a single embryo. Embryos are collected by combing the plate surface with a brush under a strong water stream. Wash extensively (5 to 10 min) with cold tap water, first to push all embryos to the last sieve and then to remove yeast debris carried over the last sieve (yeast flow through the last sieve). Embryos look like sand on the last sieve. *Note:* in places where tap water is of low quality, use distilled water for all steps.

2. Wash the collected embryos with a few milliliters of freshly prepared isotonic EW solution (0.7% NaCl, 0.04% Triton X-100) at room temperature. Transfer the embryos to a 500 ml cylinder containing EW solution. Allow the embryos to settle and wash twice with 500 ml of EW solution. Floating embryos can also be used for the extract preparation. For small amounts of embryos, washing steps can be carried out in Falcon tubes and embryos can be collected by centrifugation at 1000 rpm for 5 min at 4°.

3. Fill the cylinder with 200 ml of EW solution and vigorously agitate with a magnetic stirrer. For dechlorination, add 60 ml of 13% bleach (3% sodium

hypochlorite final) and incubate for 3 min at room temperature under agitation. Quickly transfer the dechlorinated embryos to the sieve and wash extensively with a strong stream of tap water for about 5 min.

4. Transfer the embryos to a 100 ml cylinder and wash twice with 100 ml of DE buffer (10 mM HEPES, pH 7.4, 5 mM DTT). Many embryos will float. These embryos can also be used for the preparation. As in step 2, small amounts of embryos can be collected and washed in Falcon tubes.

5. Remove the buffer and measure the volume of remaining embryos. Add one volume of DEI buffer (DE buffer supplemented with 1× Complete protease inhibitor cocktail from Roche, see also step 4 of the previous section) and homogenize at 4° by 20 strokes of a Potter-Elvehjem homogenizer at about 1500 rpm. Keep the homogenate on ice.

6. Spin the homogenate in a tabletop ultracentrifuge at 24,000 rpm (40,000×g) in a TLS-55 rotor at 4° for 20 min. Small volumes of homogenate can be spun in a microcentrifuge at 14,000 rpm for 15 min at 4°.

7. Collect the clear cytoplasmic interphase by puncturing the tube with a syringe. Add glycerol to a final concentration of 10%, aliquot, and flash freeze in liquid nitrogen. Store at −80°.

4. THE TRANSLATION ASSAY

The translational efficiency of an extract is subject to some batch-to-batch variation and is influenced by the assay conditions. Cell-free translation using embryo and ovary extracts is sensitive to variations in the concentration of Mg^{2+} and K^+. While the optimal concentration of K^+ ranges between 40 and 80 mM for most mRNAs, the response to Mg^{2+} varies greatly (Fig. 2.2). In addition, spermidine and DTT may affect the translational efficiency of some mRNAs. These parameters should be optimized for each mRNA tested. A typical translation reaction together with the optimal range of concentrations for various critical parameters are shown in Table 2.1.

mRNAs to be compared in a translation reaction should be assayed under the same conditions (in parallel or, when possible, internally controlled) and, optimally, also synthesized in the same batch. Importantly, a translation curve with increasing amounts of mRNA should be performed to select a concentration in the linear range of translation.

As mentioned above, the mRNA should contain a cap structure to improve its translational efficiency and stability. Because the cap analog and GTP compete for incorporation at the 5' end of the mRNA, optimal capping during mRNA synthesis is obtained by preincubating the reaction mix in the absence of GTP for 5 min at 37° to allow for cap incorporation and subsequently adding the GTP to allow for RNA synthesis. In addition,

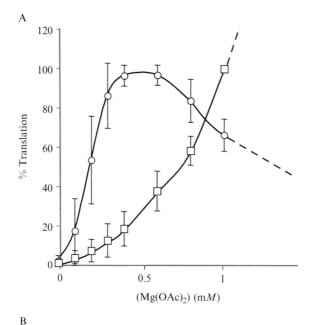

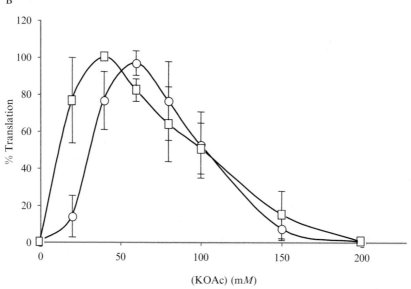

Figure 2.2   Effect of Mg^{2+} and K^+ concentrations on translation. The translational efficiencies of Firefly (open circles) and *Renilla* (open squares) luciferase mRNAs were measured in response to increasing concentrations of magnesium (A) and potassium (B) acetate. Reaction conditions were chosen following the procedure described in Table 2.1. Reactions contained 80 mM KOAc when Mg^{2+} was tested, and 0.6 mM Mg (OAc)$_2$ when K^+ was tested, and lacked spermidine and DTT. The translational efficiency was determined by measuring the luciferase activity, which was plotted as the percentage of the maximal activity obtained in each experiment. Each data set represents the average of at least four experiments.

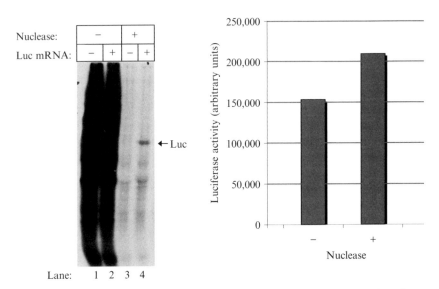

Figure 2.3 Micrococcal nuclease treatment. Embryo extract was treated with 0.15 units/μl micrococcal nuclease in the presence of 1 mM CaCl$_2$ for 2 min at 20°, or mock treated. The reaction was stopped with 2 mM EGTA, and the extract was used to translate Firefly luciferase mRNA in the presence of [^{35}S]methionine. Part of the reaction was loaded in an SDS-polyacrylamide gel, and the protein products visualized by autoradiography (left panel). The other part was used to measure the luciferase activity (right panel). Nuclease treatment effectively eliminated the background translation from endogenous mRNAs and did not decrease exogenous luciferase mRNA translation.

modified cap analogs with improved geometry for appropriate incorporation at the 5′ end of the mRNA have been developed (Stepinski *et al.*, 2001). Usually, translation improves if the mRNA contains a poly(A) tail longer than 31 residues. Optimally, the poly(A) tail is encoded in the DNA construct used to synthesize RNA. Poly(A) tails of more than 20 residues are unstable in most bacteria. Thus, we recommend to keep plasmids containing poly(A) stretches in the bacterial strain XL1-Blue. Long poly(A) tails can also be added after mRNA synthesis by the use of yeast poly(A) polymerase (yPAP, Amersham). When required, cordycepin (3′ deoxyadenosine) can be incorporated at the 3′ end of the mRNA using yPAP to prevent further adenylation by activities present in the translation extract.

Translation requires an ATP regenerating system (creatine phosphate and creatine kinase). Creatine phosphate is prepared in water, aliquoted, and stored at −80°, and remainders of aliquots should be discarded after thawing. Creatine kinase is prepared as a stock solution at a concentration of 10 mg/ml in 20 mM HEPES, pH 7.4, 50% glycerol, and is stable at −20° for up to a year. Contrary to some cell-free systems, translation in embryo

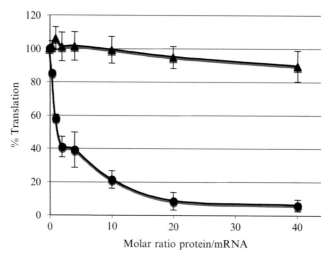

Figure 2.4 Translation inhibition by the RNA-binding protein Sex-lethal (SXL). Increasing amounts of recombinant SXL (circles) or a control protein (mRBD, triangles) were incubated in translation reactions containing both an mRNA target for SXL (Firefly luciferase fused to the untranslated regions of msl-2 mRNA; Gebauer *et al.*, 1999) and *Renilla* mRNA as an internal control. The translational efficiency was determined by measuring the respective luciferase activities. The Firefly luciferase values were corrected for *Renilla* luciferase expression and plotted as percentages against the molar ratio protein:mRNA. The activity obtained in the absence of added protein was taken as 100%.

extracts is not stimulated by the addition of GTP. In addition, we have observed that *de novo* translation is largely inactive in extracts that have been preincubated at 25° for at least 15 min in the presence of an energy-generating system and, thus, have already initiated translation (F. Gebauer and M. W. Hentze, unpublished).

The use of luciferase and other reporter systems whose activities can be measured enzymatically has circumvented the need to eliminate the endogenous mRNAs, which would otherwise generate a background that prevents the detection of the exogenous translation product (Fig. 2.3, lanes 1 and 2). Indeed, to study regulatory mechanisms it is often more convenient to preserve the complement of endogenous mRNAs to maximally mimic physiological conditions. If needed, endogenous mRNAs can be eliminated efficiently without significant loss of translation activity by treating the extract with 0.15 units/μl micrococcal nuclease for 2 to 4 min at 20° after adjusting the extract to 1 mM CaCl$_2$ (see Fig. 2.3). Micrococcal nuclease treatment is stopped by adding EGTA to a final concentration of 2 mM.

Buffers such as Tris or PBS may inhibit the translation reaction. Thus, when testing the effect of recombinant proteins on translation, we recommend dialyzing the proteins against HEPES-based buffers in the absence of salts. We usually dialyze proteins against a buffer containing 20 mM HEPES, pH 8.0, 0.2 mM EDTA, 1 mM DTT, 0.01% NP40, and 20% glycerol. Ideally, different concentrations of recombinant protein are tested in reactions that contain two exogenously added mRNAs: the mRNA under study and a control mRNA that serves normalization purposes. An example of the effect of adding a recombinant regulatory protein on translation is shown in Fig. 2.4.

ACKNOWLEDGMENTS

F. Gebauer was supported by Grants SGR05/00669 from DURSI and BFU2006–01874/BMC from the Spanish Ministry of Education and Science. M. W. Hentze acknowledges support by multiple sources, especially the European Molecular Biology Laboratory and the Deutsche Forschungsgemeinschaft.

REFERENCES

Beckmann, K., Grskovic, M., Gebauer, F., and Hentze, M. W. (2005). A dual inhibitory mechanism restricts msl-2 mRNA translation for dosage compensation in *Drosophila*. *Cell* **122**, 529–540. Erratum in: *Cell* **123**, 171.

Castagnetti, S., Hentze, M. W., Ephrussi, A., and Gebauer, F. (2000). Control of oskar mRNA translation by Bruno in a novel cell-free system from *Drosophila* ovaries. *Development* **127**, 1063–1068.

Chekulaeva, M., Hentze, M. W., and Ephrussi, A. (2006). Bruno acts as a dual repressor of oskar translation, promoting mRNA oligomerization and formation of silencing particles. *Cell* **124**, 521–533.

Clark, I. E., Wyckoff, D., and Gavis, E. R. (2000). Synthesis of the posterior determinant Nanos is spatially restricted by a novel cotranslational regulatory mechanism. *Curr. Biol.* **10**, 1311–1314.

Gebauer, F., Corona, D. F., Preiss, T., Becker, P. B., and Hentze, M. W. (1999). Translational control of dosage compensation in *Drosophila* by Sex-lethal: Cooperative silencing via the 5′ and 3′ UTRs of msl-2 mRNA is independent of the poly(A) tail. *EMBO J.* **18**, 6146–6154.

Hernandez, G., Vazquez-Pianzola, P., Sierra, J. M., and Rivera-Pomar, R. (2004). Internal ribosome entry site drives cap-independent translation of reaper and heat shock protein 70 mRNAs in *Drosophila* embryos. *RNA* **10**, 1783–1797.

Jeske, M., Meyer, S., Temme, C., Freudenreich, D., and Wahle, E. (2006). Rapid ATP-dependent deadenylation of nanos mRNA in a cell-free system from *Drosophila* embryos. *J. Biol. Chem.* **281**, 25124–25133.

Lie, Y. S., and Macdonald, P. M. (2000). *In vitro* translation extracts prepared from *Drosophila* ovaries and embryos. *Biochem. Biophys. Res. Commun.* **270**, 473–481.

Maroto, F. G., and Sierra, J. M. (1988). Translational control in heat-shocked *Drosophila* embryos: Evidence for the inactivation of initiation factor(s) involved in the recognition of mRNA cap structure. *J. Biol. Chem.* **263**, 15720–15725.

Scott, M. P., Storti, R. V., Pardue, M. L., and Rich, A. (1979). Cell-free protein synthesis in lysates of *Drosophila melanogaster* cells. *Biochemistry* **18**, 1588–1594.

Stepinski, J., Waddell, C., Stolarski, R., Darzynkiewicz, E., and Rhoads, R. E. (2001). Synthesis and properties of mRNAs containing the novel "anti-reverse" cap analogs 7-methyl(3′-O-methyl)GpppG and 7-methyl (3′-deoxy)GpppG. *RNA* **7**, 1486–1495.

Storti, R. V., Scott, M. P., Rich, A., and Pardue, M. L. (1980). Translational control of protein synthesis in response to heat shock in *D. melanogaster* cells. *Cell* **22**, 825–834.

Tuschl, T., Zamore, P. D., Lehmann, R., Bartel, D. P., and Sharp, P. A. (1999). Targeted mRNA degradation by double-stranded RNA *in vitro*. *Genes Dev.* **13**, 3191–3197.

Vazquez-Pianzola, P., Hernandez, G., Suter, B., and Rivera-Pomar, R. (2006). Different modes of translation for hid, grim and sickle mRNAs in *Drosophila*. *Cell Death Differ.* **14**(2), 286–295.

Zapata, J. M., Maroto, F. G., and Sierra, J. M. (1991). Inactivation of mRNA cap-binding protein complex in *Drosophila melanogaster* embryos under heat shock. *J. Biol. Chem.* **266**, 16007–16014.

CHAPTER THREE

USE OF *IN VITRO* TRANSLATION EXTRACT DEPLETED IN SPECIFIC INITIATION FACTORS FOR THE INVESTIGATION OF TRANSLATIONAL REGULATION

Daniel R. Gallie

Contents

Abstract

Regulation of gene expression often involves the control of translation mediated through one or more initiation factors that are required for the translation of eukaryotic mRNAs. Genetic and molecular biological approaches can be highly useful in the initial identification of translational regulation, but the use of *in vitro* translation lysates can be essential in elucidating the details of translational regulatory mechanisms. Wheat germ lysate has long been used for *in vitro* translation studies. The noncompetitive conditions that prevail in this lysate as it is normally produced, however, preclude the translational regulatory analysis of many mRNAs involving the preferential recruitment of initiation factors. The development of lysate depleted in specific translation initiation factors converts wheat germ lysate from a noncompetitive system to one that is competitive in a fast and simple procedure that enables it to be used in the analysis of many more translational regulatory mechanisms than is currently possible with unfractionated lysate.

Department of Biochemistry, University of California, Riverside, California

Methods in Enzymology, Volume 429
ISSN 0076-6879, DOI: 10.1016/S0076-6879(07)29003-2

1. Introduction

Eukaryotic translation initiation differs from that in bacteria in its increased number and complexity of factors that are involved in protein synthesis. In some cases, these factors represent regulatory proteins that target specific mRNAs to promote or repress their translation. Translation in eukaryotes also requires a larger number of initiation factors to assemble a ribosome at the appropriate initiation codon. The regulatory role that initiation factors play in determining the level of expression at a genome-wide level is only now receiving attention. Elucidating the contribution that they make will be essential in understanding how the translational machinery influences the composition of the proteome in a given cell, tissue, or organ. The use of plants as a model for translation and translational regulation in higher eukaryotes has several advantages. Plants provide a ready and inexpensive source of material for analysis, mutations can be easily generated, and they possess signaling pathways and stress responses that are conserved in many instances with those of other eukaryotes. In addition, a translation lysate derived from wheat germ has long been used to study protein synthesis. The wheat embryo, from which wheat germ lysate is made, is rich in the factors required for protein synthesis and low in endogenous mRNAs. While this has the advantage of producing an active translation system, it also means that translation is carried out under conditions that are noncompetitive, i.e., an excess of translational machinery for the mRNA being translated. This can make the study of translational regulation difficult or impossible as the high level of translation factors obscures those features of an mRNA that contribute to controlling expression at the translational level. Because most regulation of translation occurs during the initiation phase of protein synthesis, depleting the lysate of specific initiation factors can convert this noncompetitive system into a competitive one in which regulatory features of an mRNA can be revealed. In this chapter, we describe the preparation of fractionated lysates depleted for specific initiation factors and their use in the study of translational regulation.

2. Factors Involved in Translation Initiation

Early in initiation, the $5'$-cap structure (m^7GpppN, where N represents any nucleotide) and the $3'$-terminal poly(A) tail cooperate to recruit those translation initiation factors critical for the early steps that lead to binding of an 40S ribosomal subunit to an mRNA (Gallie, 2002a). The $5'$-cap structure serves as the binding site for the eukaryotic initiation factor (eIF) 4F that is composed of three subunits: eIF4E, eIF4A, and eIF4G. eIF4E functions as the cap-binding subunit, eIF4A possesses RNA helicase activity required to remove secondary structure within the 5′ leader sequence that would otherwise inhibit scanning of the 40S ribosomal subunit, and eIF4G is a large subunit that binds eIF4E and eIF4A through direct protein–protein interactions. eIF4G also recruits other proteins involved in stimulating 40S ribosomal subunit binding to an mRNA such as eIF3 and the poly(A)-binding protein (PABP). The interaction between PABP and eIF4G is conserved in plants, yeast, and animals and serves to stabilize the binding of eIF4F to the 5′-cap (Wei *et al.*, 1998). In plants and animals, PABP also interacts with eIF4B, a factor that assists the activities of eIF4A and eIF4F (Bushell *et al.*, 2001; Le *et al.*, 1997, 2000). The 5′-cap and poly(A) tail, therefore, serve to recruit eIF4G to the mRNA through the proteins that bind each mRNA element, i.e., eIF4E and PABP, respectively. Two related but distinct eIF4G proteins are expressed in plants, animals, and yeast (Browning *et al.*, 1992; Goyer *et al.*, 1993; Gradi *et al.*, 1998). The two plant eIF4G proteins, referred to as eIF4G and eIFiso4G, differ in size (165 kDa and 86 kDa, respectively) and share only 30% identity.

3. Experimental Methods to Generate and Use Fractionated Translation Extracts

3.1. *In vitro* RNA synthesis

T7-based monocistronic and dicistronic luciferase constructs have been described previously (Gallie *et al.*, 1989, 1991, 2000). The polyadenylated, monocistronic and dicistronic luciferase constructs that contain the 5′-leader sequence from tobacco etch virus (TEV), the 5′-leader sequence from tobacco mosaic virus (TMV) that is referred to as Ω, or control sequences have been described previously (Gallie, 2002b; Niepel and Gallie, 1999). Following linearization downstream of the poly(A)$_{50}$ tract, the DNA concentration is quantitated spectrophotometrically and brought to 0.5 mg/ml. *In vitro* transcription is carried out for 2 h as described previously (Yisraeli and Melton, 1989) using 40 mM Tris–HCl, pH 7.5, 6 mM MgCl$_2$, 100 μg/ml bovine serum albumin (BSA), 0.5 mM each of ATP, CTP, UTP, and GTP, 10 mM dithiothreitol (DTT), 0.3 units/μl RNasin (Promega), and 0.5 units/μl T7 RNA polymerase. Capped RNAs are synthesized using 3 μg of template in the same reaction mix except GTP is used at 160 μM and 1 mM of either GpppG or m^{7}GpppG is included. Under these conditions more than 95% of the mRNA is capped.

3.2. Preparation of fractionated lysates

To generate an eIF4F/eIFiso4F or PABP-dependent lysate, 200 μl of commercial wheat germ extract (Promega) is thawed on ice. Once thawing begins, the tube is briefly hand mixed to facilitate complete thawing rapidly. To prepare eIF4F/eIFiso4F-dependent lysate, 300 μl of m^{7}GTP-Sepharose (Pharmacia) is equilibrated in 1 ml N$'$ buffer (20 mM HEPES-KOH, pH 7.6, 1 mM DTT, 0.1 mM EDTA, 10% glycerol) for 40 min, washed twice in one volume N$'$ buffer, and the supernatant is removed. Then 200 μl of wheat germ extract is added to the m^{7}GTP-Sepharose resin and incubated with rotation at 4$°$ for 15 min. The lysate is collected by centrifugation (800$\times g$ for 1 min) through a spin column (Promega) and used immediately. To prepare PABP-dependent lysate, 100 μl of poly(A)-agarose (Sigma) is equilibrated in 0.5 ml N$'$ buffer for 40 min, washed twice in one volume of N$'$ buffer, and the supernatant is removed. Then 200 μl of wheat germ extract is added to the poly(A)-agarose and incubated with rotation at 4$°$ for 15 min. The lysate is collected by centrifugation (800$\times g$ for 1 min) through a spin column (Promega) and used immediately.

The depletion of initiation factors such as eIF4G, eIF4E, eIFiso4G, eIFiso4E, eIF4A, eIF4B, eIF3, eEF2, or PABP is confirmed by Western analysis following resolution of the lysate by sodium dodecyl sulfate polyacrylamide gel electrophoresis (SDS–PAGE). Because eIF4G, eIFiso4G, and eIF4B also bind poly(A) RNA, albeit with considerably lower affinity than does PABP, and PABP is known to physically interact with eIF4G, eIFiso4G, and eIF4B (Le *et al.*, 1997, 2000), which in turn can interact with eIF4A and eIF3, the incubation of wheat germ lysate with poly(A)-agarose is effective in reducing the level of eIF4G and eIFiso4G in addition to the depletion of PABP, whereas no reduction was observed for the heat shock protein, HSP101, that was used as a control (Fig. 3.1) (Gallie, 2001; Gallie and Browning, 2001).

The translational characteristics of the eIF4F/eIFiso4F or PABP-dependent lysates are reproducible when lysates are prepared in a similar fashion. It is particularly important for reproducible results among experiments that the same volumes of lysate and m^{7}GTP-Sepharose resin [or poly(A)-agarose resin] are used each time and the incubation time of the resin with the lysate is not varied. As the activity of lysate is lost over time following thawing, it is important to perform the fractionation protocol in as short a time as possible and maintain all components on ice at all times. Translation should be performed immediately following preparation of the eIF4F/eIFiso4F or PABP-dependent lysate. For the greatest reproducibility between experiments, a batch of fractionated lysate can be prepared and immediately frozen as aliquots, each of which can be used once for translation. Multiple freeze/thaw cycles of the lysate should be avoided. Moreover, to minimize any variability in the unfractionated lysate used to prepare fractionated lysate, a

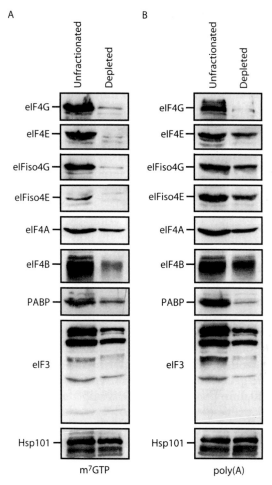

Figure 3.1 Depletion of eIF4F and eIFiso4F from wheat germ lysate. Wheat germ lysate was incubated with (A) m⁷GTP-Sepharose or (B) poly(A)-Sepharose for 30 min. Western analysis was performed to determine the level of eIF4G, eIF4E, eIFiso4G, eIFiso4E, eIF4A, eIF4B, eIF3, and PABP relative to the unfractionated lysate. Western analysis of the heat shock protein, HSP101, was performed as a control. (Reproduced with permission from Gallie, 2001.)

sufficient quantity of lysate from the same lot should be purchased (or prepared in the laboratory) prior to the initiation of the analysis.

3.3. SDS–PAGE analysis

Proteins are fractionated using thin (0.75 mm) SDS-polyacrylamide gels prepared as described (Laemmli, 1970; Sambrook *et al.*, 1989). Wearing gloves, the gel casting parts are cleaned with 95% ethanol and assembled.

Supplementation of the lysates with native or recombinant initiation factors is done prior to the addition of the mRNA to be translated. The purification of recombinant initiation factors or factors purified from wheat germ extract has been described for PABP (Le *et al.*, 1997), eIF4F and eIFiso4F (Browning *et al.*, 1992), eIF4B (Browning *et al.*, 1987), eIF4A (Lax *et al.*, 1986), and recombinant eIFiso4G and eIFiso4E (van Heerden and Browning, 1994). These can be added singly or in any combination to analyze the contribution of each to the translational regulation.

3.5.1. Luciferase assay

Two-microliter aliquots of the wheat germ translation lysate are removed from each reaction and added to 100 μl luciferase assay buffer (25 mM tricine, pH 8, 5 mM MgCl$_2$, 0.1 mM EDTA, supplemented with 33.3 mM DTT, 270 μM coenzyme A, and 500 μM ATP) at room temperature and mixed briefly. Luciferase activity is measured following injection of 100 μl of 0.5 mM luciferin using a Monolight 2010 Luminometer (Analytical Luminescence Laboratory). Each translation reaction is assayed in duplicate and the average and standard deviation for the triplicate translation reactions are determined. It is important that the volume of the luciferase assay be constant for all samples as luciferase uses oxygen in the luciferase reaction and the use of equal volumes for all samples eliminates any differences in the surface-to-volume ratio that would otherwise affect the rate of oxygen diffusion into the reaction. If it is desirable to perform a time course of translation to determine the rate of translation, 2-μl aliquots of lysate can be removed at specific time points during the reaction and assayed for luciferase activity. The slope of the curve representing the increase in luciferase activity over time can be determined and this serves as a measure of the rate of translation.

3.5.2. Examples

Analysis of the cap dependency of translation of an mRNA   To examine the extent to which a cap stimulates translation from a given mRNA, eIF4F-dependent, eIFiso4F-dependent lysate, generated by depleting wheat germ lysate of eIF4F (composed of eIF4G and eIF4E) and eIFiso4F (composed of eIFiso4G and eIFiso4E) through their binding to m^{7}GTP-Sepharose resin, can be used to translate the mRNA in its capped and uncapped forms. Reduction of the level of eIF4E, eIF4G, eIFiso4E, and eIFiso4G up to 90 to 95% can be achieved (see Fig. 3.1). This reduces the amount of these factors to a point that competitive conditions are created, but allows a low level of the factors to remain to permit the 5′-cap to recruit the factors to an mRNA. The eIF4F or eIFiso4F dependency of the fractionated lysate can be measured by translating capped-*luc*-A$_{50}$ mRNA in the fractionated lysate supplemented with increasing amounts of purified eIF4F or eIFiso4F. The extent to which the reporter mRNA is translated is determined by measuring luciferase activity. Other reporters can be assayed accordingly or the amount of protein produced determined using SDS–PAGE/fluorography for radiolabeled proteins or Western analysis if antiserum to the protein is available. Reduction in the level of eIF4F and eIFiso4F can reduce translation more than 95% (Gallie and Browning, 2001). Residual translational activity of the fractionated lysate may be the result of the low level of either eIF4G and eIFiso4G remaining in the lysate (see Fig. 3.1). Supplementation with eIF4F (up to 16 nM, the highest concentration tested) increased reporter mRNA translation nearly 10-fold in the fractionated lysate, but did not affect translation of the same mRNA in unfractionated lysate (Gallie and Browning, 2001). Supplementation with eIFiso4F also increased reporter mRNA translation in the fractionated lysate but not in the unfractionated lysate.

Wheat germ lysate is highly message dependent because of a low concentration of endogenous transcripts and the high level of unengaged translational machinery. As a consequence, those features that increase the competitiveness of an mRNA, such as a 5'-cap structure, would not be expected to provide a translational advantage under the noncompetitive conditions that prevail in normal lysate and would only do so in fractionated lysate where translation is competitive. Each preparation of eIF4F-dependent, eIFiso4F-dependent lysate (or the PABP-dependent lysate) is programmed with capped or uncapped mRNAs to determine the degree to which translation is cap dependent. The presence of the cap has been shown to increase translation 3-fold in the eIF4F-dependent, eIFiso4F-dependent lysate and 10-fold in the PABP-dependent lysate, but has little effect in normal lysate (Gallie and Browning, 2001). Supplementation of the PABP-dependent lysate (which was also reduced in the level of eIF4F and eIFiso4F, see Fig. 3.1) with increasing amounts of eIF4F or eIFiso4F reduced the cap dependency of translation, indicating that the depletion of PABP in combination with the partial reduction of eIF4F and eIFiso4F increases cap-dependent translation to a greater extent than does a reduction in eIF4F and eIFiso4F alone. Interestingly, the basal level of translational activity in the PABP-dependent lysate was substantially lower than that observed in the eIF4F/eIFiso4F-dependent or complete lysates (Gallie and Browning, 2001), suggesting that in addition to eIF4F and eIFiso4F, PABP contributes to the overall translational activity of the lysate.

Just as a 5'-cap provides a translational advantage in fractionated lysates, those features that reduce the competitiveness of an mRNA, such as moderate secondary structure within the 5'-leader, would be expected to be revealed in the fractionated lysate where the lower level of initiation factors required for removing secondary structure would reduce the ability to melt moderate secondary structure that impairs translation. Translation of *luc* reporter mRNAs containing a stable stem–loop structure in the 5'-leader in eIF4F/eIFiso4F-dependent revealed that translation is sensitive to the presence of even moderate secondary structure (e.g., −4.5 kcal/mol) at an

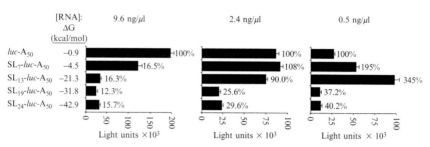

Figure 3.2 5′-Proximal secondary structure is inhibitory to translation in eIF4F/eIFiso4-F-reduced lysate. A stem–loop (SL) with a 7, 13, 19, or 24 base pair stem was introduced 4 nt downstream of the 5′-terminus of luc-A$_{50}$ mRNA to result in SL$_7$-luc-A$_{50}$, SL$_{13}$-luc-A$_{50}$, SL$_{19}$-luc-A$_{50}$, and SL$_{24}$-luc-A$_{50}$, respectively. The free energy (ΔG) of the control leader (i.e., present in luc-A$_{50}$) and each stem–loop construct is indicated. The luc mRNA constructs were synthesized $in\ vitro$ as capped, polyadenylated mRNAs and translated in eIF4F/eIFiso4F-dependent wheat germ lysate at three concentrations: 9.6 ng/μl (left), 2.4 ng/μl (middle), and 0.5 ng/μl (right). Each mRNA construct was translated in triplicate and the average value and standard deviation of the absolute level of expression from each construct are reported as a histogram. Luciferase expression is also indicated as a percentage (indicated to the right of each histogram) of the SL$_{24}$-luc-A$_{50}$ mRNA. (Reproduced with permission from Gallie and Browning, 2001.)

RNA concentration of 9.6 ng/μl (Fig. 3.2) (Gallie and Browning, 2001). A substantially more stable secondary structure in the 5′-leader was required to inhibit translation in an unfractionated lysate (Gallie and Browning, 2001). Moderately stable secondary structures inhibited translation increasingly in the fractionated lysate as the concentration of the input mRNA increased (see Fig. 3.2), indicating that higher concentrations of mRNA act to titrate RNA helicase activity (e.g., from eIF4A and eIF4F). When a low concentration of RNA (i.e., 0.5 ng/μl) was used in the fractionated lysate, expression from mRNAs containing moderate secondary structure was actually higher than an mRNA containing little secondary structure in its 5′-leader (see Fig. 3.2). This may have been a result of the sequence introduced to generate the secondary structure increasing the length of the 5′-leader, a factor known to increase translation (Gallie and Walbot, 1992). However, the advantage conferred by the additional sequence at a low RNA concentration was lost at higher RNA concentrations and the inhibitory effect of the moderate secondary structure was revealed (see Fig. 3.2). The lack of an inhibitory effect from the presence of moderate secondary structure at a low RNA concentration in a fractionated lysate may be a result of limited RNA helicase activity remaining in the lysate. Increasing the RNA concentration in the fractionated lysate would increase the demand for RNA helicase activity. For those mRNAs containing moderate secondary structure, the advantage of the increased length of the 5′-leader conferred at a low concentration of RNA is lost at higher RNA concentrations when the secondary structure can no longer be removed by the limited amount of RNA helicase activity, resulting in translation inhibition. These data illustrate that the translation of mRNAs in a fractionated lysate at the appropriate RNA concentration can reveal the effect of even moderate secondary structure on protein synthesis. Therefore, the extent to which any mRNA leader sequence may regulate translation through the presence of secondary structure can be measured with eIF4F/eIFiso4F-dependent or PABP-dependent lysates.

Translation of *luc* reporter mRNAs with an unstructured leader or with a moderately stable secondary structure in eIF4F/eIFiso4F-dependent lysate that was supplemented with either eIF4F or eIFiso4F revealed that eIF4F increased translation from structured mRNA whereas eIFiso4F did not (Gallie and Browning, 2001). These results show how fractionated lysate can be used to reveal functional differences between the eIF4F isoforms. Other depleted initiation factors, such as eIF4A or eIF4B, can be added to the fractionated lysates to examine their impact on translation under these competitive conditions. The fractionated lysates can also be used to examine the effect of specific regulatory proteins that are not standard initiation factors to examine their regulatory function during competitive translation.

Analysis of cap-dependent viral translational enhancer The 68 nt 5'-leader (called Ω) of tobacco mosaic virus (TMV), a single-strand, positive-sense RNA virus, functions as a translational enhancer (Gallie and Walbot, 1992; Gallie et al., 1987, 1988). Ω promotes release of the genomic mRNA from the virion particle while enhancing translation of the 5'-proximal cistron encoding the replicase through a cotranslation disassembly process. Host cell 40S ribosomal subunits are recruited by Ω to which coat protein only loosely binds (Mundry et al., 1991). During translation elongation, ribosomes synthesize replicase protein from the 5'-cistron and simultaneously strip the coat protein from the viral RNA. Although the genomic RNA is capped, it is an unusual mRNA in that it does not terminate with a poly(A) tail but instead contains a 204 nt 3'-untranslated region (3'-UTR). Although the TMV 3'-UTR also functions as a translation enhancer (Gallie and Walbot, 1990; Leathers et al., 1993), the 3'-end of the virion particle does not undergo disassembly until after replicase protein is synthesized (Wu et al., 1994), suggesting that the 3' translational enhancer cannot participate in the first round of translation. Ω enhances the translation of reporter mRNAs in the absence of other viral sequences or viral proteins, including the TMV 3'-UTR. Thus, Ω facilitates ribosome recruitment to a virion particle in the absence of the participation of the 3' translational enhancer, enabling it to promote translation initiation of encapsidated RNA without assistance from any 3'-terminal regulatory element (Gallie and Walbot, 1990, 1992; Gallie et al., 1987, 1988; Leathers et al., 1993). The heat shock protein, HSP101, binds to the translational enhancer within Ω and is sufficient to mediate the translational enhancement associated with Ω (Wells et al., 1998).

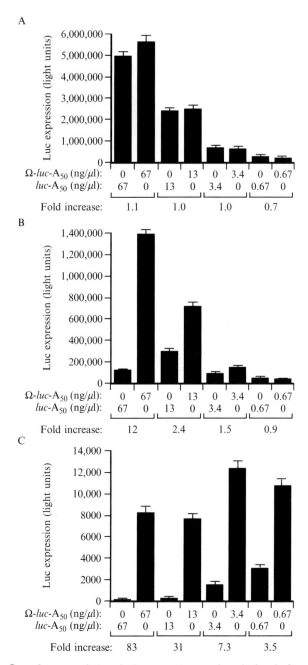

Figure 3.3 Ω confers a translational advantage *in vitro* when the level of eIF4F, eIFiso4F, or PABP is limiting. Unfractionated (A), eIF4F/eIFiso4F-dependent (B), or PABP-dependent (C) wheat germ lysate was programmed with capped Ω-*luc*-A$_{50}$ or *luc*-A$_{50}$

Genetic analysis suggested that the translational activity of HSP101 requires eIF4G and eIF3 (Wells *et al.*, 1998).

Because of the excess of unengaged translational machinery and low level of endogenous transcripts in wheat germ lysate, Ω conferred little translational advantage at any RNA concentration in the unfractionated lysate (Fig. 3.3A) (Gallie, 2002b), making a detailed analysis of this translational enhancer difficult to achieve *in vitro*. Therefore, fractionated wheat germ lysate was used to determine whether Ω would provide a translational advantage under the competitive conditions that prevail following the removal of the excess translational machinery. Capped-*luc*-A_{50} and Ω-*luc*-A_{50} mRNAs were translated at different RNA concentrations in eIF4F/eIFiso4F-dependent lysate. At low mRNA concentrations, Ω failed to stimulate translation substantially as translation remained noncompetitive when a low mRNA concentration was used (Fig. 3.3B). However, Ω stimulated translation up to 12-fold higher when the mRNA concentration was increased (see Fig. 3.3B), demonstrating that the stimulatory effect of this well-characterized translational enhancer could be revealed best when fractionated wheat germ lysate was used.

Lysate depleted of PABP by incubating lysate with poly(A)-agarose was also used in the analysis of function of Ω in promoting translation. Because PABP interacts with eIF4G and eIFiso4G (Le *et al.*, 1997, 2000), the incubation of lysate with poly(A)-agarose is effective in reducing the level of eIF4G and eIFiso4G in addition to the depletion of PABP, whereas no reduction was observed for HSP101 or other components of the translational machinery (see Fig. 3.1) (Gallie, 2001; Gallie and Browning, 2001). In the PABP-dependent lysate, Ω conferred a translational advantage at all RNA concentrations (Fig. 3.3C) (Gallie, 2002b). The stimulatory effect of Ω increased with RNA concentration, demonstrating that the analysis of translational regulatory elements that confer a competitive advantage to an mRNA is best revealed in fractionated lysates at a range of RNA concentration that achieves a competitive condition.

Fractionated wheat germ lysate could also be used to examine the initiation factor requirements for the function of Ω. The ability of Ω to improve the translation of an mRNA in PABP-dependent or eIF4F/eIFiso4F-dependent lysate suggests recruitment of a limited factor required for translation initiation. Restoring the factor in question through its supplementation to

---

mRNAs at the concentration indicated below the histograms. The degree to which each mRNA was translated was determined by luciferase assays. Luciferase activity is indicated as the average (from 2 μl of lysate) of three translation reactions with the standard deviation for each construct shown. The degree to which the presence of Ω increased translation relative to the control (i.e., fold increase) is indicated below each pair of mRNAs for each concentration tested. (Reproduced with permission from Gallie, 2002b.)

the lysate would be expected to reduce the translational advantage conferred by Ω. By determining whether a factor can reduce the translational advantage conferred by an RNA element, the requirement for that factor can be established. As genetic analysis had suggested that eIF4G was required for the HSP101-mediated function of Ω (Wells *et al.*, 1998), the requirement of eIF4F or eIFiso4F could be examined using fractionated lysates. Capped-*luc*-A_{50} and Ω-*luc*-A_{50} mRNAs were translated in PABP-dependent lysate (which was reduced in eIF4F/eIFiso4F) at an RNA concentration that provided a substantial degree of translational enhancement. The lysate was supplemented with either eIF4F or eIFiso4F and their effect on translation and the translational advantage conferred by Ω were determined. eIF4F and eIFiso4F purified from wheat do not contain eIF4A, therefore, the purified eIF4F and eIFiso4F were supplemented with eIF4A. The translational advantage conferred by Ω was reduced following supplementation with eIF4F but not with eIFiso4F (Gallie, 2002b), suggesting that Ω functions by recruiting eIF4F when the factor is present in limiting amounts, but that the translational advantage conferred by Ω is lost when the concentration of eIF4F is no longer limiting.

Analysis of cap-independent translation conferred by a viral 5′-leader   The 5′-leader of tobacco etch virus (TEV), a potyvirus whose genomic mRNA is polyadenylated but naturally lacks a 5′ cap structure, confers cap-independent translation to an mRNA and exhibits internal ribosome entry site (IRES) activity when present in the intercistronic region of a dicistronic mRNA (Carrington and Freed, 1990; Gallie, 2001; Gallie *et al.*, 1995; Niepel and Gallie, 1999). The TEV 5′-leader stimulated cap-independent translation up to 73-fold in eIF4F/eIFiso4F-dependent lysate, whereas it had little to no effect in unfractionated lysate (Gallie, 2001). Similar results were obtained for IRES activity (Fig. 3.4), indicating that these mechanisms do function *in vitro* and that the TEV 5′-leader confers a translational advantage only under conditions of competitive translation. These results demonstrate that the cap-independent translation and IRES activity conferred by the TEV 5′-leader that are not observed in normal lysate can be easily revealed when fractionated lysate is employed.

   The translational advantage conferred by the TEV IRES under these conditions was lost when the fractionated lysate was supplemented with eIF4F (or to a lesser extent, eIFiso4F), but not when supplemented with eIF4E, eIFiso4E, eIF4A, or eIF4B (Gallie, 2001). Supplementation of the lysate with eIF4G, the large subunit of eIF4F, specifically reduced the competitive advantage conferred by the TEV IRES, demonstrating that this subunit of eIF4F was required. Addition of either eIF4A or eIF4B reduced the translational advantage conferred by the TEV IRES only to a small extent, but substantially improved the ability of eIF4F to reduce the translational advantage conferred by the TEV IRES (Gallie, 2001). Thus, the

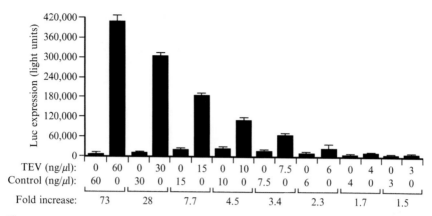

Figure 3.4 The TEV IRES directs internal initiation *in vitro* under competitive conditions. eIF4F/eIFiso4F-reduced wheat germ lysate was programmed with uncapped control (i.e., GUS-SL-Con$_{144}$-*luc*-A$_{50}$) or uncapped TEV IRES-containing (i.e., GUS-SL-TEV-*luc*-A$_{50}$) dicistronic constructs at the concentration indicated below the histograms. Luciferase activity is reported as the average (from 2 μl of lysate) of three translation reactions with the standard deviation for each construct shown. The degree to which the presence of the TEV 5′-leader increased translation relative to the control (i.e., fold increase) is indicated below each pair of mRNAs for each concentration tested. (Reproduced with permission from Gallie, 2001.)

combinatorial supplementation of fractionated lysate can be used to examine the functional interaction between initiation factors during translation.

Because PABP interacts with eIF4G and eIFiso4G (Le *et al.*, 1997, 2000), PABP-dependent lysate was used to determine whether PABP affected the extent to which the TEV 5′-leader sequence functions to stimulate cap-independent translation. Depletion of PABP from the lysate following incubation with poly(A)-agarose was confirmed by Western, which also revealed some reduction in eIF4G, eIFiso4G, eIF4A, eIF3, and eIF4B (see Fig. 3.1). To examine the activity of the TEV 5′-leader in the PABP-reduced lysate, monocistronic TEV-*luc*-A$_{50}$ mRNA and a control mRNA that contained a leader of similar length but unrelated in sequence were translated over a range of RNA concentrations. The TEV 5′-leader stimulated translation up to 95-fold when the lysate was programmed with a high level of RNA (Gallie, 2001). As was observed in the eIF4F/eIFiso4F-reduced lysate, the degree to which the TEV 5′-leader stimulated cap-independent translation increased with an increase in RNA concentration. Similar results were obtained when the TEV 5′-leader was tested as part of the intercistronic region of a dicistronic mRNA construct. In the PABP-reduced lysate, the TEV IRES increased internal initiation up to 79-fold relative to the control mRNA. These data demonstrate that PABP-dependent lysate, like eIF4F/eIFiso4F-dependent lysate, can be used to reveal and investigate IRES activity and the requirement for specific *trans*-acting factors.

REFERENCES

Browning, K. S., Maia, D. M., Lax, S. R., and Ravel, J. M. (1987). Identification of a new protein synthesis initiation factor from wheat germ. *J. Biol. Chem.* **262,** 538–541.

Browning, K. S., Webster, C., Roberts, J. K., and Ravel, J. M. (1992). Identification of an isozyme form of protein synthesis initiation factor 4F in plants. *J. Biol. Chem.* **267,** 10096–10100.

Bushell, M., Wood, W., Clemens, M. J., and Morley, S. J. (2001). Disruption of the interaction of mammalian protein synthesis eukaryotic initiation factor 4B with the poly(A)-binding protein by caspase- and viral protease-mediated cleavages. *Eur. J. Biochem.* **267,** 1083–1091.

Carrington, J. C., and Freed, D. D. (1990). Cap-independent enhancement of translation by a plant potyvirus 5′ nontranslated region. *J. Virol.* **64,** 1590–1597.

Gallie, D. R. (2001). Cap-independent translation conferred by the 5′-leader of tobacco etch virus is eIF4G-dependent. *J. Virol.* **75,** 12141–12152.

Gallie, D. R. (2002a). Protein-protein interactions required during translation. *Plant Mol. Biol.* **50,** 949–970.

Gallie, D. R. (2002b). The 5′-leader of tobacco mosaic virus promotes translation through enhanced recruitment of eIF4F. *Nucl. Acids Res.* **30,** 3401–3411.

Gallie, D. R., and Browning, K. S. (2001). eIF4G functionally differs from eIFiso4G in promoting internal initiation, cap-independent translation, and translation of structured mRNAs. *J. Biol. Chem.* **276,** 36951–36960.

Gallie, D. R., Feder, J. N., Schimke, R. T., and Walbot, V. (1991). Post-transcriptional regulation in higher eukaryotes: The role of the reporter gene in controlling expression. *Mol. Gen. Genet.* **228,** 258–264.

Gallie, D. R., Lucas, W. J., and Walbot, V. (1989). Visualizing mRNA expression in plant protoplasts: Factors influencing efficient mRNA uptake and translation. *Plant Cell* **1,** 301–311.

Gallie, D. R., and Walbot, V. (1990). RNA pseudoknot domain of tobacco mosaic virus can functionally substitute for a poly(A) tail in plant and animal cells. *Genes Dev.* **4,** 1149–1157.

Gallie, D. R., and Walbot, V. (1992). Identification of the motifs within the tobacco mosaic virus 5′ leader responsible for enhancing translation. *Nucl. Acids Res.* **20,** 4631–4638.

Gallie, D. R., Ling, J., Niepel, M., Morley, S. J., and Pain, V. M. (2000). The role of 5′-leader length, secondary structure and PABP concentration on cap and poly(A) tail function during translation in *Xenopus* oocytes. *Nucl. Acids Res.* **28,** 2943–2953.

Gallie, D. R., Sleat, D. E., Watts, J. W., Turner, P. C., and Wilson, T. M. A. (1987). The 5′-leader sequence of tobacco mosaic virus RNA enhances the expression of foreign gene transcripts *in vitro* and *in vivo. Nucl. Acids Res.* **15,** 3257–3273.

Gallie, D. R., Sleat, D. E., Watts, J. W., Turner, P. C., and Wilson, T. M. A. (1988). Mutational analysis of the tobacco mosaic virus 5′-leader for altered ability to enhance translation. *Nucl. Acids Res.* **16,** 883–893.

Gallie, D. R., Tanguay, R. L., and Leathers, V. (1995). The tobacco etch viral 5′ leader and poly(A) tail are synergistic regulators of translation. *Gene* **165,** 233–238.

Goyer, C., Altmann, M., Lee, H. S., Blanc, A., Deshmukh, M., Woolford, J. L., Trachsel, H., and Sonenberg, N. (1993). Tif4631 and Tif4632—Two yeast genes encoding the high-molecular-weight subunits of the cap-binding protein complex (eukaryotic initiation factor-4F) contain an RNA recognition motif-like sequence and carry out an essential function. *Mol. Cell. Biol.* **13,** 4860–4874.

Gradi, A., Imataka, H., Svitkin, Y. V., Rom, E., Raught, B., Morino, S., and Sonenberg, N. (1998). A novel functional human eukaryotic translation initiation factor 4G. *Mol. Cell. Biol.* **18,** 334–342.

Laemmli, U. K. (1970). Cleavage of structural proteins during the assembly of the head of bacteriophage T4. *Nature* **227**, 680–685.

Lax, S. R., Lauer, S. J., Browning, K. S., and Ravel, J. M. (1986). Purification and properties of protein synthesis initiation and elongation factors from wheat germ. *Methods Enzymol.* **118**, 109–128.

Le, H., Tanguay, R. L., Balasta, M. L., Wei, C.-C., Browning, K. S., Metz, A. M., Goss, D. J., and Gallie, D. R. (1997). Translation initiation factors eIF-iso4G and eIF-4B interact with the poly(A)-binding protein and increase its RNA binding activity. *J. Biol. Chem.* **272**, 16247–16255.

Le, H., Browning, K. S., and Gallie, D. R. (2000). The phosphorylation state of poly(A)-binding protein specifies its binding to poly(A) RNA and its interaction with eukaryotic initiation factor (eIF) 4F, eIFiso4F, and eIF4B. *J. Biol. Chem.* **275**, 17452–17462.

Leathers, V., Tanguay, R., Kobayashi, M., and Gallie, D. R. (1993). A phylogenetically conserved sequence within viral 3′ untranslated RNA pseudoknots regulates translation. *Mol. Cell. Biol.* **13**, 5331–5347.

Mundry, K. W., Watkins, P. A., Ashfield, T., Plaskitt, K. A., Eisele-Walter, S., and Wilson, T. M. (1991). Complete uncoating of the 5′ leader sequence of tobacco mosaic virus RNA occurs rapidly and is required to initiate cotranslational virus disassembly *in vitro. J. Gen. Virol.* **72**, 769–777.

Niepel, M., and Gallie, D. R. (1999). Identification and characterization of the functional elements within the tobacco etch viral 5′-leader required for cap-independent translation. *J. Virol.* **73**, 9080–9088.

Sambrook, J., Fritsch, E. F., and Maniatis, T. (1989). "Molecular Cloning." Cold Spring Harbor Press, Cold Spring Harbor, NY.

van Heerden, A., and Browning, K. S. (1994). Expression in *Escherichia coli* of the two subunits of the isozyme form of wheat germ protein synthesis initiation factor 4F. Purification of the subunits and formation of an enzymatically active complex. *J. Biol. Chem.* **269**, 17454–17457.

Wei, C-C., Balasta, M. L., Ren, J., and Goss, D. J. (1998). Wheat germ poly(A) binding protein enhances the binding affinity of eukaryotic initiation factor 4F and (iso)4F for cap analogues. *Biochemistry* **37**, 1910–1916.

Wells, D. R., Tanguay, R. L., Le, H., and Gallie, D. R. (1998). HSP101 functions as a specific translational regulatory protein whose activity is regulated by nutrient status. *Genes Devel.* **12**, 3236–3251.

Wu, X., Xu, Z., and Shaw, J. G. (1994). Uncoating of tobacco mosaic virus RNA in protoplasts. *Virology* **200**, 256–262.

Yisraeli, J. K., and Melton, D. A. (1989). Synthesis of long, capped transcripts *in vitro* by SP6 and T7 RNA polymerases. *Methods Enzymol.* **180**, 42–50.

CHAPTER FOUR

A Highly Efficient and Robust *In Vitro* Translation System for Expression of Picornavirus and Hepatitis C Virus RNA Genomes

Yuri V. Svitkin* *and* Nahum Sonenberg[†]

Contents

* Department of Biochemistry, McGill University, Montreal, Quebec, Canada
[†] Department of Biochemistry and McGill Cancer Center, McGill University, Montreal, Quebec, Canada

Methods in Enzymology, Volume 429
ISSN 0076-6879, DOI: 10.1016/S0076-6879(07)29004-4

Abstract

A Krebs-2 cell-free extract that efficiently translates encephalomyocarditis virus (EMCV) RNA and extensively processes the viral polyprotein is also capable of supporting complete infectious EMCV replication. The system displays high RNA synthesis activity and *de novo* synthesis of virus up to titers of 2×10^7 to 6×10^7 plaque-forming units (pfu)/ml. The preparation of Krebs-2 cell extract and methods of analysis of EMCV-specific processes *in vitro* are described. We also demonstrate that the Krebs-2 cell-free system translates the entire open reading frame of the hepatitis C virus (HCV) RNA and properly processes the viral polyprotein when supplemented with canine microsomal membranes. In addition to processing, other posttranslational modifications of HCV proteins take place *in vitro*, such as the N-terminal glycosylation of the E1 and the E2 precursor (E2-p7) and phosphorylation of NS5A. The HCV RNA-programmed Krebs-2 cell-free extract should prove very useful as a novel screen for drugs that inhibit NS3-mediated processing. The use of this system should help fill the gap in understanding the regulation of synthesis and maturation of HCV proteins. With further optimization of cell-free conditions, the entire reconstitution of infectious HCV synthesis *in vitro* might become feasible.

1. INTRODUCTION

The translation of the genomes of positive strand RNA viruses in cell-free systems is an important means for elucidation of the mechanisms of virus gene expression. Initial reports demonstrating that encephalomyocarditis virus (EMCV) RNA can stimulate amino acid incorporation in an extract of mammalian cells date back to the late 1960s (Aviv *et al.*, 1971; Kerr *et al.*, 1966; Mathews and Korner, 1970). The translation of EMCV RNA in these systems was not complete and did not yield mature viral proteins. A drastic improvement in translation efficiency of mRNAs *in vitro* was achieved by employing extracts in which endogenous incorporation had been decreased by digestion of cellular mRNAs with micrococcal nuclease rather than by preincubation at 37° to achieve ribosomal runoff (Pelham and Jackson, 1976). Nuclease-treated Krebs-2 cell extract (hereafter referred to as Krebs-2 S10 extract) and rabbit reticulocyte lysate (RRL) translate EMCV RNA efficiently and accurately with the formation of almost all virus-specific proteins (Pelham, 1978; Svitkin and Agol, 1978). Analyses of EMCV polyprotein processing *in vitro* by different techniques made it possible to delineate the organization of the picornavirus genome and identify viral proteases responsible for cleavages (Gorbalenya *et al.*, 1979; Jackson, 1986; Palmenberg *et al.*, 1979; Svitkin *et al.*, 1979). Efficient translation of poliovirus (PV) RNA was achieved in nuclease-treated HeLa cell extracts (Molla *et al.*, 1991). Moreover, conditions were found under which PV translation, RNA replication, and RNA encapsidation occur in succession in the same test tube to produce infectious virus particles (Barton *et al.*, 1996; Molla *et al.*, 1991).

We found that programming Krebs-2 S10 extracts with EMCV RNA also yields infectious virus (Svitkin and Sonenberg, 2003). This chapter describes the preparation of Krebs-2 S10 extracts and the composition of the reaction mixture that was optimized for coupled translation–replication of EMCV RNA and generation of infectious virus. In addition, we describe a Krebs-2 cell extract-based system for translation of hepatitis C virus (HCV) RNA (Svitkin *et al.*, 2005b). This system has made it possible for the first time to translate the entire open reading frame of HCV *in vitro* and to reconstitute processing and other posttranslational modifications of HCV proteins.

2. Cell-Free Model for EMCV Replication

2.1. Picornavirus replication: An overview

The *Picornaviridae* RNA viruses include PV, human rhinovirus, foot-and-mouth disease virus, and EMCV. The genome of picornaviruses consists of a 7- to 9-kb-long positive-strand RNA with a small viral protein (VPg) and a poly(A) tail present at the 5′ and 3′ terminus, respectively (Agol, 2002). Within cells, the viral RNA sequentially directs virus-specific translation and negative strand RNA synthesis. The minus strand then serves as a template for the synthesis of new plus-strand RNA molecules. At a late stage of infection, the amplified copies of the plus-strand RNA are incorporated into viral capsid intermediate structures to produce virions.

Picornaviruses utilize a cap-independent internal ribosome entry site (IRES)-mediated mechanism for ribosome binding (Jang *et al.*, 1988; Pelletier and Sonenberg, 1988). Translation proceeds from a single initiation site, and the expression of multiple virus genes occurs via processing of the viral polyprotein. This consists of a series of cotranslational and posttranslational cleavages, which in the case of EMCV are accomplished by the viral protease 3C[pro] and its precursor 3ABC[pro] (Gorbalenya *et al.*, 1979; Jackson, 1986; Palmenberg *et al.*, 1979). RNA replication involves negative and positive

RNA synthesis and requires functions of several viral proteins (including the RNA polymerase 3D^{pol} and the genome-linked protein VPg) and cellular factors (Paul, 2002).

2.2. Experimental results

We developed a system for complete replication of EMCV *in vitro* by using an extract derived from Krebs-2 cells (Svitkin and Sonenberg, 2003). The optimization of this system involved studying the effects of many variables (such as salt, EMCV RNA and other component concentrations, temperature, and time of incubation) on viral translation, RNA replication, and virus yields.

In the absence of exogenous mRNA, the nuclease-treated Krebs-2 S10 extract exhibits a very low level of incorporation of [^{35}S]methionine into proteins. When EMCV RNA is added, incorporation of [^{35}S]methionine is robust, and after appropriate incubation, translation yields all the known virus proteins (Fig. 4.1A). The kinetics of the appearance of P3, 3CD, and 3D (the C-terminal portion of the polyprotein) is consistent with the complete translation of the viral genome over 0.5 h of incubation time. (Based on this value, the average elongation rate of the EMCV polyprotein can be estimated to be not less than 1.2 amino acids per second per ribosome. Because the time of the appearance of these polypeptides also includes the time periods required for the initiation of translation and for processing of the respective precursor polypeptides, the real rate of elongation is likely to be higher.) Subsequent incubation results in extensive cleavages of the L-P1–2A and P3 polypeptides into mature viral proteins by the viral proteases *de novo* (3C^{pro}/3ABC^{pro}).

In addition to the virus-specific proteases, the translation of EMCV RNA yields an active RNA polymerase and other nonstructural viral proteins that are required for RNA replication. This could be demonstrated by pulse labeling of *in vitro* translation–RNA replication reactions with [α-^{32}P]CTP and analysis of the newly synthesized RNAs by electrophoresis (Barton *et al.*, 1996). The major RNA product synthesized between 4 and 6 h of incubation comigrates with the full-length EMCV RNA in an agarose gel, although some slowly migrating RNA species, a putative double-stranded replicative-form RNA (Barton *et al.*, 1995), is also apparent (Fig. 4.1B). The preferential synthesis of plus-strand RNA is consistent with the known asymmetry of picornaviral replication *in vivo* (Giachetti and Semler, 1991; Novak and Kirkegaard, 1991). The time-course analysis of plus-strand RNA synthesis reveals the maximum rate of RNA replication at ~4 h of incubation (Fig. 4.1B).

A small amount of the capsid protein 1B was evident in an EMCV RNA translation–replication reaction mixture after a 4-h incubation period, indicative of the final cleavage of 1AB, which occurs at a late stage of the maturation of virions (see Fig. 4.1A). At this time, generation of infectious

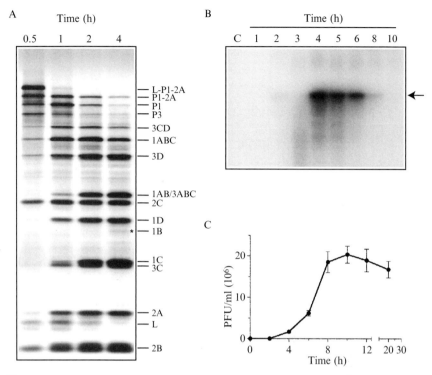

Figure 4.1 EMCV translation, RNA replication, and virus synthesis in EMCV RNA-programmed Krebs-2 S10 extract. (A) Time course of synthesis and processing of EMCV-specific proteins. EMCV RNA (20 μg/ml) was translated in the presence of [^{35}S] methionine under standard reaction conditions. At the times indicated, 5-μl aliquots of the reaction mixture were withdrawn and analyzed by sodium dodecyl sulfate (SDS)–15% polyacrylamide gel electrophoresis (PAGE). An autoradiogram of the dried gel is shown. The assignment of polypeptides is based on their comparison with proteins synthesized in EMCV-infected BHK-21 cells (Svitkin *et al.*, 1998). An asterisk indicates the position of 1B. (B) Kinetics of EMCV RNA replication. EMCV RNA (10 μg/ml)-programmed reaction mixtures were pulse labeled with [α-^{32}P]CTP for 1 h (the label was added 30 min before the times indicated on the figure). Products of RNA synthesis were isolated and analyzed by native 1% agarose gel electrophoresis and autoradiography. The arrow indicates the position of single-stranded EMCV RNA. (C) Time course of the infectivity titer of *in vitro*-synthesized EMCV. Reactions were programmed with EMCV RNA (10 μg/ml) at 32° for the indicated periods of time, treated with RNase A/T1, and assayed for infectivity following serial dilutions. The data are averages (with standard deviations from the means) of three independent titer determinations. (Reprinted from Svitkin and Sonenberg, 2003, with permission from the American Society for Microbiology.)

virus particles is first detected (Fig. 4.1C). During subsequent incubation, the virus titer is raised exponentially, typically to 2×10^7 plaque-forming units (pfu)/ml. For unknown reasons, optimal RNA and virus syntheses occurred at significantly lower concentrations of input EMCV RNA than translation (i.e., 10 μg/ml compared to > 30 μg/ml) (Svitkin and Sonenberg, 2003). In general, the kinetics of viral functions depicted in Fig. 4.1 is consistent with the sequential occurrence of viral translation, RNA replication, and virus assembly *in vitro*.

3. Materials for Cell-Free Synthesis of EMCV

3.1. Animals, cells, virus, and viral RNA

1. Female white mice (BALB/c, 6 to 8 weeks old).
2. Krebs-2 ascites carcinoma cells. These cells are available from several laboratories (Aviv *et al.*, 1971; Bordeleau *et al.*, 2006; Svitkin and Agol, 1978; Svitkin and Sonenberg, 2003; Villa-Komaroff *et al.*, 1974). The original cell line is kept at the Imperial Cancer Research Institute (Mill Hill, London).
3. EMCV, K2 strain (Burness, 1969). The preparation and purification of EMCV have been described (Kerr and Martin, 1972; Martin *et al.*, 1961; Svitkin *et al.*, 1998).
4. EMCV RNA, prepared by phenol/chloroform/isoamyl alcohol extraction of the purified EMCV. The viral RNA is purified by a 2 M LiCl precipitation and CHROMA SPIN-1000 column chromatography as recommended by the manufacturer (BD Biosciences, Mississauga, ON, Canada). RNA integrity is confirmed by agarose gel electrophoresis under denaturing conditions (Sambrook *et al.*, 1989).

3.2. Media and solutions

All solutions used for cell-free translation experiments should be prepared using analytical grade reagents and glass-distilled deionized water.

1. Earle's balanced salt solution (EBSS) without calcium and magnesium (Invitrogen Corp., Carlsbad, CA).
2. Dimethyl sulfoxide (DMSO).
3. Fetal bovine serum, dialyzed (Invitrogen).
4. Methionine-free Dulbecco's modified Eagle medium (DMEM) with 4.5 g/liter D-glucose, without L-glutamine and L-methionine (MP Biomedicals, Inc., Solon, OH). The medium (500 ml) is supplemented with 10 ml of 200 mM L-glutamine, 2 ml of 7.5% sodium bicarbonate, 1 ml penicillin (10,000 U/ml)–streptomycin (10,000 μg/ml) solution, and 10 ml dialyzed fetal bovine serum.
5. Micrococcal nuclease (nuclease S7; Roche Diagnostics, Laval, QC, Canada): 15,000 U/ml (~1 mg/ml). The lyophilized nuclease (15,000 U) is reconstituted by adding water (1 ml) and is stored at −20° in small aliquots.

6. CaCl$_2$: 75 mM.
7. Ethylene glycol-bis(β-aminoethyl ether)-N,N,N',N'-tetraacetic acid (EGTA): 200 mM. A suspension of the free acid in water is neutralized with KOH to pH ~7.3.
8. Dithiothreitol (DTT): 1 M, stored at $-20°$ in 1-ml aliquots.
9. Nucleoside 5′-triphosphate solutions: 100 mM. The solutions of each ATP (disodium salt), GTP (dilithium salt), CTP (disodium salt), and UTP (trisodium salt) (Roche Diagnostics) are prepared; their pH is then adjusted to 6.0 to 8.0 with 9 to 15 μl of 50% (w/v) KOH per 1 ml of the solutions (pH indicator paper is used to monitor the pH), stored at $-70°$.
10. Creatine phosphate (dipotassium salt, Calbiochem, San Diego, CA): 1 M, stored at $-70°$.
11. Creatine phosphokinase (rabbit skeletal muscle, Calbiochem): 20 mg/ml. The lyophilized enzyme (20 mg) is reconstituted by adding 1 ml of buffer containing 25 mM HEPES-KOH, pH 7.3, 1 mM DTT, and 10% (w/v) glycerol, stored at $-70°$ in small aliquots.
12. Total L-amino acid mix (lacking L-methionine): 1 mM of each amino acid. The mixture is commercially available (Promega Corp., Madison, WI) or may be prepared using an amino acid kit (Sigma-Aldrich, St. Louis, MO; this prepared mixture should be filter sterilized), stored at $-70°$ in 1-ml aliquots.
13. L-Methionine (Sigma-Aldrich): 0.4 mM solution in 1 mM DTT, stored at $-70°$.
14. Sigmacote (Sigma-Aldrich).
15. HEPES-KOH: 1 M, pH 7.7. The stock solution should be sterilized by filtration through a 0.22-μm filter; it results in pH 7.3 after dilution.
16. Buffer A (10× stock): 350 mM HEPES-KOH, pH 7.3, 1.46 M NaCl, 110 mM D-glucose, filter sterilized through a 0.22-μm filter under vacuum and stored at 4°. The buffer is diluted to 1× with water as required.
17. Buffer B: 25 mM HEPES-KOH, pH 7.3, 50 mM KCl, 1.5 mM MgCl$_2$, 1 mM DTT, prepared fresh as required.
18. Buffer C: 25 mM HEPES-KOH, pH 7.3, 1 M potassium acetate, 30 mM MgCl$_2$, 30 mM DTT, prepared fresh as required.
19. Master mix: 10 mM ATP, 2 mM GTP, 2 mM CTP, 2 mM UTP, 100 mM creatine phosphate, 1 mg/ml creatine phosphokinase, 19 unlabeled L-amino acids (lacking L-methionine, 0.2 mM each), 125 mM HEPES-KOH, pH 7.3. The mix is prepared using the stock solutions above and stored at $-70°$ in 50-μl aliquots.
20. Salt mix (EMCV): 750 mM potassium acetate, 10 mM MgCl$_2$, and 2.5 mM spermidine (trihydrochloride), stored at $-20°$.
21. Proteinase K stock solution (Roche Diagnostics): 20 mg/ml in water, stored at $-70°$ in small aliquots.

22. tRNA (*Escherichia coli*, Roche Diagnostics): 20 mg/ml in water, stored at $-70°$ in small aliquots.
23. Sodium dodecyl sulfate (SDS): 10% (w/v) stock solution.
24. Ethylenediaminetetraacetic acid (EDTA): 200 mM stock solution. A suspension of the disodium salt in water is neutralized with NaOH to pH $\sim$8.0.
25. TNES buffer (2×): 40 mM Tris-HCl, pH 8.0, 200 mM NaCl, 2 mM EDTA, and 2% (w/v) SDS.
26. Deproteinization solution: 0.4 mg/ml proteinase K and 50 μg/ml tRNA in 1× TNES buffer, prepared fresh as required using the stock solutions above.
27. RNase A/T1: 0.2 mg/ml RNase A and 1000 U/ml RNase T1 in water, stored at $-20°$ in small aliquots.
28. Trichloroacetic acid (TCA): 15%, 10%, and 5% (w/v) solutions (the 10% TCA solution is supplemented with 0.1% D,L-methionine), stored at $4°$.
29. Phosphate-buffered saline (PBS): 140 mM NaCl, 2.7 mM KCl, 10 mM Na_2HPO_4, and 1.8 mM KH_2PO_4, pH 7.3.
30. Phenol/chloroform/isoamyl alcohol (50:48:2) mixture, prepared using phenol saturated with 100 mM Tris-HCl, pH 8.0. The mixture should be stored protected from light at $4°$.
31. Ammonium acetate: 10 M, stored at $-20°$.
32. 100% Ethanol, stored at $-20°$.
33. 70% Ethanol, containing 100 mM ammonium acetate, stored at $-20°$.
34. SDS sample buffer (1.5×): 75 mM Tris–HCl, pH 6.8, 7.5% (v/v) 2-mercaptoethanol, 15% (v/v) glycerol, 3% (w/v) SDS, and 0.15% (w/v) bromophenol blue.
35. Reagents and solutions for SDS–polyacrylamide gel electrophoresis (PAGE) (Sambrook *et al.*, 1989) including 30% acrylamide stock solutions (with 29.7:0.3 and 29:1 acrylamide to N',N'-methylene bisacrylamide ratios, for preparing separating and stacking gels, respectively).
36. EN^{3}HANCE (PerkinElmer Life Sciences, Inc., Boston, MA).
37. Loading dye solution (6×): 10 mM Tris–HCl, pH 7.6, 60 mM EDTA, 60% glycerol, 0.03% bromophenol blue, and 0.03% xylene cyanol FF; this solution is commercially available (Fermentas International Inc., Burlington, ON, Canada).
38. Agarose (Biotechnology Grade).
39. TBE buffer (10×): 900 mM Tris base, 900 mM boric acid, and 20 mM EDTA, pH $\sim$8.4.

3.3. Other materials

1. Filter papers, Whatman no. 1 and 3MM (Whatman, Hillsboro, OR).
2. Dounce glass homogenizer (40 ml) with a tight-fitting pestle (Kontes, Vineland, NJ).

3. CHROMA SPIN-1000 columns (BD Biosciences).
4. Standard equipment for PAGE and agarose gel electrophoresis (Bio-Rad, Richmond, CA), gel dryer, plastic wrap, and cellophane.

3.4. Isotopes

1. [^{35}S]Methionine, translational grade: 1200 Ci/mmol, 10 mCi/ml (PerkinElmer Life Sciences).
2. [α-^{32}P]CTP: 3000 Ci/mmol, 10 mCi/ml (PerkinElmer Life Sciences).

4. METHODS FOR CELL-FREE SYNTHESIS OF EMCV

4.1. Propagation and storage of Krebs-2 cells

Krebs-2 ascites tumor cells are maintained using passages in the peritoneal cavity of mice (Villa-Komaroff *et al.*, 1974).

1. Select a mouse that had been injected 8 days previously and exhibits abdominal swelling. Euthanize the mouse by cervical dislocation or another approved method (sedating the animal by CO_2 prior to sacrificing is recommended). Pin the limbs to a Styrofoam support exposing the stomach. Saturate the skin with 70% alcohol (denatured).
2. Withdraw ascites fluid using a 10-ml syringe and a needle that allows for good flow (18 gauge).
3. Change to a 26-gauge needle and inject 0.25 ml per mouse in the peritoneal cavity. After 8 days the mice will develop a large quantity (typically 4 to 8 ml) of ascites fluid containing $\sim 10^8$ cells/ml, and their abdomens will be swollen.
4. If cell extract isolation is intended, conduct one or two additional cell passages, so that 10 to 15 mice with well-developed tumors are obtained.
5. If cells are being frozen, add an equal volume of ice-cold EBBS containing 20% DMSO to the ascites fluid. Deliver 1 ml of cell suspension (approx 5×10^7 cells) into each freezing vial. Place vials into Nalgene Cryofreezer. Incubate overnight at $-80°$, then transfer the vials into a liquid nitrogen storage tank.
6. To expand cells from a frozen stock, you should have one or two mice ready for injection. Thaw frozen cells by briefly placing the vial into a $20°$ water bath. Disinfect the outside of the vial with 70% alcohol (denatured). Once thawed, the cell suspension should be immediately used for injection (0.5 ml per mouse). Timing is important, since even brief storage of ascites at room temperature could result in its clotting and cell death.

4.2. Preparation of Krebs-2 S10 extract

For efficient incorporation of [^{35}S]methionine into proteins *in vitro*, it is necessary to deplete the endogenous methionine pool. Krebs-2 S10 extracts described here are neither dialyzed nor subjected to Sephadex G-25 chromatography. Instead, the endogenous methionine is depleted by incubation of cells in methionine-free medium as described below.

1. Harvest 8-day-old liquid tumors from 10 to 15 mice into two 250-ml conical Corning tubes containing ice-cold EBSS (~200 ml). Avoid collecting bloody tumors. Mix the cell suspension after each transfer of ascites to the tubes.
2. Collect the cells by centrifugation (120×g for 8 min at 4°) and resuspend them with ice-cold EBSS. Wash the cells by centrifugation through EBSS one more time as above. Note the cell volume after the second centrifugation and estimate the cell number (the pellet contains ~3 × 10^8 cells/ml).
3. Suspend the pellet in methionine-free DMEM supplemented with fetal bovine serum and other ingredients at 10^7 cells/ml and dispense 200 to 250 ml into each of two to four 1-liter Erlenmeyer flasks (treated with Sigmacote as recommended by the manufacturer). Seal the flasks with rubber stoppers.
4. Incubate the cells for 2 h at 37° under gentle (100 rpm) agitation on a rotary shaker.
5. Chill the cell suspension on ice and filter it through two layers of cheesecloth. Collect the cells by centrifugation as above and wash them with buffer A (twice as above, and once with centrifugation at 750×g for 8 min).
6. Carefully remove the supernatant by aspiration. Resuspend the cells in two packed-cell volumes of buffer B and allow them to swell for 20 min.
7. Break the cells with 15 to 30 strokes of a precooled tight-fitting Dounce homogenizer. Avoid generating bubbles by keeping the head of the pestle beneath the surface of the liquid. A decrease in viscosity and frothing of the suspension indicate cell lysis. To confirm cell lysis, stain an aliquot of the homogenate with 0.04% trypan blue and inspect it under a microscope. (*Note*: You should avoid excessive disruption of cells, as this causes damage to the nuclei and leakage of components that inhibit the activity of the extract; Villa-Komaroff *et al.*, 1975.)
8. Add a one-ninth volume of buffer C. Pour the homogenate into 30-ml Corex tubes and centrifuge at 18,000×g (e.g., Sorvall SS-34 rotor, 12,000 rpm) for 20 min at 4°. Carefully collect the supernatant with a pipette (avoid collecting the upper lipid layer). Dispense the supernatant into 200-μl aliquots. Flash-freeze the aliquots on dry ice or liquid nitrogen and store at −70°. The extract remains active for several years. Repeating freezing and thawing of the extract decreases its activity and is not recommended. The OD$_{260}$ of the extract should be 45 to 60 units per ml.

4.3. Nuclease treatment of the extract

To destroy endogenous mRNA and thus reduce background translation, the extract is treated with micrococcal nuclease in the presence of $CaCl_2$. This treatment should be carried out just before setting up translation reactions as follows.

1. Remove a 200-μl aliquot of Krebs-2 cell extract from storage and allow it to thaw in a 20° water bath. Quickly chill the extract on ice.
2. Add 2 μl micrococcal nuclease (15,000 U/ml) and 2 μl of 75 mM $CaCl_2$ per 200 μl extract. Mix and incubate at 20° for 20 min.
3. Add 3 μl of 200 mM EGTA (3 mM final concentration) to stop the reaction. Chill the extract on ice. (*Note:* The extract may become turbid during nuclease treatment. This precipitate readily dissolves when the extract is supplemented with the salt mix and other components as described below.)

4.4. Translation protocol

Translation reaction mixtures (20 μl) contain, by volume, 50% Krebs–2 S10 extract, 10% master mix, 10% salt mix (EMCV), 5% [^{35}S]methionine (or 0.4 mM L-methionine, where indicated), and 25% mRNA solution and water. Before assembling individual reaction mixtures, combine ingredients that are common to all samples. For example, to set up 19 reactions, proceed as follows.

1. To 200 μl Krebs-2 S10 extract, add 40 μl master mix, 40 μl of salt mix (EMCV), and 20 μl of [^{35}S]methionine.
2. Dispense 15-μl aliquots of this mixture to precooled plastic tubes.
3. If necessary, add components whose effects on translation are being investigated.
4. Add an appropriate amount of EMCV RNA (e.g., to 10 μg/ml final concentration). As a control, assemble a reaction mixture without EMCV RNA. Bring the reaction volume to 20 μl with water. Gently mix each reaction mixture.
5. Incubate the tubes at 32° as required (use Fig. 4.1A as a guide for the timing of incubation).
6. Stop reactions by adding 40 μl of 1.5× SDS sample buffer.

4.5. Analyzing translation

For determining the incorporation of [^{35}S]methionine into TCA-insoluble material, withdraw 3-μl aliquots from the samples and spot them onto squares of no. 1 Whatman filter paper (mark the squares with a pencil before use). Fix the proteins with cold 10% TCA containing methionine.

Wash the filters first with 5% TCA (two times at room temperature and once at 90°) and then with 100% ethanol. Determine TCA-insoluble radioactivity by liquid scintillation counting. For more details about assaying TCA-insoluble radioactivity, see Svitkin and Sonenberg (2004).

To characterize translation products by PAGE, denature proteins in the rest of the samples by heating (95° for 3 min). Resolve proteins by SDS–15% PAGE. (*Note:* For good resolution, we recommend using 12-cm or longer separating gels with 29.7:0.3 acrylamide to N',N'-metyhylene bisacrylamide ratio.) A greater proportion of N',N'-metyhylene bisacrylamide in the separating gel should be avoided as this causes the gel to become brittle and prone to breakage during drying. For more information on the preparation of SDS–polyacrylamide gels and separation of proteins by PAGE, refer to Sambrook *et al.* (1989). Following PAGE, fix the gel with a methanol/acetic acid solution (Sambrook *et al.*, 1989) for at least 1 h. For fluorography, treat the gel first with EN^{3}HANCE and then with water (45 min each treatment). Place the gel onto two sheets of Whatman 3MM paper, cover it with wet cellophane, and dry at 80° for 2 h in a gel-drying apparatus. Expose the gel to X-ray film at $-70°$ (Bonner, 1983). To precisely quantify radioactivity in individual protein bands, use the BAS-2000 analyzer (FUJI Medical Systems U.S.A., Inc.) or a similar instrument. Individual EMCV proteins *in vitro* can also be detected by Western immunoblotting as described previously (Svitkin *et al.*, 2005a).

4.6. EMCV RNA replication protocol

RNA synthesis in EMCV RNA translation–replication reactions is assayed as described previously (Barton *et al.*, 1996; Svitkin and Sonenberg, 2003).

1. Set up reactions in a 40-μl total volume without [^{35}S]methionine (substitute L-methionine for [^{35}S]methionine). Use a final concentration of $\sim$10 μg/ml of EMCV RNA. [*Note:* Excess input RNA inhibits RNA replication (Svitkin and Sonenberg, 2003). We recommend that EMCV RNA be titrated for each extract preparation to determine its optimal concentration.] As a negative control, use the reaction that does not contain EMCV RNA.
2. Incubate the reaction mixtures at 32° for 4 h.
3. Add 1 μl [α-^{32}P]CTP to the reaction mixtures and continue the incubation at 32° for 1 h.
4. Stop the reactions by adding 200 μl of deproteinization solution. Incubate the samples at 37° for 15 min.
5. Add 240 μl of the phenol/chloroform/isoamyl alcohol mixture. Vortex for 30 sec and centrifuge at 16,000$\times g$ for 1 min. Carefully recover the aqueous phases (withdraw 200 μl from each sample).
6. Precipitate the RNA with 20 μl of 10 M ammonium acetate and 2 vol of 100% ethanol. Store samples at $-20°$ (overnight).

4.7. Analyzing RNA replication

The labeled products of RNA synthesis are unstable. Therefore, we recommend that this analysis be carried out no later than on the next day after labeling of RNA.

1. Recover RNA by centrifugation at $16,000 \times g$ for 15 min at $4°$ (carefully remove supernatants with a pipette). Wash the pellets with 70% ethanol by centrifugation (two times).
2. Following thorough removal of the supernatants from the last wash, dry the pellets (this may be accomplished by leaving the microcentrifuge tubes open for approximately half an hour at room temperature).
3. Resuspend the RNA pellets in 40 μl $0.5\times$ TBE containing 0.1% SDS.
4. To 15-μl aliquots of the samples add 3 μl of the $6\times$ loading dye solution. Analyze RNA by electrophoresis using a native 1% agarose gel in $1\times$ TBE (without ethidium bromide). Marker RNAs of known sizes, such as EMCV RNA or ribosomal RNAs, should be loaded into a slot on the side of the gel. We recommend freshly poured $\sim$5-mm-thick gels, voltage gradient of 10 V/cm, and running the bromphenol blue $\sim$8 cm into the gel for good resolution.
5. Following electrophoresis, cut the side lane off and use it to determine the positions of the markers by staining with ethidium bromide (Sambrook *et al.*, 1989).
6. Fix the gel with cold 15% TCA for at least 1 h with gentle agitation on a rotary shaker.
7. Briefly wash the gel with distilled water. (*Note*: To avoid the loss of RNA from the gel during blotting, the washing time should not exceed 2 min.)
8. For drying, lay the gel flat on a piece of the plastic wrap with slots pointing up. Cut five sheets of Whatman 3MM paper to a size slightly exceeding that of the gel and place them on top of the gel. Stack precut paper towels on top of the Whatman 3MM paper to a height of $\sim$4 cm. Lay a glass plate and a weight on top of the stack. Leave this overnight; the gel should turn to a film during drying. If completely dry gel is desired, accomplish drying of the gel under vacuum in a conventional gel dryer (2 h at $50°$).
9. Cover the dry gel with plastic wrap, and expose it to X-ray film. Use the fluorescent labels to allow for the alignment of the film against the gel. (*Note*: To increase the speed of detection, perform the autoradiography at $-70°$ with the use of an intensifying screen.)

4.8. Virus synthesis protocol

1. Assemble the reaction components in 40-μl reaction volumes as described for assaying RNA replication (radioactive amino acids or nucleoside triphosphates are not added to these reaction mixtures at

any time). (*Note*: Use a 5 to 10 μg/ml final concentration of EMCV RNA. However, for maximal virus yields, you should determine the optimal concentration of the input RNA for each extract preparation.) Include a control reaction containing no added EMCV RNA.

2. Incubate the reaction mixtures at 32° for 8 to 20 h (use Fig. 4.1C as a guide for the timing of incubation).
3. Add 4 μl of RNase A/T1 and incubate the reaction mixtures at room temperature for 30 min.
4. Dilute the reaction mixtures 5-fold with PBS (add 176 μl).
5. Flash-freeze the reaction mixtures on dry ice and store them at −70° until titer analysis is performed.

4.9. Plaque assay for infectivity

The numbers of plaque-forming units are measured in serially diluted *in vitro* reaction mixtures by standard methods (Rueckert and Pallansch, 1981). Briefly, BHK-21 cells are grown to confluency on 60-mm petri dishes, and the serial dilutions of virus are made up with DMEM containing 2% fetal bovine serum. The cell monolayers are exposed to virus (0.25 ml) for 30 min at room temperature. The cells are then covered with 2.5 ml of medium P5 (Rueckert and Pallansch, 1981) containing 0.6% Noble agar (the agar overlay is maintained in the 44° water bath). When the overlay hardens, the liquid medium P5 overlay (2.5 ml) is added, and the plaques are allowed to develop at 37° for 26 h under 5% of CO_2. The agar is then removed, and cell monolayers are stained with 0.1% crystal violet in 20% ethanol. After removing the stain and washing the dishes with water, the plaques appear as transparent spots (typically 1 to 2 mm diameter) on a blue background of adherent cells.

5. *IN VITRO* TRANSLATION OF HCV RNA

5.1. HCV replication: An overview

Hepatitis C virus (HCV) is a leading cause of chronic hepatitis and liver cirrhosis in the developed world. HCV is an enveloped virus belonging to the genus *Hepacivirus* in the family *Flaviviridae*. The virus possesses a ∼9.6-kb-long positive-strand RNA that is translated into a polyprotein of ∼3010 amino acids (Reed and Rice, 2000). Akin to picornaviruses and pestiviruses, the translation of HCV RNA is IRES mediated (Pestova *et al.*, 1998; Tsukiyama-Kohara *et al.*, 1992).

HCV was found difficult to study owing to the lack of inexpensive animal models and inefficient propagation of most virus strains in cell culture. Nevertheless, data obtained through the use of transient expression systems and the replicon systems have provided a general picture about HCV genome organization and polyprotein processing (reviewed in Bartenschlager *et al.*, 2004; Lindenbach and Rice, 2001). These studies have suggested that the viral polyprotein is cleaved cotranslationally and posttranslationally into at least 10 polypeptides: (NH_2) C-E1-E2-p7-NS2-NS3-NS4A-NS4B-NS5A-NS5B (COOH). Cleavages within the structural region and at the p7/NS2 junction are mediated by host cell signal peptidase(s) located in the lumen of the endoplasmic reticulum (ER). The core protein (C) is the major component of the nucleocapsid. Envelope proteins E1 and E2 are type I transmembrane glycoproteins. The NS region is processed by two overlapping viral proteases. The NS2–3 autoproteinase cleaves the polyprotein at the NS2/3 site. The NS3 serine proteinase with the assistance of the NS4A cofactor cleaves the polyprotein at all sites downstream of the NS3 carboxy terminus. NS3 is also an RNA helicase, which, together with RNA-dependent RNA polymerase NS5B, other NS proteins, and host factors, forms a membrane-associated RNA replication complex (Moradpour *et al.*, 2004).

5.2. General characteristics of HCV RNA-directed translation and polyprotein processing *in vitro*

Elucidation of the mechanisms of HCV infection requires developing diverse methodological approaches. One promising way of studying HCV replication would be the modeling of this process in a test tube, a method that proved to work for PV and EMCV. A major obstacle toward achieving this goal had been that the systems for translation of HCV RNA *in vitro* were by and large inefficient, yielding only structural proteins and aberrant products. No NS5B was evident, precluding the occurrence of RNA replication. We recently demonstrated that Krebs-2 S10 extracts translate HCV RNA completely and accurately when supplemented with canine pancreatic microsomal membranes (CMMs) (Svitkin *et al.*, 2005b). CMMs are known to mediate processing, such as signal peptide cleavage, membrane insertion, translocation, and core glycosylation of proteins (Walter and Blobel, 1983). However, we reported for the first time that in HCV RNA-programmed Krebs-2 extract, CMMs support most NS3 protease-mediated cleavages and also stabilize HCV mRNA during translation.

To establish conditions that would allow translation of the entire open reading frame of HCV RNA *in vitro*, the RNA transcribed from the infectious H77 HCV cDNA clone was used to program nuclease-treated cytoplasmic extracts from different sources, such as Krebs-2, Huh7, and HeLa cell extracts, as well as commercially available RRL (Svitkin *et al.*, 2005b). CMMs were included in all the systems in order to facilitate processing and maturation of proteins. The expression of HCV proteins was most efficient with Krebs-2 and Huh7 S10 extracts (Svitkin *et al.*, 2005b).

Because the Krebs-2 S10 extract performed more reliably and was available in larger quantities, as compared to Huh7 S10 extract, it had been chosen for further optimization. The translation of HCV RNA exhibited an unusually high potassium salt optimum for an uncapped mRNA (i.e., 160 mM), and was much more accurate with KCl than potassium acetate (data not shown). The advantage of the use of KCl rather than potassium acetate for the synthesis of authentic as opposed to abnormal products has also been demonstrated in the systems translating other viral RNAs (Jackson, 1991; Svitkin et al., 1981). HCV RNA did not differ from most mRNA with respect of magnesium salt concentration optimum for translation (2.5 mM).

Under optimal ionic conditions, the CMMs-supplemented Krebs-2 S10 extract is highly efficient in the translation of HCV (H77C) RNA (Fig. 4.2, left panel). The products in vitro are similar in size to authentic HCV proteins (i.e., the core protein [C], E1, NS2, NS3, NS4B, NS5A, and NS5B). We confirmed the identity of the products in vitro by Western blot analyses using antibodies against C, E1, E2, NS3, NS5A, and NS5B (Svitkin et al., 2005b). A time-course experiment demonstrated that NS5B, which corresponds to the C-terminal portion of the polyprotein, first appeared at 90 min. Such a late appearance of NS5B is surprising, given the fact that in EMCV RNA-programmed translation extracts the C-terminal polypeptide P3 and its cleavage products appear by 30 min (the coding region of EMCV RNA is shorter than that of HCV RNA by only 24%). Thus, the translation rate of HCV RNA in vitro may be significantly slower than that of EMCV RNA. Alternatively, slow HCV polyprotein processing, in particular cleavage at the NS5A/5B junction, may be responsible for the relatively late appearance of NS5B. Importantly, the RNA of the JFH1 genotype 2a HCV isolate (Wakita et al., 2005) also directed, in a CMM-dependent manner, the synthesis of mature viral proteins (Fig. 4.2, right panel). Thus, the system described is applicable for the expression of proteins of different HCV genotypes.

6. MATERIALS FOR *IN VITRO* TRANSLATION OF HCV RNA

1. Infectious HCV cDNA clones, pCV-H77C (Yanagi et al., 1997) or pJFH1 (Kato et al., 2001), were used. pCV-H77C was a kind gift from Jens Bukh and Robert Purcell (National Institutes of Health, Bethesda, MD). pJFH1 was kindly provided by Takaji Wakita (Tokyo Metropolitan Institute for Neuroscience, Tokyo, Japan).
2. Uncapped HCV RNA was transcribed from pCV-H77C or pJFH1 using RiboMAX large-scale RNA production protocol (Promega

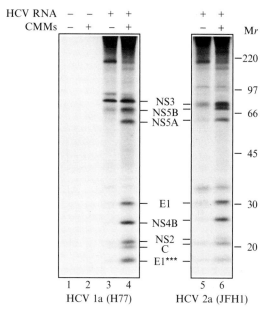

Figure 4.2 Products of HCV RNA translation in Krebs-2 S10 extract. The assays were performed with genotype 1a, strain H77 (lanes 3 and 4) or genotype 2a, strain JFH1 (lanes 5 and 6) HCV RNA (20 μg/ml) in the absence (− lanes 3 and 5) or presence (+ lanes 4 and 6) of CMMs. Incubation in the presence of [^{35}S]methionine was at 32° for 3 h. Translation products were resolved by SDS–15% PAGE and detected by fluorography. Analyses of reaction mixtures that did not contain mRNA are shown in lanes 1 and 2. The assignment of genotype 1a (H77) HCV polypeptides is based on their reactivity with corresponding antibodies in Western blotting (Svitkin *et al.*, 2005b). E1 * * * is the form of E1 that lacks carbohydrate (see below). E2-p7 glycoprotein is highly heterogeneous and requires a longer exposure of the gel to the film for detection (data not shown). NS4A (~6-kDa polypeptide) is expressed in the presence of CMMs, and could be resolved on a higher percentage polyacrylamide gel (18% acrylamide; data not shown). The positions of the ^{14}C-methylated protein molecular weight markers (GE Healthcare) are shown at the right.

Corp.). Prior to transcription, the plasmids were linearized with *Xba*I and treated with mung bean nuclease to remove the unpaired vector-derived nucleotides. HCV RNA is purified by 2 *M* LiCl precipitation and CHROMA SPIN-1000 column chromatography as recommended by the manufacturer. The integrity of HCV RNA should be confirmed by denaturing agarose gel electrophoresis (Sambrook *et al.*, 1989).

3. CMMs, available commercially (Promega Corp.), are stored at −70°.
4. CMM storage buffer: 50 m*M* triethanolamine, 2 m*M* DTT, and 250 m*M* sucrose, stored at −70°.
5. Salt mix (HCV): 1 *M* KCl, 5 m*M* MgCl$_2$, and 2.5 m*M* spermidine (trihydrochloride), stored at −20°. (*Note*: Potassium and magnesium concentrations in this mixture may need to be optimized.)

6. RNase/EDTA solution (10×): 5 mg/ml RNase A and 100 mM EDTA, pH 8.0, stored at −20°.
7. PNGase F (N-glycosidase F, recombinant, lyophilisate; Roche Diagnostics). The content of the bottle (100 U) is dissolved in 0.1 ml water and stored at 4°.
8. Nonidet P-40 (NP-40): 10% solution.
9. PNGase F buffer 1: 5 mM EDTA, 2% 2-mercaptoethanol, and 2% SDS, prepared fresh as required.
10. PNGase F buffer 2: 50 mM sodium phosphate, pH 7.5, 5 mM EDTA, and 0.1% SDS, prepared fresh as required.
11. PNGase F buffer 3: 50 mM sodium phosphate, pH 7.5, 5 mM EDTA, 1%-mercaptoethanol, 2% NP-40, and complete protease inhibitor cocktail (Roche Diagnostics), prepared fresh as required.
12. Proteinase K solution (0.3 mg/ml), prepared fresh as required by diluting the proteinase K stock solution with water.
13. CaCl$_2$: 200 mM.
14. Phenylmethylsulfonyl fluoride (PMSF): 10 mg/ml in isopropanol, prepared fresh as required.
15. Western Lightning Chemiluminescence kit (PerkinElmer Life Sciences).
16. CHROMA SPIN-10 columns (BD Biosciences).
17. 100% Methanol (or 70% acetone), stored at −20°.

Other materials are the same as used for EMCV synthesis.

7. Methods and Applications of *In Vitro* Translation of HCV RNA

7.1. HCV RNA *in vitro* translation protocol

Conditions for the translation of HCV RNA *in vitro* are similar to those described for EMCV RNA, except a higher KCl concentration is used and CMMs are included in the reaction cocktail. Final reaction mixtures contain 9 μl of Krebs-2 S10 extract, 1 μl of CMMs (or CMM storage buffer, where appropriate), 2 μl of master mix, 2 μl of salt mix (HCV), 1 μl of [^{35}S] methionine, and 0.4 μg of HCV RNA. (*Note*: To optimize the processing of HCV polyprotein, the amount of CMMs used in the reaction mixture may need to be titrated.) The reaction mixtures are reconstituted to the final volume of 20 μl with water. Incubation is at 32° for 3 h. Reactions are stopped by the addition of 40 μl of 1.5× SDS-sample buffer. [^{35}S]Methionine incorporation is assayed in 3-μl aliquots of the samples as detailed above. The gel banding pattern of translation products is visualized by fluorography after SDS–15% PAGE of the samples. (*Note*: Upon incubation with [^{35}S]methionine, some Krebs-2 S10 preparations produce a broad background band of 20 to 30 kDa. This band, presumably representing [^{35}S] methionyl-tRNA, can be greatly reduced or eliminated if reactions are terminated by adding 2 μl of RNase/EDTA solution followed by incubation at 32° for 5 min. If detection of particular virus proteins is desired, Western blotting as detailed previously can be used (Svitkin *et al.*, 2005b).

7.2. Characterizing NS3 protease inhibitors

We previously evaluated the use of the HCV RNA-programmed Krebs-2 S10 extract for characterizing the specificity of NS3 serine protease inhibitors, potential therapeutic agents of HCV infection (Svitkin *et al.*, 2005b). The NS3 protease is known to be sensitive to inhibition by specific pentapeptides and hexapeptides derived from the amino terminal NS3 cleavage products (Tan *et al.*, 2002). One highly specific inhibitor of NS3 activity (compound A) was shown to dramatically reduce replication of a subgenomic HCV RNA in Huh7 cells (Pause *et al.*, 2003; Tsantrizos *et al.*, 2003). This compound, kindly provided by Daniel Lamarre and Michael G. Cordingley [Boehringer Ingelheim (Canada) Ltd., Laval, QC, Canada], was assayed for its ability to inhibit HCV polyprotein processing *in vitro*. At 0.25 to 1000 nM concentrations, compound A did not have an effect on [^{35}S] methionine incorporation into protein in HCV RNA-programmed translation reactions (Svitkin *et al.*, 2005b). However, it decreased the accumulation of the mature viral proteins derived from the NS3-NS5B portion of the polyprotein (Fig. 4.3). At 50 mM and higher concentrations of compound A, the NS4B and NS5A bands disappeared and the intensities of the NS3, NS2–3, and NS5B bands were significantly reduced. Concomitantly, the enhanced accumulation of the high-molecular-weight precursor polypeptides was evident. As expected, compound A did not inhibit the cleavages carried out by a cellular signal peptidase(s), i.e., appearance of proteins C, E1 (E1 * and E1 * * * glycosylation forms), and NS2. These results validate the HCV RNA *in vitro* translation assay for testing the specificities and potencies of NS3 inhibitors. To characterize an NS3 inhibitor by this method, proceed as follows.

1. Prepare the serial dilutions of the compound. (*Note*: Many compounds are only slightly soluble in water and their stock solutions may have to be made up with 100% DMSO. Because DMSO inhibits *in vitro* translation, it should be present in the serial dilutions at 1% or lower concentrations.)
2. Add 1 μl of each dilution to a 20-μl reaction mixture containing CMMs, [^{35}S]methionine, HCV RNA, and other components as specified above. Include a control HCV RNA translation reaction supplemented with the solvent alone.
3. Incubate at 32° for 3 h.
4. Stop the reactions by adding 40 μl of 1.5× SDS sample buffer.

4. Withdraw three 16-μl aliquots into fresh microcentrifuge tubes.
5. To one aliquot add 2 μl of proteinase K solution (0.3 mg/ml) and 2 μl of water. To another aliquot add 2 μl of each proteinase K solution and 10% NP40. Supplement the third (control) aliquot with water alone (4 μl).
6. Incubate at 0° for 60 min.
7. Stop digestion by adding 2 μl of 10 mg/ml PMSF and 2 μl of 200 mM EGTA (add these components to all the samples).
8. Add 48 μl of 1.5× SDS sample buffer.
9. Denature the proteins by heating them at 95° for 15 min. Quickly microcentrifuge the samples.
10. Analyze the samples by SDS–15% PAGE and fluorography. Alternatively, transfer the proteins to a nitrocellulose membrane for Western blot analysis.

This technique is also applicable for analysis of E2-p7 topology. However, for reliable detection, the E2-p7 glycoprotein should be rendered homogeneous by deglycosylation. Thus, the analysis of E2-p7 localization would require conducting steps 3 to 10 in the previous section after step 7 in this section.

8. Perspectives and Future Applications

Although any cell and tissue could potentially be used as a source of translational extract, RRL is a system of choice for many researchers due to its superior translation rate and commercial availability (Jackson and Hunt, 1983). However, RRL tends to use spurious initiation sites on some mRNAs, such as PV RNA, an adverse feature not characteristic of HeLa or Krebs-2 S10 extracts (Dorner et al., 1984). The added disadvantage of RRL is its relatively low cap and poly(A) tail dependence (Munroe and Jacobson, 1990; Svitkin et al., 1996). Finally, RRL does not support picornavirus replication to any measurable extent (data not shown). Because Krebs-2 cells are highly permissive for EMCV growth, we chose these cells as a source of an extract for modeling of EMCV replication in vitro. We also favored the Krebs-2 S10 extract over other translation systems because of its reproducible performance, low cost, and ease to prepare in large quantities.

Similar to the HeLa cell-free system for de novo poliovirus replication (Barton et al., 1996; Molla et al., 1991), the Krebs-2 S10 extract supports the coordinated functions necessary for synthesis of EMCV and mengovirus (Fata-Hartley and Palmenberg, 2005; Svitkin and Sonenberg, 2003). The expression of virus genomes in this system has proved to be useful for studying translational regulation of EMCV (Svitkin et al., 2005a) and the mechanisms of action of anticardiovirus drugs (Fata-Hartley and Palmenberg, 2005). A pertinent question is whether or not the Krebs-2 S10 extract is capable of supporting replication of picornaviruses not belonging to the *Cardiovirus* genus. We found that the Krebs-2 S10 extract synthesizes infectious PV when programmed with PV RNA; however, this synthesis was ~10 to 20 times less robust than that carried out by the HeLa S10 extract (data not shown). Conversely, EMCV replicated more efficiently in Krebs-2 than in HeLa S10 extract. Thus, extracts from cells to which a particular virus is adapted seem to be best suited for this virus synthesis.

An obvious inconvenience of the viral RNA-programmed translation–replication assays is the need to grow infectious viruses for the purpose of template RNA preparation. However, recent studies have validated synthetic picornavirus RNAs with a ribozyme at the 5′-end as mRNAs for cell-free reactions (Fata-Hartley and Palmenberg, 2005; Herold and Andino, 2000). The hammerhead ribozyme catalyzes the removal of 5′ nonviral nucleotides that strongly inhibit plus-strand RNA synthesis. Importantly, with the use of this technique it has become possible to uncouple reactions involved in minus- and plus-strand RNA synthesis and conduct mutational analysis of picornaviruses *in vitro* (Fata-Hartley and Palmenberg, 2005; Herold and Andino, 2000).

Several features distinguish virus synthesis in a cell-free environment from that *in vivo* (Barton *et al.*, 2002). One is that *in vitro*, the input RNA is present in a close-to-saturating concentration and the competition from cellular mRNAs for translation is lacking. Under these conditions, viral translation proceeds with the maximum rate already in the beginning of incubation. In contrast *in vivo*, virus protein synthesis is heavily dependent on the synthesis of new viral mRNAs taking place in the middle of the infection cycle. We recently emphasized the importance of using a low concentration of the input EMCV RNA and an extract that was not nuclease treated for recapitulating the regulation of EMCV translation by 4E-BPs (eIF4E-binding proteins) observed *in vivo* (Svitkin *et al.*, 2005a).

An apparent limitation for virus RNA synthesis *in vitro* is a short supply of nucleoside triphosphates at a time when this synthesis is occurring. Raising the concentration of creatine phosphate, which was also shown to increase by about 3-fold maximal titers of EMCV *in vitro* (Svitkin and Sonenberg, 2003), provides a partial solution to this problem. A much more radical solution is the substitution of a dialysis or continuous-flow cell-free system for the conventional batch system (Mikami *et al.*, 2006b). Under the latter conditions, not only are the NTPs maintained at constant levels during RNA synthesis, but also the waist products are continuously removed from the incubation mixture. That RNA synthesis after EMCV RNA translation proceeds more efficiently in the dialysis system than in the system utilizing the batch protocol has recently been demonstrated (H. Imataka, personal communication).

The Krebs-2 S10 extract also completely translates HCV RNA, provided that it is supplemented with CMMs. No products resulting from initiation at spurious internal sites on HCV RNA have been detected, attesting to the high fidelity of translation initiation in this system (Svitkin et al., 2005b). The appearance of almost all mature virus proteins is consistent with efficient processing of the viral polyprotein in vitro by host signal peptidase(s) and virus-specific proteases, NS3 and NS2–3. Realization of cotranslational and posttranslational modifications of newly synthesized HCV proteins is another important asset of the system. In the presence of CMMs, four products differing in the extent of glycosylation of E1 are synthesized. A similar glycosylation pattern of E1 was revealed in cells transiently expressing HCV glycoproteins (Dubuisson et al., 1994). The absence of the discrete E2 band is attributed to incomplete N-linked glycosylation and the lack of cotranslational cleavage at the E2-p7 junction. Why these E2 maturation events cannot be recapitulated in vitro with complete fidelity is not known. However, the release of E2 from the N-terminus of p7 is also inefficient in vivo (Lin et al., 1994; Selby et al., 1994). We also noted that band NS5A shifted slightly upward upon chase or prolonged incubation (Svitkin et al., 2005b). The treatment with phosphatase, on the other hand, increased the electrophoretic mobility of NS5A but not of other HCV proteins. This result provides evidence for phosphorylation of NS5A in vitro.

It should be noted that even under optimal conditions, there is a cessation of HCV RNA translation in vitro after $\sim$60 min of incubation (Svitkin et al., 2005b). The decay of HCV mRNA cannot fully account for this effect. In fact, we found that in the CMM-supplemented Krebs-2 S10 extract, almost 90% of HCV RNA remains intact over the 60-min incubation period. A likely possibility is that eIF2α and eIF2α-kinases become phosphorylated in the presence of ATP and the ATP-regenerating system and that these modifications cause the inhibition of translation initiation (Kaufman, 2000; Mikami et al., 2006a). Obviously, eIF2α phosphorylation would preclude the occurrence of new rounds of translation initiation in vitro irrespective of the nature of mRNA in use.

Despite efficient expression of NS5B and other components of the HCV RNA replication complex in the Krebs-2 S10 incubation mixture, our attempts to detect [α-^{32}P]CTP incorporation into minus- and plus-strand HCV RNAs in this system were not successful. In this regard, the Huh-7.5 cell line that can be productively infected with the JFH1 HCV isolate (Lindenbach et al., 2005; Wakita et al., 2005) seems superior to Krebs-2 cells as a potential source of HCV RNA replication-competent extracts. Future studies would seek to identify components and conditions essential for infectious HCV synthesis in vitro.

ACKNOWLEDGMENTS

We thank Jens Bukh, Robert Purcell, and Takaji Wakita for the infectious HCV cDNA clones, Ralf Bartenschlager, Michinori Kohara, Darius Moradpour, Jane A. McKeating, and Michael Houghton for the antibodies against HCV proteins, Daniel Lamarre and Michael G. Cordingley for compound A and helpful discussions, and Sandra Perreault and Colin Lister for excellent technical assistance. We are also grateful to Vadim I. Agol for advice on the preparation of the Krebs-2 cell-free translation system and Hiroaki Imataka for communicating his unpublished results to us. This work was supported by grants from the Canadian Institute of Health Research (CIHR) to N.S. N.S. is a CIHR Distinguished Scientist and a Howard Hughes Medical Institute International Scholar.

REFERENCES

Agol, V. I. (2002). Picornavirus genome: An overview. *In* "Molecular Biology of Picornaviruses" (B. L. Semler and E. Wimmer, eds.), pp. 127–148. ASM Press, Washington, DC.

Aviv, H., Boime, I., and Leder, P. (1971). Protein synthesis directed by encephalomyocarditis virus RNA: Properties of a transfer RNA-dependent system. *Proc. Natl. Acad. Sci. USA* **68,** 2303–2307.

Bartenschlager, R., Frese, M., and Pietschmann, T. (2004). Novel insights into hepatitis C virus replication and persistence. *Adv. Virus Res.* **63,** 71–180.

Barton, D. J., Black, E. P., and Flanegan, J. B. (1995). Complete replication of poliovirus *in vitro:* Preinitiation RNA replication complexes require soluble cellular factors for the synthesis of VPg-linked RNA. *J. Virol.* **69,** 5516–5527.

Barton, D. J., Morasco, B. J., and Flanegan, J. B. (1996). Assays for poliovirus polymerase, 3D^{Pol}, and authentic RNA replication in HeLa S10 extracts. *Methods Enzymol.* **275,** 35–57.

Barton, D. J., Morasco, B. J., Smerage, L. E., and Flanegan, J. B. (2002). Poliovirus RNA replication and genetic complementation in cell-free reactions. *In* "Molecular Biology of Picornaviruses" (B. L. Semler and E. Wimmer, eds.), pp. 461–469. ASM Press, Washington, DC.

Bonner, W. M. (1983). Use of fluorography for sensitive isotope detection in polyacrylamide gel electrophoresis and related techniques. *Methods Enzymol.* **96,** 215–222.

Bordeleau, M. E., Mori, A., Oberer, M., Lindqvist, L., Chard, L. S., Higa, T., Belsham, G. J., Wagner, G., Tanaka, J., and Pelletier, J. (2006). Functional characterization of IRESes by an inhibitor of the RNA helicase eIF4A. *Nat. Chem. Biol.* **2,** 213–220.

Burness, A. T. (1969). Purification of encephalomyocarditis virus. *J. Gen. Virol.* **5,** 291–303.

Dorner, A. J., Semler, B. L., Jackson, R. J., Hanecak, R., Duprey, E., and Wimmer, E. (1984). *In vitro* translation of poliovirus RNA: Utilization of internal initiation sites in reticulocyte lysate. *J. Virol.* **50,** 507–514.

Dubuisson, J., Hsu, H. H., Cheung, R. C., Greenberg, H. B., Russell, D. G., and Rice, C. M. (1994). Formation and intracellular localization of hepatitis C virus envelope glycoprotein complexes expressed by recombinant vaccinia and Sindbis viruses. *J. Virol.* **68,** 6147–6160.

Fata-Hartley, C. L., and Palmenberg, A. C. (2005). Dipyridamole reversibly inhibits mengovirus RNA replication. *J. Virol.* **79,** 11062–11070.

Giachetti, C., and Semler, B. L. (1991). Role of a viral membrane polypeptide in strand-specific initiation of poliovirus RNA synthesis. *J. Virol.* **65,** 2647–2654.

Gorbalenya, A. E., Svitkin, Y. V., Kazachkov, Y. A., and Agol, V. I. (1979). Encephalo-myocarditis virus-specific polypeptide p22 is involved in the processing of the viral precursor polypeptides. *FEBS Lett.* **108**, 1–5.

Herold, J., and Andino, R. (2000). Poliovirus requires a precise 5′ end for efficient positive-strand RNA synthesis. *J. Virol.* **74**, 6394–6400.

Jackson, R. J. (1986). A detailed kinetic analysis of the *in vitro* synthesis and processing of encephalomyocarditis virus products. *Virology* **149**, 114–127.

Jackson, R. J. (1991). Potassium salts influence the fidelity of mRNA translation initiation in rabbit reticulocyte lysates: Unique features of encephalomyocarditis virus RNA transla-tion. *Biochim. Biophys. Acta* **1088**, 345–358.

Jackson, R. J., and Hunt, T. (1983). Preparation and use of nuclease-treated rabbit reticulo-cyte lysates for the translation of eukaryotic messenger RNA. *Methods Enzymol.* **96**, 50–74.

Jang, S. K., Kräusslich, H. G., Nicklin, M. J., Duke, G. M., Palmenberg, A. C., and Wimmer, E. (1988). A segment of the 5′ nontranslated region of encephalomyocarditis virus RNA directs internal entry of ribosomes during *in vitro* translation. *J. Virol.* **62**, 2636–2643.

Kato, T., Furusaka, A., Miyamoto, M., Date, T., Yasui, K., Hiramoto, J., Nagayama, K., Tanaka, T., and Wakita, T. (2001). Sequence analysis of hepatitis C virus isolated from a fulminant hepatitis patient. *J. Med. Virol.* **64**, 334–339.

Kaufman, R. J. (2000). The double-stranded RNA-activated protein kinase PKR. *In* "Translational Control of Gene Expression" (N. Sonenberg, J. W. B. Hershey, and M. B. Mathews, eds.), pp. 503–527. Cold Spring Harbor Laboratory Press, Cold Spring Harbor, NY.

Kerr, I. M., and Martin, E. M. (1972). Simple method for the isolation of encephalomyo-carditis virus ribonucleic acid. *J. Virol.* **9**, 559–561.

Kerr, I. M., Cohen, N., and Work, T. S. (1966). Factors controlling amino acid incorpora-tion by ribosomes from Krebs II mouse ascites-tumour cells. *Biochem. J.* **98**, 826–835.

Lin, C., Lindenbach, B. D., Pragai, B. M., McCourt, D. W., and Rice, C. M. (1994). Processing in the hepatitis C virus E2-NS2 region: Identification of p7 and two distinct E2-specific products with different C termini. *J. Virol.* **68**, 5063–5073.

Lindenbach, B. D., and Rice, C. M. (2001). Flaviviridae: the viruses and their replication. *In* "Fields Virology" (D. M. Knipe and P. M. Howley, eds.), Vol. 1, pp. 991–1042. Lippincott Williams & Wilkins, Philadelphia, PA.

Lindenbach, B. D., Evans, M. J., Syder, A. J., Wolk, B., Tellinghuisen, T. L., Liu, C. C., Maruyama, T., Hynes, R. O., Burton, D. R., McKeating, J. A., and Rice, C. M. (2005). Complete replication of hepatitis C virus in cell culture. *Science* **309**, 623–626.

Maley, F., Trimble, R. B., Tarentino, A. L., and Plummer, T. H., Jr. (1989). Characteriza-tion of glycoproteins and their associated oligosaccharides through the use of endogly-cosidases. *Anal. Biochem.* **180**, 195–204.

Martin, E. M., Malec, J., Sved, S., and Work, T. S. (1961). Studies on protein and nucleic acid metabolism in virus-infected mammalian cells. 1. Encephalomyocarditis virus in Krebs II mouse-ascites-tumour cells. *Biochem. J.* **80**, 585–597.

Mathews, M., and Korner, A. (1970). Mammalian cell-free protein synthesis directed by viral ribonucleic acid. *Eur. J. Biochem.* **17**, 328–338.

Mikami, S., Kobayashi, T., Yokoyama, S., and Imataka, H. (2006a). A hybridoma-based *in vitro* translation system that efficiently synthesizes glycoproteins. *J. Biotechnol.* **127**, 65–78.

Mikami, S., Masutani, M., Sonenberg, N., Yokoyama, S., and Imataka, H. (2006b). An efficient mammalian cell-free translation system supplemented with translation factors. *Protein Expr. Purif.* **46**, 348–357.

Molla, A., Paul, A. V., and Wimmer, E. (1991). Cell-free, *de novo* synthesis of poliovirus. *Science* **254,** 1647–1651.

Moradpour, D., Brass, V., Bieck, E., Friebe, P., Gosert, R., Blum, H. E., Bartenschlager, R., Penin, F., and Lohmann, V. (2004). Membrane association of the RNA-dependent RNA polymerase is essential for hepatitis C virus RNA replication. *J. Virol.* **78,** 13278–13284.

Munroe, D., and Jacobson, A. (1990). mRNA poly(A) tail, a 3′ enhancer of translational initiation. *Mol. Cell. Biol.* **10,** 3441–3455.

Novak, J. E., and Kirkegaard, K. (1991). Improved method for detecting poliovirus negative strands used to demonstrate specificity of positive-strand encapsidation and the ratio of positive to negative strands in infected cells. *J. Virol.* **65,** 3384–3387.

Palmenberg, A. C., Pallansch, M. A., and Rueckert, R. R. (1979). Protease required for processing picornaviral coat protein resides in the viral replicase gene. *J. Virol.* **32,** 770–778.

Paul, A. V. (2002). Possible unifying mechanism of picornavirus genome replication. *In* "Molecular Biology of Picornaviruses" (B. L. Semler and E. Wimmer, eds.), pp. 227–246. ASM Press, Washington, DC.

Pause, A., Kukolj, G., Bailey, M., Brault, M., Do, F., Halmos, T., Lagace, L., Maurice, R., Marquis, M., McKercher, G., Pellerin, C., Pilote, L., *et al.* (2003). An NS3 serine protease inhibitor abrogates replication of subgenomic hepatitis C virus RNA. *J. Biol. Chem.* **278,** 20374–20380.

Pelham, H. R. B. (1978). Translation of encephalomyocarditis virus RNA *in vitro* yields an active proteolytic processing enzyme. *Eur. J. Biochem.* **85,** 457–462.

Pelham, H. R., and Jackson, R. J. (1976). An efficient mRNA-dependent translation system from reticulocyte lysates. *Eur. J. Biochem.* **67,** 247–256.

Pelletier, J., and Sonenberg, N. (1988). Internal initiation of translation of eukaryotic mRNA directed by a sequence derived from poliovirus RNA. *Nature* **334,** 320–325.

Pestova, T. V., Shatsky, I. N., Fletcher, S. P., Jackson, R. J., and Hellen, C. U. (1998). A prokaryotic-like mode of cytoplasmic eukaryotic ribosome binding to the initiation codon during internal translation initiation of hepatitis C and classical swine fever virus RNAs. *Genes Dev.* **12,** 67–83.

Reed, K. E., and Rice, C. M. (2000). Overview of hepatitis C virus genome structure, polyprotein processing, and protein properties. *Curr. Top. Microbiol. Immunol.* **242,** 55–84.

Rueckert, R. R., and Pallansch, M. A. (1981). Preparation and characterization of encephalomyocarditis (EMC) virus. *Methods Enzymol.* **78,** 315–325.

Sambrook, J., Fritsch, E. F., and Maniatis, T. (1989). "Molecular Cloning: A Laboratory Manual," 2nd ed. Cold Spring Harbor Laboratory Press, Cold Spring Harbor, NY.

Schmidt-Rose, T., and Jentsch, T. J. (1997). Transmembrane topology of a CLC chloride channel. *Proc. Natl. Acad. Sci. USA* **94,** 7633–7638.

Selby, M. J., Glazer, E., Masiarz, F., and Houghton, M. (1994). Complex processing and protein:protein interactions in the E2:NS2 region of HCV. *Virology* **204,** 114–122.

Svitkin, Y. V., and Agol, V. I. (1978). Complete translation of encephalomyocarditis virus RNA and faithful cleavage of virus-specific proteins in a cell-free system from Krebs-2 cells. *FEBS Lett.* **87,** 7–11.

Svitkin, Y. V., and Sonenberg, N. (2003). Cell-free synthesis of encephalomyocarditis virus. *J. Virol.* **77,** 6551–6555.

Svitkin, Y. V., and Sonenberg, N. (2004). An efficient system for cap- and poly(A)-dependent translation *in vitro*. *Methods Mol. Biol.* **257,** 155–170.

Svitkin, Y. V., Gorbalenya, A. E., Kazachkov, Y. A., and Agol, V. I. (1979). Encephalomyocarditis virus-specific polypeptide p22 possessing a proteolytic activity: Preliminary mapping on the viral genome. *FEBS Lett.* **108,** 6–9.

Svitkin, Y. V., Ugarova, T. Y., Chernovskaya, T. V., Lyapustin, V. N., Lashkevich, V. A., and Agol, V. I. (1981). Translation of tick-borne encephalitis virus (flavivirus) genome *in vitro:* Synthesis of two structural polypeptides. *Virology* **110**, 26–34.

Svitkin, Y. V., Ovchinnikov, L. P., Dreyfuss, G., and Sonenberg, N. (1996). General RNA binding proteins render translation cap dependent. *EMBO J.* **15**, 7147–7155.

Svitkin, Y. V., Hahn, H., Gingras, A. C., Palmenberg, A. C., and Sonenberg, N. (1998). Rapamycin and wortmannin enhance replication of a defective encephalomyocarditis virus. *J. Virol.* **72**, 5811–5819.

Svitkin, Y. V., Herdy, B., Costa-Mattioli, M., Gingras, A. C., Raught, B., and Sonenberg, N. (2005a). Eukaryotic translation initiation factor 4E availability controls the switch between cap-dependent and internal ribosomal entry site-mediated translation. *Mol. Cell. Biol.* **25**, 10556–10565.

Svitkin, Y. V., Pause, A., Lopez-Lastra, M., Perreault, S., and Sonenberg, N. (2005b). Complete translation of the hepatitis C virus genome *in vitro:* Membranes play a critical role in the maturation of all virus proteins except for NS3. *J. Virol.* **79**, 6868–6881.

Tan, S. L., Pause, A., Shi, Y., and Sonenberg, N. (2002). Hepatitis C therapeutics: Current status and emerging strategies. *Nat. Rev. Drug Discov.* **1**, 867–881.

Tsukiyama-Kohara, K., Iizuka, N., Kohara, M., and Nomoto, A. (1992). Internal ribosome entry site within hepatitis C virus RNA. *J. Virol.* **66**, 1476–1483.

Tsantrizos, Y. S., Bolger, G., Bonneau, P., Cameron, D. R., Goudreau, N., Kukolj, G., LaPlante, S. R., Llinas-Brunet, M., Nar, H., and Lamarre, D. (2003). Macrocyclic inhibitors of the NS3 protease as potential therapeutic agents of hepatitis C virus infection. *Angew. Chem. Int. Ed.* **42**, 1356–1360.

Villa-Komaroff, L., McDowell, M., Baltimore, D., and Lodish, H. F. (1974). Translation of reovirus mRNA, poliovirus RNA and bacteriophage Qb RNA in cell-free extracts of mammalian cells. *Methods Enzymol.* **30**, 709–723.

Villa-Komaroff, L., Guttman, N., Baltimore, D., and Lodish, H. F. (1975). Complete translation of poliovirus RNA in a eukaryotic cell-free system. *Proc. Natl. Acad. Sci. USA* **72**, 4157–4161.

Wakita, T., Pietschmann, T., Kato, T., Date, T., Miyamoto, M., Zhao, Z., Murthy, K., Habermann, A., Krausslich, H. G., Mizokami, M., Bartenschlager, R., and Liang, T. J. (2005). Production of infectious hepatitis C virus in tissue culture from a cloned viral genome. *Nat. Med.* **11**, 791–796.

Walter, P., and Blobel, G. (1983). Preparation of microsomal membranes for cotranslational protein translocation. *Methods Enzymol.* **96**, 84–93.

Yanagi, M., Purcell, R. H., Emerson, S. U., and Bukh, J. (1997). Transcripts from a single full-length cDNA clone of hepatitis C virus are infectious when directly transfected into the liver of a chimpanzee. *Proc. Natl. Acad. Sci. USA* **94**, 8738–8743.

CHAPTER FIVE

A Practical Approach to Isolate 48S Complexes: Affinity Purification and Analyses

Nicolas Locker *and* Peter J. Lukavsky

Contents

MRC Laboratory of Molecular Biology, Cambridge, United Kingdom

Methods in Enzymology, Volume 429
ISSN 0076-6879, DOI: 10.1016/S0076-6879(07)29005-6

Abstract

In vitro assembly of eukaryotic translation initiation complexes requires purification of ribosomal subunits, eukaryotic initiation factors, and initiator tRNA
from natural sources and therefore yields only limited material for functional and structural studies. In this chapter, we describe a robust, affinity
chromatography-based method for the isolation of eukaryotic 48S initiation
complexes from rabbit reticulocyte lysate (RRL). Both canonical and internal
ribosome entry site (IRES)-containing mRNAs labeled with a streptomycin
aptamer sequence at the 3′ end can be used to purify milligram quantities of
48S particles in a simple, two-step procedure. The 48S complexes purified
with this method are properly assembled at the initiation codon, contain the
expected RNA and protein components in a 1:1 stoichiometry, and are functional
intermediates along the initiation pathway.

1. INTRODUCTION

Initiation of translation in higher eukaryotes can occur in two major
modes: canonical initiation, which requires the full set of eukaryotic initiation factors (eIFs), Met-tRNA$_i^{Met}$ and a 5′-capped mRNA to assemble the
80S ribosome at the authentic AUG start codon, and internal ribosome
entry site (IRES)-mediated initiation, which uses only a subset of eIFs and
a highly structured 5′ UTR (Hellen and Sarnow, 2001; Kapp and Lorsch,
2004b; Merrick, 2004; Pestova *et al.*, 2001; Sachs *et al.*, 1997). In the canonical initiation mode, the pathway commences with the recruitment of a
43S particle, comprising the 40S subunit, eIF1, 1A, 3, and the eIF2/GTP/
Met-tRNA$_i^{Met}$ ternary complex, to the 5′-cap structure of the mRNA
mediated through interactions with the 5′-cap-binding complex, eIF4F.
This ribosomal complex then scans the 5′ UTR to locate the AUG start
codon. When the 43S components are assembled at the initiation codon, a
48S complex is formed. Proper start codon selection is controlled by eIF1
and 1A, which are believed to modulate 40S subunit conformation (Maag
et al., 2005; Pestova *et al.*, 1998a). Upon codon–anticodon base pairing
between the mRNA and Met-tRNA$_i^{Met}$ in the ribosomal P site, a conformational change is proposed to occur, which releases eIF1 and triggers
eIF5-mediated hydrolysis of eIF2-bound GTP, P$_i$ release, and subsequent
dissociation of eIF2/GDP from the 48S complex (Algire *et al.*, 2005;
Unbehaun *et al.*, 2004). Finally, eIF5B mediates release of the remaining

eIFs and the joining of the 60S subunit in another GTP-dependent process to form 80S ribosomes (Pestova *et al.*, 2000; Unbehaun *et al.*, 2004).

In IRES-mediated initiation of translation, the requirements for eIFs are greatly reduced. The genomic RNAs of many viruses are incompatible with a 5′ end-dependent, scanning-based mechanism of AUG start codon selection because they do not bear a 5′ cap and have a highly structured 5′ UTR. Instead, translation initiation on these mRNAs occurs via IRESs located in the 5′ UTR, which recruit ribosomal subunits directly at or near the initiation codon (Hellen and Sarnow, 2001; Sachs *et al.*, 1997). Viral IRES RNAs differ both in nucleotide length and predicted secondary structure, and require different subsets of canonical eIFs to form a 48S complex. At least three distinct mechanisms for the formation of initiation complexes have been identified so far (Pestova *et al.*, 2001). In the mechanism used by encephalomyocarditis virus (EMCV) and EMCV-like IRESs from other picornaviruses, components of the eIF4F complex bind specifically to the IRES and thereby recruit the 43S particle near the AUG start codon (Pestova *et al.*, 1996). In contrast, hepatitis C virus (HCV) and other HCV-like IRESs can form stable binary IRES–40S complexes with the initiation codon placed in the ribosomal P site; only eIF3 and the ternary complex are required to form a functional 48S complex (Pestova *et al.*, 1998b). A third, distinct mechanism is employed by the intergenic IRES of the cricket paralysis virus (CrPV) and other CrPV-like IRESs, which assemble 80S ribosomes without any eIFs or Met-tRNA$_i^{Met}$ through binding to the ribosomal P and E sites, thereby initiating translation from the A site (Wilson *et al.*, 2000).

Formerly, both the canonical and IRES-mediated initiation pathways have been studied by reconstituting 48S complexes *in vitro* using ribosomal subunits, Met-tRNA$_i^{Met}$, and individual eIFs purified from natural sources (Algire *et al.*, 2002; Asano *et al.*, 2002; Benne *et al.*, 1979; Pestova *et al.*, 1996). The reconstituted initiation complexes up to the 48S stage have revealed the distinct eIF requirements of different IRESs and have helped to elucidate the role of several eIFs during canonical initiation (for reviews, see Hershey and Merrick, 2000; Pestova and Hellen, 2000; Pestova *et al.*, 2001). While reconstitution gives a precise control over the composition of the particles, the purification of the eIFs and Met-tRNA$_i^{Met}$ is rather labor intensive and their yield often poor. More recently, alternative methods were presented to study aspects of initiation complex formation. These novel methods are all based on affinity chromatography and focus on isolating intermediate complexes along the initiation pathway (Boehringer *et al.*, 2005; Ji *et al.*, 2004; Locker *et al.*, 2006). Ji and coworkers isolated initiation complexes formed onto mutant HCV IRES RNAs to study 48S and 80S assembly defects (Ji *et al.*, 2004). Boehringer and coworkers isolated HCV 80S complexes for cryoelectron microscopy (EM) studies using tagged IRES RNA (Boehringer *et al.*, 2005). We have developed a protocol to prepare milligram quantities of canonical and IRES 48S complexes for structural and functional studies (Locker *et al.*, 2006). The attractiveness of isolation methods using affinity chromatography is the possibility of finding new cofactors, intermediates, or subcomplexes at large scale suitable for biochemical and structural studies. That being said, isolated complexes also require strict characterization of composition and stoichiometry to ensure the authenticity of the complexes. Here we provide detailed protocols for our affinity chromatography-based method for the isolation of 48S complexes from rabbit reticulocyte lysate (RRL). In addition, we describe experiments we used to analyze the composition of the isolated particles and their activity in subunit joining experiments.

2. DESIGN OF STREPTO-TAGGED mRNAs FOR AFFINITY PURIFICATION OF 48S COMPLEXES

RNA oligonucleotides used for affinity purification consist of a $5'$ UTR (canonical or IRES element) followed by approximately 85 nucleotides (nt) of the authentic open reading frame (ORF), a toeprint primer-binding site (Pestova *et al.*, 1996), a short uracil-cytosine linker followed by the StreptoTag (Bachler *et al.*, 1999) (Fig. 5.1A). The StreptoTag is a small RNA aptamer, less than 50 nt in length (Wallace and Schroeder, 1998), which had been applied previously as an affinity tag to isolate RNA–protein complexes from cellular extracts (Bachler *et al.*, 1999). The StreptoTag was attached at the $3'$ end of the mRNA to enable both the internal ribosome entry and scanning mode of initiation to occur freely on the $5'$ end of mRNAs and to ensure that only "full-length" mRNAs are bound to the affinity column. The aptamer binds reversibly to dihydrostreptomycin coupled to a Sepharose 6B matrix with micromolar affinity in a magnesium-dependent manner and the assembled complexes can be eluted with a buffer containing free streptomycin (10 μM). The length of the ORF is best kept around 85 nt to ensure sufficient spacing between the 48S particle and the StreptoTag. Shorter spacing seems to interfere with efficient resin binding of the tagged mRNAs and therefore decreases the yield of the purified complexes. Furthermore, this length is also ideal to perform toeprinting experiments using the toeprint primer-binding site inserted downstream of the ORF (Pestova *et al.*, 1996). Two restriction sites, *Hind*III and *Xba*I, are used for subcloning of different mRNAs ($5'$ UTR and 85 nt ORF, see Fig. 5.1A). DNA inserts encoding the different mRNA elements were prepared from overlapping primers using standard polymerase chain reaction (PCR) and cloning techniques as described (Locker *et al.*, 2006). Tagged mRNAs are transcribed and purified using established protocols (Lukavsky and Puglisi, 2004) and stored at −20° in 10 mM Tris–HCl solution (pH 7.4) at a concentration of about 1.0 A_{260} units/μl.

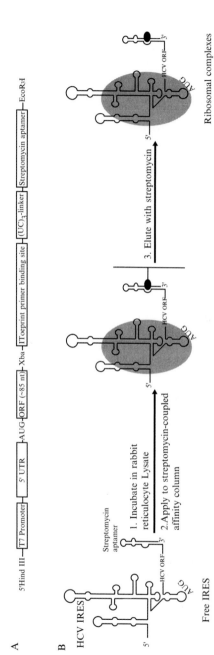

Figure 5.1 Design of streptomycin aptamer–based affinity purification. (A) Schematic showing the design of tagged RNAs used for affinity purification and analysis of initiation complexes. (B) Strategy for affinity purification of HCV IRES 48S complexes from RRL. (Reprinted with permission from Locker *et al.*, 2006.)

3. Affinity Purification of 48S Complexes

3.1. Preparation of the dihydrostreptomycin-coupled sepharose 6B matrix

This recipe yields enough dihydrostreptomycin-coupled Sepharose 6B matrix for two 8-ml glass Econo-columns (Bio-Rad), which are used for affinity purification. The preparation of the matrix is straightforward and is done over 3 days. All the solutions are freshly prepared and water refers to sterile water (Millipore).

3.2. Protocol 1

1. Rehydrate a total of 3 g of epoxy-activated Sepharose 6B (Amersham) matrix in 50 ml of water in a 50-ml Falcon tube and place it on a roller for 10 min at room temperature (RT) to ensure complete resuspension.
2. Decant the whole suspension into a sintered glass filter (Schott, 500 ml capacity) and wash with 600 ml of water. After washing, transfer the resin to a fresh 50-ml Falcon tube.
3. Rinse the filter with 10 ml of coupling solution [10 mM NaOH and 3 mM dihydrostreptomycin (Sigma) in water] and combine with resin. Add additional coupling solution to the resin so that the final buffer volume is 35 to 40 ml. Place the tube on a roller at RT for 2 h, then shake in a shaker incubator in the dark at 250 rpm at 37° overnight.
4. Decant the resin and buffer onto the sintered glass filter and wash with 200 ml of 10 mM NaOH solution. Transfer the resin to a fresh 50-ml Falcon tube.
5. Rinse the filter with 10 ml of ethanolamine solution (Sigma, 6% in water) and combine with resin. Add additional ethanolamine solution so that the final buffer volume is 35 to 40 ml. Place the tube on a roller at RT for 10 min and then shake in a shaker incubator in the dark at 250 rpm at 42° overnight.
6. Decant the resin and buffer onto the sintered glass filter. Apply three rounds of washes with alternating pH to the resin starting with 50 ml of low pH buffer (0.1 M NaOAc and 0.5 M NaCl in water, pH 4) and then 50 ml high pH buffer (0.1 M Tris–HCl and 0.5 M NaCl in water, pH 7.4), etc. Finally, wash the resin with 200 ml of water and then transfer to a fresh 50-ml Falcon tube.
7. Resuspend the resin in 16 ml of storage solution (10 mM Tris–HCl and 10 mM NaN$_3$ in water, pH 7.6) and store in the dark at 4° for up to 3 weeks.

3.3. 48S assembly in RRL and isolation of 48S complexes by affinity chromatography

The overall strategy of the affinity purification is outlined in Fig. 5.1B. First, the tagged mRNA is incubated in RRL to assemble ribosomal complexes. The lysate mixture is then applied to the dihydrostreptomycin-coupled affinity column and is eluted after several washes by competition with free streptomycin. For the assembly of ribosomal particles, untreated RRL (Green Hectares, USA) is used, which is usually shipped in 50-ml aliquots. Since a typical 48S assembly reaction requires only 1 to 4 ml of RRL and since RRL should not be defrosted more than twice, it is advisable to defrost 50 ml of RRL initially and to dispense it into smaller aliquots and refreeze them for subsequent usage. The assembly of ribosomal complexes is performed mainly at 37°, but binding to the affinity column and subsequent washes are done at 4°. Elution from the column, on the other hand, occurs only at RT. Therefore, the chromatographic apparatus (peristaltic pump, UV monitor, and column) should be assembled on a sturdy tray, which will allow easy transfer of the entire setup from 4° to RT for the elution step.

3.4. Protocol 2

1. Incubate 4 ml of freshly thawed, untreated RRL with 12 ml of binding buffer [20 mM Tris–HCl, 10 mM MgCl$_2$, 120 mM KCl, 8% sucrose, 2 mM dithiothreitol (DTT), pH 7.6] containing 2.5 μl of ribonuclease inhibitor (Promega) and half a tablet of protease inhibitor cocktail (EDTA-free, Roche) for 10 min at 37°.
2. Add puromycin (Sigma) to the lysate mixture (final concentration of 1 mM) and incubate for 10 min on ice and then 10 min at 37°.
3. Add GMPPNP (Sigma) to the lysate mixture (final concentration of 0.2 to 2.0 mM) and incubate for 5 min at 37°.
4. Add Strepto-tagged mRNA to the lysate mixture (final concentration of 1 μM). Incubate for 10 min at 37° to allow for assembly of ribosomal complexes. If scanning is required for the assembly of 48S complexes, ATP should be included (1 mM final concentration) at this stage.
5. Assemble a low-pressure liquid chromatography apparatus on a sturdy tray, including an 8 ml glass Econo-column (Bio-Rad), a peristaltic pump, and a UV monitor (UV-1, Amersham), and connect the latter to a chart recorder (Rec 112, Amersham).
6. Place 8 ml of the dihydrostreptomycin-coupled resin suspension (protocol 1) into an 8-ml Econo-column (Bio-Rad) in a cold room (4°). Equilibrate the resin with three column volumes of binding buffer, prechilled to 4°.

7. Load the 48S assembly reaction (step 4) slowly (0.8 ml/min) onto the column while monitoring the UV absorbance at 280 nm. After complete loading, wash with binding buffer at the same flow rate and monitor UV absorbance until a stable baseline is reached, usually after five column volumes of binding buffer (Fig. 5.2A).

8. Place the entire chromatography setup to RT for elution and wash the column with two or three more column volumes of binding buffer, preequilibrated to RT, at 0.8 ml/min until the UV monitor is equilibrated to RT and a stable baseline is reached.

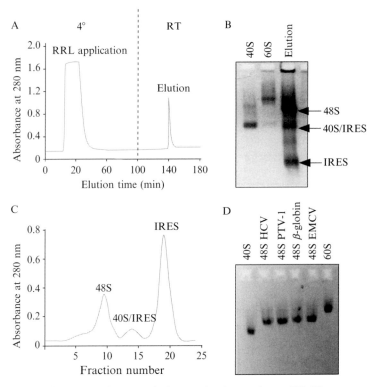

Figure 5.2 Affinity purification of ribosomal 48S complexes. (A) Chromatographic profile of a 48S complex assembled in RRL applied to a streptomycin affinity column. Plots of absorbance at 280 nm versus elution time are shown. (B) Visualization of the eluted fraction by 1% native agarose gel. Lanes are loaded as noted on the gel using purified 40S and 60S subunits as size markers. Bands corresponding to free IRES, IRES/40S binary, and 48S complexes are indicated. (C) Sucrose density gradient profile of affinity purified ribosomal complexes assembled onto an IRES RNA. Plots of absorbance at 260 nm versus sucrose density gradient fractions are shown; peaks corresponding to free IRES, IRES/40S binary, and 48S complexes are indicated. (D) Analysis of pooled and concentrated 48S fractions assembled onto various tagged mRNAs by 1% native agarose gel. Lanes are loaded as indicated on the gel. (Reprinted with permission from Locker *et al.*, 2006.) (See color insert.)

9. Elute the complexes with binding buffer containing 10 μM streptomycin (Sigma) at an increased flow rate (4 ml/min) into a 50-ml Falcon tube on ice (see Fig. 5.2A). Save 50 μl of the eluate (typically 10 ml total volume) for gel analysis.

10. Immediately transfer the eluted ribosomal complexes to a prechilled Ti90 tube (Beckman) and pellet by centrifugation at 45,000 rpm in a Ti90 rotor (Beckman) for 16 h at 4°. The resulting pellet can be stored at −20° or immediately resuspended for the final purification step (protocol 3).

11. In the meantime, analyze the elution on a 1% agarose gel with tris–borate as running buffer (both containing 0.5 μg/ml ethidium bromide). Use purified 40S and 60S subunits as size markers (Fig. 5.2B). Electrophoresis is performed at 100 V for 30 min at 4° and complexes are visualized by shortwave UV light.

3.5. Final cleanup of 48S complexes by sucrose density gradient centrifugation

Since the eluted fraction (see Fig. 5.2B) usually also contains free tagged mRNA and binary mRNA–40S complexes in addition to the desired 48S complexes, a final sucrose density gradient centrifugation step (Fig. 5.2C) is required to obtain pure 48S particles shown in Fig 5.2D.

3.6. Protocol 3

1. Resuspend the ribosomal pellet (step 10, protocol 2) in 200 μl of resuspension buffer (20 mM Tris–HCl, 100 mM KOAc, 200 mM KCl, 2.5 mM MgCl$_2$, 2 mM DTT, pH 7.6) at 4°. Carefully layer the resuspended mixture onto a 15% to 40% sucrose density gradient prepared in a Beckman Ultra-Clear tube (25 × 89 mm) with the same buffer. Centrifuge at 22,000 rpm in an SW28 rotor (Beckman) for 16 h at 4°.

2. Fractionate the gradient into 1.5 ml fractions from the bottom of the tube to the top. Monitor absorbance at 260 nm (see Fig. 5.2C).

3. Analyze each fraction on a 1% agarose gel as before (step 11, protocol 2).

4. Pool fractions containing pure 48S complexes and save an aliquot for native agarose gel analysis (see Fig. 5.2D). Concentrate the pooled fractions in a centricon device (YM-50, Millipore) to a final concentration of 10 A_{260} units/ml and store at −20°. Alternatively, particles can also be pelleted as before (step 10, protocol 2) and stored at −20°.

Each dihydrostreptomycin-coupled affinity column can be used up to three times for the isolation of 48S complexes from RRL: twice with about the same efficiency (1.0 to 1.5 A_{260} units/ml of RRL) and then one more time with 50% of the initial efficiency. This rapid decay of the resin is not observed when used with other lysates, which could be because of oxidative damage to the resin caused by components in the RRL. From the extinction coefficient of the three RNA components of the particle (18S rRNA, Met-tRNA$_i^{Met}$, and tagged mRNA) determined with the biopolymer calculator (http://paris.chem.yale.edu/extinct.html), we calculated that one A_{260} unit corresponds to 43 pmol of the 48S complex (about 2.5 MDa) and therefore the final yield (1.0 to 1.5 A_{260} units/ml of RRL) after sucrose density gradient centrifugation corresponds to 110 to 160 μg complex for each milliliter of RRL. The integrity of the complexes throughout the purification procedures was tested by immunoblotting all fractions of the individual purification steps for the presence of eIF2 and eIF3 with antibodies against the eIF2α or eIF3d subunits, respectively (Fig. 5.3A). In this example, 48S complexes were assembled onto tagged HCV IRES and purified using our protocols. An estimated 20% of eIF2 and 35% of eIF3

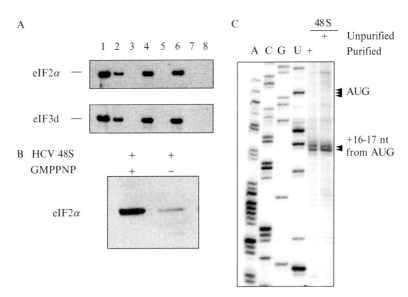

Figure 5.3 Analysis of the efficiency of the purification protocol. (A) The presence of eIF2 and eIF3 is followed throughout the purification steps by immunoblotting. Lanes are indicated as follows: (1) initial assembly reaction applied to the affinity column, (2) flow through of the application fraction, (3) wash fraction, (4) elution, (5) supernatant of the centrifugation step after elution, (6) 48S peak from the sucrose density gradient in Fig. 5.2C, (7) 40S/IRES peak from the same gradient, and (8) pooled top fractions from the same gradient. (B) Immunoblotting analysis of HCV IRES 48S complexes assembled in the absence and presence of GMPPNP using anti-eIF2α antibodies. (C) Toeprinting assay of both purified and unpurified 48S HCV initiation complexes. Arrows denote positions of the initiation AUG codon and toeprinting stops. A dideoxynucleotide sequence generated with the same primer and tagged RNA (shown on the left) was run in parallel. (Reprinted with permission from Locker *et al.*, 2006.)

present in the RRL was not incorporated into assembled HCV 48S particles and therefore detected in the flowthrough of the affinity column (see Fig. 5.3A, lanes 1 and 2). This result was not altered by increasing the incubation time of the tagged RNA in the RRL (see protocol 2). We therefore assume that the eIFs found in the flowthrough of the affinity column are part of ribosomal complexes assembled onto endogenous, untagged mRNAs in the untreated RRL. Neither the wash of the affinity column nor the supernatant of the pelleting steps nor the top fractions of the sucrose density gradient contained eIF2 and eIF3, indicating that the initially assembled particle stayed intact during the affinity purification and subsequent sucrose density gradient centrifugation (see Fig. 5.3A, lanes 3, 5, and 8). Correspondingly, about 80% and 65% of eIF2 and eIF3 initially present in the RRL could be detected in the elution fraction of the affinity column containing the 48S complexes and the pooled 48S fractions from the final sucrose density gradient (see Fig. 5.3A, lanes 4 and 6).

To obtain 48S particles, GMPPNP, a nonhydrolyzable analogue of GTP that stalls ribosome assembly at the 48S stage (Merrick, 1979), must be added to the assembly reaction to replace eIF2-bound GTP. Incorporation of GMPPNP in the complex blocks the subsequent GTP hydrolysis and eIF2/GDP release (Locker et al., 2006). It is therefore important to ensure that an efficient exchange of GTP for GMPPNP occurs. Considering that eIF2-Met-tRNA$_i^{Met}$ binds GTP with a K_d of 0.2 μM (Kapp and Lorsch, 2004a), we initially assembled 48S complexes in the presence of 2 mM GMPPNP (Locker et al., 2006). This concentration gave a large (160,000-fold) excess of GMPPNP over the GTP bound to 43S, 48S, or ternary complexes, which were present in the assembly reaction at a concentration of about 10 to 15 nM (estimated from the final yield). We soon realized that such a large excess of this rather expensive chemical, GMPPNP, is not necessary and that concentrations as low as 0.2 mM (16,000-fold excess) work as efficiently. In contrast, a significant amount of 48S complexes was found lacking eIF2 and Met-tRNA$_i^{Met}$ when the GMPPNP concentration was further lowered to 20 μM or omitted all together (Fig. 5.3B) (Locker et al., 2006).

During the entire purification protocol, care was also taken to keep the 48S particles at a sufficiently high concentration to avoid dissociation of components. While eluting the complex from the affinity column, the flow rate was increased to 4 ml/min, yielding a sharp elution profile (see Fig. 5.2A) corresponding to a total elution volume of 10 ml and a particle concentration of about 20 nM. At this concentration, neither eIF2 nor eIF3 dissociates from the 48S particles as judged from the immunoblotting analysis (see Fig. 5.3A, lanes 5 and 6). During all subsequent steps of the purification, the 48S particle concentration was kept at least at 50 nM.

The following section describes methods that we used to analyze proper 48S assembly at the AUG start codon and the stoichiometry of the particle components and to assess whether the purified 48S complexes are functional intermediates along the 80S assembly pathway competent in subunit joining experiments.

 ## 4. Analysis of the Purified 48S Complexes

4.1. Toeprinting of affinity purified 48S complexes

Inhibition of primer extension of reverse transcriptase, so-called toeprinting, is a powerful assay to monitor mRNA positioning on the 40S ribosomal subunit (Hartz et al., 1988). In a properly assembled 48S complex, arrest of reverse transcription leads to characteristic bands 16 to 18 nt downstream of the adenine of the AUG initiation codon on a sequencing gel. Such toeprints indicate that the AUG start codon is placed in the P site of the small ribosomal subunit and base paired to Met-tRNA$_i^{Met}$ (Hartz et al., 1988). To perform a toeprinting experiment, a primer has to be annealed to an mRNA under native conditions about 80 to 100 nt downstream of the AUG start codon. All our tagged mRNAs therefore contain a toeprint primer-binding site downstream of the ORF (Pestova et al., 1996) (see Fig. 5.1A). In the following protocol, affinity-purified 48S particles are compared to in vitro assembled, unpurified 48S complexes (using the same tagged mRNA) by toeprinting analysis.

4.2. Protocol 4

1. Dilute 20 μl of affinity-purified 48S particles (stock = 10 A_{260} units/ml) with 20 μl of toeprinting buffer [20 mM Tris–HCl, 100 mM KOAc, 2.5 mM Mg(OAc)$_2$, 5% sucrose, 2 mM DTT, 0.1 mM GMPPNP, and 0.25 mM spermidine, pH 7.6] at 4°. Transfer to a polycarbonate tube (13 × 51 mm, Beckman) and pellet the complexes by centrifugation for 1 h at 65,000 rpm at 4° in a TLA100.3 rotor (Beckman).
2. Resuspend the pellet in 40 μl of toeprinting buffer at 4° and pellet again to complete the buffer exchange.
3. Resuspend the ribosomal 48S complexes in 40 μl toeprinting buffer, incubate for 3 min at 30°, and then add 5 pmol of toeprint primer 5'-GGGATTTCTGATCTCGGCG-3' (Pestova et al., 1996). The reaction is then placed on ice for 10 min to anneal the primer.
4. Add 1 mM dNTPs, 5 mM Mg(OAc)$_2$ (both final concentration), 1 μl [α-^{32}P]ATP (3000 Ci/mmol; Amersham), and 0.7 U of avian myeloblastosis virus reverse transcriptase (Promega, 24 U/ml) to the reaction to a total volume of 50 μl and allow the primer extension reaction to occur for 45 min at 30°.

5. For comparison, perform a second toeprinting reaction on initiation complexes *in vitro* assembled in RRL using the same tagged mRNA, following previously published procedures (Wilson *et al.*, 2000). Incubate a master mix containing 15 μl RRL, 0.15 μl of ribonuclease inhibitor (Promega), and GMPPNP (1 mM final concentration) for 5 min at 30°, then add 0.5 μg of tagged mRNA and incubate for another 5 min at the same temperature. Dilute the reaction mixture to 40 μl with the toeprinting buffer and then follow steps 3 and 4 in this protocol.
6. Carefully remove proteins from the toeprinting reactions by phenol extraction, and then precipitate cDNAs with ethanol at −20° overnight. Analyze the cDNA products on a standard 6% sequencing gel comparing with appropriate dideoxynucleotide sequence ladders performed on the unbound mRNA using the same primer.

Toeprinting of both the assembled, but unpurified, as well as the purified HCV 48S complexes yielded the same toeprints at position +16 and +17 nt downstream of the AUG start codon (Fig. 5.3C). This shows that affinity-purified HCV 48S complexes are correctly assembled with the initiation codon placed in the ribosomal P site and that the purification process does not alter the arrangement of components in the particle. To further test the quality of the isolated particles and especially the stoichiometry of the 48S components, more quantitative assays are required, which are described in the next section.

4.3. Quantitative immunoblotting of 48S complexes

Our goal here is to show the relative stoichiometry of eIF2, eIF3, and eIF5 within purified 48S complexes. The quantities of eIF2α, eIF3d, and eIF5 were assessed against individual standard curves using purified, recombinant proteins. All the quantification was done by chemiluminescence detection on film using standard immunoblotting techniques. We will briefly describe the detection method we use for quantification and then focus more on the construction of standard curves and optimization of 48S loading using HCV 48S complexes as an example (Fig. 5.4A–C).

Recombinant eIF2α, eIF3d, and eIF5 are purified as previously described (Locker *et al.*, 2006; Pestova *et al.*, 2000). The concentrations of these proteins are determined using extinction coefficients predicted from primary sequence and absorbance measurements at 280 nm of the proteins under native and/or denaturing conditions. The linear response ranges for each protein were found to be 90 to 450 fmol for eIF2α, 6 to 90 pmol for eIF3d, and 10 to 1200 fmol for eIF5, respectively. The middle of the linear range of each protein established the optimal amount of 48S complex used for comparison: 255 fmol of 48S particles (1 A_{260} unit = 43 pmol) for eIF2α, 30 pmol for eIF3d, and 500 fmol for eIF5, respectively. The optical density

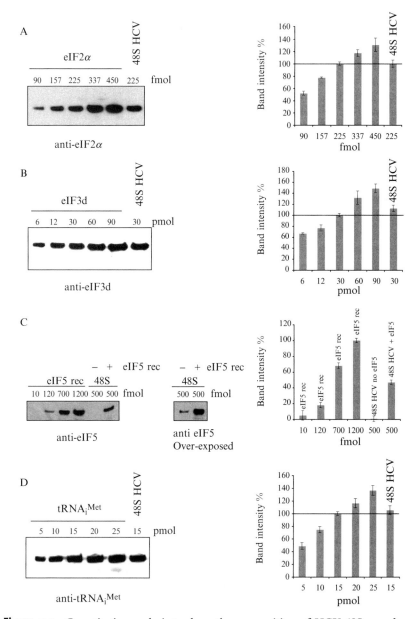

Figure 5.4 Quantitative analysis to show the composition of HCV 48S complexes. (A) Quantitative immunoblotting analysis of purified HCV 48S complexes using antibodies specific to eIF2α. Dilutions of purified recombinant eIF2α (90 to 450 fmol) and 255 fmol of purified HCV 48S complexes were loaded as indicated and resolved by 12% NuPAGE gel. Band intensities were quantified using ImageQuant software and relative levels of eIF2α were normalized to that of the 225 fmol eIF2α intensity. A graphical representation of the relative intensities is displayed on the right. All error bars are standard

of each eIF within 48S particles loaded at the above values is used to normalize the band intensity in each blot.

The serial dilutions of purified eIF2α, eIF3d, or eIF5 and the fixed amount of 48S complex are loaded on 4 to 12% NuPAGE gels (Invitrogen). Bands are resolved by electrophoresis, transferred to nitrocellulose membranes, and blocked with 5% dry milk in PBS–Tween 0.2%. Membranes are then probed for eIF2α with monoclonal (Abcam, ab5369, at 1/2000 dilution), for eIF3d with polyclonal (PTGlab, 10219-1-AP, at 1/1000 dilution), and for eIF5 with polyclonal (SantaCruz Biotechnology, sc-282, at 1/2000 dilution) antibodies. Primary antibodies are detected using appropriate HRP-coupled secondary antibodies (Abcam, ab6728 for eIF2α and ab6721 for eIF3d and eIF5, respectively; both at 1/2000 dilution), processed using enhanced chemiluminescence (ECL reagent, Amersham) and exposed onto Hyperfilm ECL (Amersham). The films are scanned at a resolution of 600 dots per inch in gray scale mode. ImageQuant software is used to convert band images to histograms. Rectangular areas are defined around each band to obtain the total optical density. The same sized rectangles are used to define the baselines, which are subsequently subtracted from the total optical density.

---

error of the mean. Using a response curve analysis, an eIF2α concentration of 225 ± 16 fmol within the HCV 48S complexes has been determined. (B) Quantitative immunoblotting analysis of purified HCV 48S complexes using antibodies specific to eIF3d. Dilutions of recombinant eIF3d (6 to 90 pmol) and 30 pmol of purified HCV 48S complexes were loaded as indicated and resolved by 12% NuPAGE gel. Band intensities were quantified using ImageQuant software and relative levels of eIF3d were normalized to that of the 30 pmol eIF3d intensity. A graphic representation of the relative intensities is displayed on the right. All error bars are standard error of the mean. Using a response curve analysis, an eIF3d concentration of 38.7 ± 4.2 pmol within the HCV 48S complexes has been determined. (C) Quantitative immunoblotting analysis of purified HCV 48S complexes assembled in the presence or absence of recombinant eIF5 using antibodies specific to eIF5. Dilutions of recombinant eIF5 (10 to 1200 fmol) and 500 fmol of purified HCV 48S complexes were loaded as indicated and resolved by 12% NuPAGE gel. Native eIF5 in 48S particles assembled without addition of recombinant eIF5 could be detected only upon overexposure of the immunoblots (on the right) and was estimated to be bound to 1% of the HCV 48S complexes. Intensities of normally exposed bands were quantified using ImageQuant software, and relative levels of eIF5 were normalized to that of the 1200 fmol eIF5 intensity. A graphic representation of the relative intensities is displayed on the right. All error bars are standard error of the mean. Using a response curve analysis, an eIF5 concentration of 470 ± 20 fmol within the HCV 48S complexes (loaded at 500 fmol) has been determined. (D) Quantitative northern blot analysis of purified 48S complexes using a probe specific to tRNA$_i^{Met}$. Dilutions of transcribed tRNA$_i^{Met}$ (5 to 25 pmol) and 15 pmol of purified HCV 48S complexes were loaded as indicated and resolved by 8% denaturing PAGE. Band intensities were quantified using ImageQuant software and relative levels of tRNA$_i^{Met}$ were normalized to that of the 15 pmol tRNA$_i^{Met}$ intensity. A graphic representation of the relative intensities is displayed below. All error bars are standard error of the mean. Using a response curve analysis, a tRNA$_i^{Met}$ concentration of 17.3 ± 1.5 pmol within the HCV 48S complexes has been determined. (Reprinted with permission from Locker *et al.*, 2006.)

The quantitative analysis of eIFs within the HCV 48S complex clearly showed that the purified particles contain both eIF2 and eIF3 at the expected 1:1 stoichiometry with 40S subunits (Locker et al., 2006) (see Fig. 54A and B). These complexes were further analyzed for the presence of eIF5, which was expected to be bound to the particles, since GTP hydrolysis is blocked by GMPPNP (see Fig. 5.4C). However, HCV IRESs contain less than 10 fmol native eIF5 corresponding to about 1% of the HCV 48S complexes (see Fig. 5.4C, lane 5), which was clearly detectable only upon overexposure of the Western blot (see Fig. 5.4C, overexposed). This seems to reflect the naturally low abundance of native eIF5 in RRL (Pestova et al., 2000). To test whether eIF5 can bind at 1:1 stoichiometry with other components, HCV 48S complexes were assembled in RRL supplemented with recombinant eIF5 (1 μM final concentration). In the 48S particle supplemented with purified eIF5, the protein could be detected as a stably bound component of the 48S complex, at 1:1 stoichiometry (see Fig. 5.4C, lane 6).

4.4. Quantitative Northern blotting of 48S complexes

Here we show how to quantify the amount of Met-tRNA$_i^{Met}$ within 48S complexes using Northern blot analysis. In principle, Met-tRNA$_i^{Met}$ should be present at 1:1 stoichiometry with eIF2 as part of the ternary complex. However, this method allows us to directly prove this stoichiometry independent of eIF2. The quantity of Met-tRNA$_i^{Met}$ was determined against a standard curve using transcribed and purified tRNA$_i^{Met}$ (Pestova and Hellen, 2001). Serial dilutions of tRNA$_i^{Met}$ ranging from 5 to 25 pmol are loaded on a 8% denaturing polyacrylamide gel (8 M urea) and resolved by electrophoresis together with 15 pmol of purified 48S complexes assembled onto tagged mRNA. The bands are transferred to a nylon membrane and hybridized with ^{32}P-labeled probe complementary to the 3' end of tRNA$_i^{Met}$ (5'-GGTAGCAGAGGATGGTTTCGATCC-3') in ExpressHyb^{TM} solution (BD Biosciences); they were then exposed onto film and the resulting bands are visualized by PhosphoImager analysis and quantified as described for the quantitative immunoblotting using ImageQuant software.

As shown in Fig. 5.4D, quantitative Northern blotting confirms that one equivalent of Met-tRNA$_i^{Met}$ is present in HCV 48S particles. The same results were obtained for 48S complexes assembled onto other IRES RNAs or canonical mRNA (Locker et al., 2006). These results also indirectly confirm the 1:1 stoichiometry of eIF2 and eIF3 in the particle, since the association of Met-tRNA$_i^{Met}$ with the 40S subunit withstands sucrose density gradient centrifugation, which was used as the final purification step, only in the presence of eIF2 (Unbehaun et al., 2004) and eIF3 is a prerequisite for stable recruitment of the ternary complex to both HCV IRES-mediated and canonical 48S particles (Ji et al., 2004; Maag et al., 2005; Otto and Puglisi, 2004).

5. FUNCTIONAL ANALYSIS OF THE PURIFIED 48S COMPLEXES

The toeprinting experiment and quantitative analysis suggest that the isolated 48S particles are properly assembled initiation complexes, but it remains to be addressed whether they represent functional intermediates along the initiation pathway or GMPPNP-stalled, dead-end complexes. The affinity purification relies on the fact that eIF2-bound GTP is exchanged to GMPPNP to block GTP hydrolysis and stall ribosome assembly at the 48S stage. Addition of GMPPNP during the ribosome assembly in RRL is absolutely crucial for obtaining homogeneous particles (Locker *et al.*, 2006). If GMPPNP is omitted, only a small amount of 48S complexes carries the ternary complex (Locker *et al.*, 2006), and therefore a large proportion of the 48S complexes is nonfunctional (see Fig. 5.3B). On the other hand, if intermediates downstream of the 48S complex or initiated 80S ribosomes are to be assembled, GMPPNP needs to be backexchanged with GTP to allow eIF5-stimulated hydrolysis of eIF2-bound GTP (Chakrabarti and Maitra, 1991; Chaudhuri *et al.*, 1994; Das *et al.*, 2001; Unbehaun *et al.*, 2004) and subsequent eIF5B-mediated, GTP hydrolysis-dependent joining of the 60S subunit (Pestova *et al.*, 2000). In the next section, we address this question and show how the purified 48S complexes can be used to study eIF release and subunit joining.

5.1. Protocol 5: GTP hydrolysis assay using purified 48s complexes

1. Incubate 2.5 pmol of purified 48S complexes in 200 μl release buffer (20 mM Tris–HCl, 100 mM KOAc, 2.5 mM MgCl$_2$, 2 mM DTT, pH 7.6) with 100 μM GTP, 50 μCi of [$\gamma-^{32}$P]GTP, and 10 pmol of recombinant eIF5 for 30 min at 37°.
2. Remove 20-μl aliquots at various time points over the 30-min period and assay for the amount of inorganic ^{32}P-P$_i$ release reflecting the amount of total GTP hydrolysis using previously published protocols (Conway and Lipmann, 1964).

5.2. Protocol 6: Analysis of eIF5-induced eIF release from purified 48S complexes

1. Incubate 25 pmol of purified 48S complexes in 150 μl of release buffer containing 1.5 mM GTP and 100 pmol of eIF5 for 30 min at 37°.
2. Remove 20-μl aliquots at various time points. Pellet ribosomal complexes by centrifugation for 1 h at 65,000 rpm at 4° in a TLA100.3 rotor (Beckman) using polycarbonate tubes (13 × 51 mm, Beckman).

3. Analyze the presence of eIF2 and/or eIF3 in pellets and supernatants by immunoblotting as described before.

5.3. Protocol 7: Analysis of eIF5B-induced eIF release from purified 48S complexes

1. Incubate 25 pmol of 48S complexes in 150 μl of release buffer containing 1.5 mM GTP, 100 pmol of eIF5, 100 pmol of ΔeIF5B, human eIF5B C-terminal domain [587 to 1220, which is fully active in subunit joining (Pestova *et al.*, 2000)], and 40 pmol of 60S subunits for 15 min at 37°.
2. Remove 20-μl aliquots at various time points. Pellet the ribosomal complexes by centrifugation for 1 h at 65,000 rpm at 4° in a TLA100.3 rotor (Beckman) using polycarbonate tubes (13 × 51 mm, Beckman).
3. Analyze the presence of eIF2 and/or eIF3 in pellets and supernatants by immunoblotting as described before.

5.4. Protocol 8: Sucrose density gradient analysis of 80S ribosome formation

1. Incubate 5 pmol of purified 48S complexes assembled onto ^{32}P-labeled, tagged mRNA in 50 μl of release buffer supplemented with 1.5 mM GTP, 20 pmol of eIF5, 20 pmol of ΔeIF5B (Pestova *et al.*, 2000), and 10 pmol of 60S subunits for 15 min at 37°.
2. Carefully layer the assembly reaction onto a 10% to 50% sucrose density gradient prepared in a Beckman Ultra-Clear tube (25 × 89 mm) with release buffer.
3. Centrifuge at 22,000 rpm in an SW28 rotor (Beckman) for 16 h at 4°C and analyze the gradient from the top to the bottom using a gradient fractionator (Brandel) as previously described (Anthony and Merrick, 1992; Pestova *et al.*, 1996; Wilson *et al.*, 2000).

In Fig. 5.5A and B, GTP hydrolysis and eIF2 release are monitored using purified HCV 48S complexes assembled in RRL (48S-eIF5$^-$) or RRL supplemented with recombinant eIF5 (48S-eIF5$^+$). A large excess of GTP was used in order to drive the back-exchange of eIF2-bound GMPPNP for GTP. Both particles were incubated with a 12,000-fold (GTP hydrolysis) or 9000-fold (eIF2 release) excess of GTP and a 4-fold excess of eIF5 to ensure that the back-exchange to GTP and binding of eIF5 are not rate limiting. In the case of 48S–eIF5$^-$ complexes, addition of both eIF5 and GTP is required to stimulate GTP hydrolysis and eIF2 release (see Fig. 5.5A and B). The 48S–eIF5$^+$ particles, which contain eIF5 at 1:1 stoichiometry with other components, in contrast, show no sign of GTP hydrolysis or eIF2 release (see Fig. 5.5A and B). This indicates that GMPPNP bound to eIF2 within 48S complexes can be exchanged with GTP only in the absence

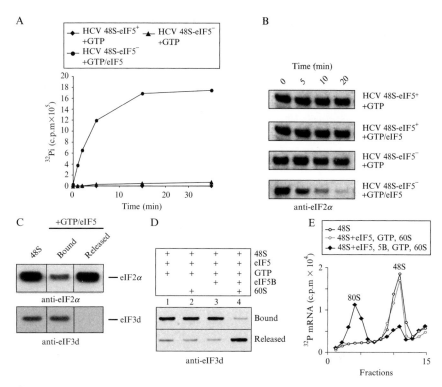

Figure 5.5 Release of eIFs from HCV 48S complexes during subunit joining. (A) Analysis of eIF5-induced GTP hydrolysis in purified HCV IRES 48S complexes assembled in the presence (HCV 48S-eIF5$^+$) or absence (HCV 48S-eIF5$^-$) of recombinant eIF5. 48S complexes were treated with [$\gamma-^{32}$P]GTP and eIF5, and ^{32}P-P$_i$ in the reaction was quantified at different time points to reflect the amount of total GTP hydrolysis. (B) Analysis of eIF5-mediated eIF2 release from purified HCV IRES 48S complexes assembled in the presence (HCV 48S-eIF5$^+$) or absence (HCV 48S-eIF5$^-$) of recombinant eIF5. Detection of eIF2 in the ribosomal pellet (not released from the complex) by immunoblotting with antibodies against eIF2α at several time points during incubation with GTP alone or GTP/eIF5 is indicated. (C) Analysis of eIF5-mediated eIF release from purified HCV IRES 48S complexes. Detection of eIF2 and eIF3 by immunoblotting with antibodies against eIF2α and eIF3d, respectively. Samples were analyzed before (48S) and after incubation of HCV 48S complexes with GTP and eIF5 (bound or released) for 15 min. (D) Analysis of eIF5B-mediated eIF3 release from purified HCV IRES 48S complexes. Detection of bound or released eIF3 by immunoblotting with antibody against eIF3d after incubation of HCV 48S complexes with different combinations of GTP, eIF5, eIF5B, and 60S subunits is as indicated for 15 min. (E) Sucrose density gradient analysis of purified HCV 48S complexes before and after incubation with GTP, eIF5, 60S subunits, and with or without eIF5B. The positions of ribosomal complexes are indicated above the appropriate peaks. The first fractions have been omitted for clarity. (Reprinted with permission from Locker *et al.*, 2007.)

of eIF5. The presence of eIF5 in the 48S particles seems to inhibit the release of GMPPNP, and as a result, its back-exchange to GTP. Therefore, 48S complexes must be assembled in the absence of supplementary eIF5, if eIF release and subunit joining are to be studied.

48S complexes assembled from purified components have revealed the requirement of eIF5 and GTP for displacement of eIF2 and eIF5, eIF5B, GTP, and 60S subunits to release eIF3 during subunit joining (Unbehaun *et al.*, 2004). Our 48S particles isolated and purified by affinity chromatography show exactly the same eIF requirements and convert into 80S ribosomes when treated with eIF5, eIF5B, and 60S subunits in the presence of GTP (Fig. 5.5C–E).

6. Conclusion

We presented detailed protocols for the affinity chromatography-based isolation of both canonical and IRES 48S initiation complexes from RRL. The proper assembly, composition, and stoichiometry of the components within the isolated particles were assessed by toeprinting and quantitative Northern and Western blot analyses. Furthermore, eIF release experiments showed that the isolated particles are functional intermediates along the initiation pathway and that they can therefore be used to study 80S ribosome assembly downstream of 48S formation. We have recently applied this method to study 48S and 80S assembly defects of mutant HCV IRES RNA lacking domain II (Locker *et al.*, 2006), which is essential for IRES function (Rijnbrand *et al.*, 1995). By analyzing eIF composition and their proper assembly within 48S complexes as a function of IRES RNA mutation, we could show that HCV IRES domain II is not required for 48S complex formation. Instead, we found that domain II functions downstream of AUG start codon recognition, and using the experiments described in this chapter, we revealed that this domain mediates eIF2 release during subunit joining (Locker *et al.*, 2006). In addition to biochemical applications, our purification scheme also provides an efficient way to isolate milligram quantities of 48S complexes, which will greatly benefit structural studies of eukaryotic initiation complexes.

ACKNOWLEDGMENTS

We thank Yoko Shibata for comments on the manuscript. N.L. is supported by a career development fellowship from MRC.

REFERENCES

Algire, M. A., Maag, D., Savio, P., Acker, M. G., Tarun, S. Z., Jr., Sachs, A. B., Asano, K., Nielsen, K. H., Olsen, D. S., Phan, L., et al. (2002). Development and characterization of a reconstituted yeast translation initiation system. RNA 8, 382–397.

Algire, M. A., Maag, D., and Lorsch, J. R. (2005). Pi release from eIF2, not GTP hydrolysis, is the step controlled by start-site selection during eukaryotic translation initiation. Mol. Cell 20, 251–262.

Anthony, D. D., and Merrick, W. C. (1992). Analysis of 40S and 80S complexes with mRNA as measured by sucrose density gradients and primer extension inhibition. J. Biol. Chem. 267, 1554–1562.

Asano, K., Phan, L., Krishnamoorthy, T., Pavitt, G. D., Gomez, E., Hannig, E. M., Nika, J., Donahue, T. F., Huang, H. K., and Hinnebusch, A. G. (2002). Analysis and reconstitution of translation initiation in vitro. Methods Enzymol. 351, 221–247.

Bachler, M., Schroeder, R., and von Ahsen, U. (1999). StreptoTag: A novel method for the isolation of RNA-binding proteins. RNA 5, 1509–1516.

Benne, R., Brown-Luedi, M. L., and Hershey, J. W. (1979). Protein synthesis initiation factors from rabbit reticulocytes: Purification, characterization, and radiochemical labeling. Methods Enzymol. 60, 15–35.

Boehringer, D., Thermann, R., Ostareck-Lederer, A., Lewis, J. D., and Stark, H. (2005). Structure of the hepatitis C virus IRES bound to the human 80S ribosome: Remodeling of the HCV IRES. Structure (Camb.) 13, 1695–1706.

Chakrabarti, A., and Maitra, U. (1991). Function of eukaryotic initiation factor 5 in the formation of an 80 S ribosomal polypeptide chain initiation complex. J. Biol. Chem. 266, 14039–14045.

Chaudhuri, J., Das, K., and Maitra, U. (1994). Purification and characterization of bacterially expressed mammalian translation initiation factor 5 (eIF-5): Demonstration that eIF-5 forms a specific complex with eIF-2. Biochemistry 33, 4794–4799.

Conway, T. W., and Lipmann, F. (1964). Characterization of a ribosome-linked guanosine triphosphatase in Escherichia coli extracts. Proc. Natl. Acad. Sci. USA 52, 1462–1469.

Das, S., Ghosh, R., and Maitra, U. (2001). Eukaryotic translation initiation factor 5 functions as a GTPase-activating protein. J. Biol. Chem. 276, 6720–6726.

Hartz, D., McPheeters, D. S., Traut, R., and Gold, L. (1988). Extension inhibition analysis of translation initiation complexes. Methods Enzymol. 164, 419–425.

Hellen, C. U., and Sarnow, P. (2001). Internal ribosome entry sites in eukaryotic mRNA molecules. Genes Dev. 15, 1593–1612.

Hershey, J. W. B., and Merrick, W. C. (2000). The pathway and mechansim of initiation of protein synthesis. In "Translation Control of Gene Expression" (N. Sonenberg, J. W. B. Hershey, and M. B. Mathews, eds.), pp. 33–88. Cold Spring Harbor Laboratory Press, Cold Spring Harbor, NY.

Ji, H., Fraser, C. S., Yu, Y., Leary, J., and Doudna, J. A. (2004). Coordinated assembly of human translation initiation complexes by the hepatitis C virus internal ribosome entry site RNA. Proc. Natl. Acad. Sci. USA 101, 16990–16995.

Kapp, L. D., and Lorsch, J. R. (2004a). GTP-dependent recognition of the methionine moiety on initiator tRNA by translation factor eIF2. J. Mol. Biol. 335, 923–936.

Kapp, L. D., and Lorsch, J. R. (2004b). The molecular mechanics of eukaryotic translation. Annu. Rev. Biochem. 73, 657–704.

Locker, N., Easton, L. E., and Lukavsky, P. J. (2006). Affinity purification of eukaryotic 48S initiation complexes. RNA 12, 683–690.

Locker, N., Easton, L. E., and Lukavsky, P. J. (2007). HCV and CSFV IRES domain II mediate eIF2 release during 80S ribosome assembly. EMBO J. 26, 795–805.

Lukavsky, P. J., and Puglisi, J. D. (2004). Large-scale preparation and purification of polyacrylamide-free RNA oligonucleotides. *RNA* **10**, 889–893.

Maag, D., Fekete, C. A., Gryczynski, Z., and Lorsch, J. R. (2005). A conformational change in the eukaryotic translation preinitiation complex and release of eIF1 signal recognition of the start codon. *Mol. Cell.* **17**, 265–275.

Merrick, W. C. (1979). Evidence that a single GTP is used in the formation of 80 S initiation complexes. *J. Biol. Chem.* **254**, 3708–3711.

Merrick, W. C. (2004). Cap-dependent and cap-independent translation in eukaryotic systems. *Gene* **332**, 1–11.

Otto, G. A., and Puglisi, J. D. (2004). The pathway of HCV IRES-mediated translation initiation. *Cell* **119**, 369–380.

Pestova, T. V., and Hellen, C. U. (2000). The structure and function of initiation factors in eukaryotic protein synthesis. *Cell. Mol. Life Sci.* **57**, 651–674.

Pestova, T. V., and Hellen, C. U. (2001). Preparation and activity of synthetic unmodified mammalian tRNAi(Met) in initiation of translation *in vitro*. *RNA* **7**, 1496–1505.

Pestova, T. V., Hellen, C. U., and Shatsky, I. N. (1996). Canonical eukaryotic initiation factors determine initiation of translation by internal ribosomal entry. *Mol. Cell. Biol.* **16**, 6859–6869.

Pestova, T. V., Borukhov, S. I., and Hellen, C. U. (1998a). Eukaryotic ribosomes require initiation factors 1 and 1A to locate initiation codons. *Nature* **394**, 854–859.

Pestova, T. V., Shatsky, I. N., Fletcher, S. P., Jackson, R. J., and Hellen, C. U. (1998b). A prokaryotic-like mode of cytoplasmic eukaryotic ribosome binding to the initiation codon during internal translation initiation of hepatitis C and classical swine fever virus RNAs. *Genes Dev.* **12**, 67–83.

Pestova, T. V., Lomakin, I. B., Lee, J. H., Choi, S. K., Dever, T. E., and Hellen, C. U. (2000). The joining of ribosomal subunits in eukaryotes requires eIF5B. *Nature* **403**, 332–335.

Pestova, T. V., Kolupaeva, V. G., Lomakin, I. B., Pilipenko, E. V., Shatsky, I. N., Agol, V. I., and Hellen, C. U. (2001). Molecular mechanisms of translation initiation in eukaryotes. *Proc. Natl. Acad. Sci. USA* **98**, 7029–7036.

Rijnbrand, R., Bredenbeek, P., van der Straaten, T., Whetter, L., Inchauspe, G., Lemon, S., and Spaan, W. (1995). Almost the entire 5' non-translated region of hepatitis C virus is required for cap-independent translation. *FEBS Lett.* **365**, 115–119.

Sachs, A. B., Sarnow, P., and Hentze, M. W. (1997). Starting at the beginning, middle, and end: Translation initiation in eukaryotes. *Cell* **89**, 831–838.

Unbehaun, A., Borukhov, S. I., Hellen, C. U., and Pestova, T. V. (2004). Release of initiation factors from 48S complexes during ribosomal subunit joining and the link between establishment of codon-anticodon base-pairing and hydrolysis of eIF2-bound GTP. *Genes Dev.* **18**, 3078–3093.

Wallace, S. T., and Schroeder, R. (1998). *In vitro* selection and characterization of streptomycin-binding RNAs: Recognition discrimination between antibiotics. *RNA* **4**, 112–123.

Wilson, J. E., Pestova, T. V., Hellen, C. U., and Sarnow, P. (2000). Initiation of protein synthesis from the A site of the ribosome. *Cell* **102**, 511–520.

> CHAPTER SIX

YEAST PHENOTYPIC ASSAYS ON TRANSLATIONAL CONTROL

Bumjun Lee, Tsuyoshi Udagawa, Chingakham Ranjit Singh, *and* Katsura Asano

Contents

Molecular, Cellular, and Developmental Biology Program, Division of Biology, Kansas State University, Manhattan, Kansas

Methods in Enzymology, Volume 429
ISSN 0076-6879, DOI: 10.1016/S0076-6879(07)29006-8

Abstract

This chapter describes phenotypic assays on specific and general aspects of translation using yeast *Saccharomyces cerevisiae* as a model eukaryote. To study the effect on start codon selection stringency, a *his4⁻* or *his4-lacZ* allele altering the first AUG to AUU is employed. Mutations relaxing the stringent selection confer the His$^+$ phenotype in the *his4⁻* strain background or increase expression from *his4-lacZ* compared to that from wild-type *HIS4-lacZ* (Sui⁻ phenotype). Translation of the Gcn4p transcription activator is strictly regulated by amino acid availability depending on upstream ORF (uORF) elements in the *GCN4* mRNA leader. Mutations reducing the eIF2/GTP/Met-tRNA$_i^{Met}$ complex level or the rate of its binding to the 40S subunit derepress *GCN4* translation by allowing ribosomes to bypass inhibitory uORFs in the absence of the starvation signal (Gcd⁻ phenotype). Mutations impairing scanning or AUG recognition generally impair translational *GCN4* induction during amino acid starvation (Gcn⁻ phenotype). Different amino acid analogs or amino acid enzyme inhibitors are used to study Gcd⁻ or Gcn⁻ phenotypes. The method of polysome profiling is also described to gain an ultimate "phenotypic" proof for translation defects.

1. INTRODUCTION

Alteration in the process of protein synthesis (translation), caused by genetic mutations or by covalent modifications induced by external stimuli, has diverse effects on cell physiology. These effects include changes in the rate of overall protein synthesis as well as translation of specific mRNA (specific translational control). If the accuracy of start codon selection by an initiating ribosome is altered, the cell produces proteins from start codons other than AUG, and some of the proteins produced in this manner can have toxic effects on cells. If alteration in translation machineries specifically affects the mRNA encoding a transcription factor, the expression pattern of genes controlled by the transcription factor is also changed, leading to diverse physiological effects (Dever, 2002). These ideas have been established using the yeast *Saccharomyces cerevisiae* as a eukaryotic model organism.

The process of translation is composed of three phases: initiation, elongation, and termination. Eukaryotic translation involves at least 10 eukaryotic translation initiation factors (eIFs 1, 1A, 2, 2B, 3, 4A, 4B, 4F, 5, and 5B), in contrast to three initiation factors (IFs 1, 2, and 3) involved in bacterial translation (Pestova et al., 2006). It is noteworthy that most of the S. cerevisiae genes encoding eIFs were isolated by genetic approaches (Hinnebusch et al., 2006). The Donahue group identified genes encoding eIF1 (SUI1), eIF5 (SUI5/TIF5), and the three ($\alpha-\gamma$) subunits of eIF2 (SUI2, SUI3, and GCD11/SUI4) as required for stringent start codon selection in yeast (Castilho-Valavicius et al., 1990; Huang et al., 1997). SUI stands for suppressor of initiation codon mutations; thus, Sui$^-$ mutations increase the frequency of translation from non-AUG codons.

Earlier genetic studies done by the groups of Fink, Hutter, and Greer employed a variety of amino acid enzyme inhibitor drugs to identify genes involved in general yeast response to amino acid starvation stress (general amino acid control response) (Hinnebusch, 1992). The Hinnebusch group characterized GCD11, GCN3, GCD7, GCD1, GCD2, and GCD6 genes as encoding the eIF2γ subunit and all five ($\alpha-\epsilon$) subunits of eIF2B, respectively. The mutants altering these genes are either able to overcome the starvation$^-$ stress independent of the key protein kinase Gcn2p (Gcd$^-$ for general control derepressed) or are unable to overcome the given stress due to failure to induce the starvation response (Gcn$^-$ for general control nonderepressible) (Hinnebusch, 2005). The ability to overcome the starvation insult is due to translational activation of the Gcn4p transcription factor, governing the general control response. On amino acid starvation, Gcn4p induces hundreds of genes including those involved in amino acid biosynthesis genes that are required to overcome the starvation stress (Natarajan et al., 2001). Amino acid starvation activates the Gcn2p kinase, which then phosphorylates eIF2, thereby changing it from the substrate to the inhibitor of the guanine nucleotide exchange factor eIF2B. When eIF2B is inhibited, the level of eIF2-GTP, and hence its ternary complex (TC) with Met-tRNA$_i^{Met}$, decreases. The decrease in TC level caused by eIF2 phosphorylation specifically changes the choice of start codons by ribosomes reinitiating translation in the GCN4 mRNA leader, such that GCN4 translation is induced under starvation conditions (see Fig. 6.1 for details). Thus, direct changes in eIF2 or eIF2B activities can induce GCN4 translation independent of eIF2 phosphorylation. This mechanism allowed isolation of Gcd$^-$ mutants altering genes encoding eIF2 or eIF2B. By contrast, one mechanism to inhibit translational activation of GCN4 under starvation conditions (Gcn$^-$ phenotype) is by altering eIF2B to become resistant to inhibition by phospho-eIF2. A second mechanism is by inhibiting GCN4 translation due to impaired preinitiation complex function (see details later).

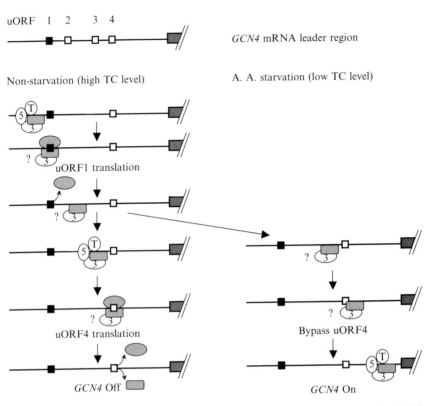

Figure 6.1 Model for *GCN4* translational control. Lines indicate the *GCN4* mRNA leader with the gray boxes to the right (followed by diagonal hashed lines) representing the *GCN4* coding region. Of the four upstream open reading frames (uORFs), shown on top, two (uORFs 1 and 4) have been shown to be necessary and sufficient for regulation of *GCN4* expression, and are depicted as filled and open squares, respectively, on the second line. The figure illustrates the ribosome movement on the leader region with the focus on its association with eIF3 (3), eIF5 (5), and eIF2 TC (T). The 40S and 60S subunits are drawn as a gray rounded rectangle and gray oval, respectively. Under nonstarvation conditions (left column), the preinitiation complex scans for and translates uORF1. Evidence suggests that eIF3 is associated with the ribosome at this stage. "?" indicates the uncertainty of this association. Following uORF1 translation, a population of 40S subunits remains associated with the mRNA and resumes scanning after reacquiring TC and other eIFs (third and fourth panels). The time it takes to scan from uORF1 to uORF4 is sufficient for all scanning ribosomes to reacquire TC before reaching uORF4, forcing them all to reinitiate at this start site (fifth to seventh panels). Under amino acid starvation conditions (right column), Gcn2p kinase is activated and phosphorylates eIF2. This phosphorylation renders eIF2 into a competitive inhibitor of GDP/GTP exchange activity (catalyzed by eIF2B; not shown), thereby reducing the level of eIF2/GTP and hence TC levels. Accordingly, a fraction of the migrating ribosomes reaches uORF4 without rebinding TC (first panel) and scan past the uORF4 AUG codon (second panel), reacquires TC in the uORF4–*GCN4* interval (third panel), and reinitiates at the *GCN4* AUG instead. Implicit in this mechanism is the fact that AUG recognition by the scanning ribosomes requires the anticodon of initiator tRNA in the TC.

Since these discoveries, the reporter constructs that were designed to detect these genetic changes have been used as convenient tools for phenotypic assays on translation initiation machineries. We used these assays to demonstrate the role for eIF4F in start codon selection (He *et al.*, 2003) and that of eIF5 for proper general control response (Singh *et al.*, 2005; Yamamoto *et al.*, 2005). This chapter describes the methods of the phenotypic assays using *HIS4* alleles as reporters for stringent AUG selection and *GCN4* alleles as reporters for eIF2 or eIF2B activities, preinitiation complex assembly, or its postassembly activities.

Besides these specific changes, the inhibition of overall translation reduces the growth rate of the cell. Defects in translation initiation result in ribosomes being shifted from polyribosomes (polysomes) engaged in translation to vacant 80S ribosomes not engaged in translation. On the other hand, defects in translation elongation result in increased abundance of the polysomes due to slower migration of ribosomes on mRNA. Thus, polysome profiling by density gradient-velocity sedimentation is an ultimate "phenotypic" proof for defects in translation activities *in vivo*.

The yeast *S. cerevisiae* also provides a convenient tool to study the biological function of foreign eukaryotic proteins. Translation machineries are strikingly similar between this yeast and other eukaryotes. Thus, the heterologous expression of proteins from these other organisms can have a significant impact on yeast translation, hence the growth or other specific phenotypes. A remarkable example is found in the study of eIF2 kinases. *S. cerevisiae* encodes Gcn2 as the sole eIF2 kinase, unlike other well-studied eukaryotic model organisms encoding multiple eIF2 kinases. Taking advantage of this, mammalian eIF2 kinases, such as PKR (protein kinase RNA-activated), were expressed in yeast *gcn2Δ* strains to study the mechanism of their activation (Dever *et al.*, 1993) and the regulation of PKR by virus-encoded translational inhibitors (Kawagishi-Kobayashi *et al.*, 1997). In this chapter, phenotypic assays on specific and general aspects of translation are described to examine the effects of mutations altering yeast translation factors or regulators and of heterologous expression of foreign proteins from various expression vectors.

2. QUANTITATIVE YEAST GROWTH ASSAY

Translation factors are, in general, essential for yeast growth. Thus, their mutations can show conditional or unconditional lethality. Conditionally lethal mutants grow normally at a permissive temperature, typically at 30°, but do not grow at a limiting temperature (for temperature-sensitive or Ts⁻ mutants, typically at 37°; for cold-sensitive or Cs⁻ mutants, typically at 18°). These can be analyzed by different growth assays, as described in this section. We will describe the assay for recessive unconditionally lethal mutations in the next section.

Yeast strains are grown in the rich YPD or minimal SD medium (Sherman *et al.*, 1974). Plasmids carrying selectable markers are introduced by transformation (Ito *et al.*, 1983). The protocol of yeast transformation described here is slightly modified from the original protocol (Schiestl and Gietz, 1989). Commonly used selectable markers are *LEU2, TRP1, HIS3, ADE2*, and *URA3* (Gietz and Sugino, 1988). Thus, the parental strain should contain one or more of the mutations *leu2, trp1, his3, ade2*, and *ura3*, which confer growth requirements for leucine, tryptophan, histidine, adenine, and uracil, respectively. Yeast transformants carrying these plasmids are grown in a selective medium, either SD containing only required compound(s) or the SC dropout medium, which is SD containing all amino acids, adenine and uracil, but lacking the compound covered by the selectable marker.

When studying Gcn⁻ mutants or strains carrying *GCN4-lacZ* reporters, care must be taken in determining the amino acid supplement to SD, because some combinations of amino acids lead to starvation for other amino acids (Niederberger *et al.*, 1981). Thus, Ile and Val must be added together with Leu, as Leu inhibits Ile/Val biosynthesis by a negative feedback loop; similarly, Trp dropout medium should be avoided because Phe and Tyr inhibit Trp biosynthesis.

The simplest method to assay yeast growth is to measure the growth rate in a liquid medium. Alternatively, the yeast strains are streaked out on a solid medium and the growth can be qualitatively evaluated by their colony size. A third, simple yet more informative growth assay is the spot assay, as described below. In this assay, fixed amounts of yeast culture are spotted on a solid medium. Some yeast mutants not only grow slowly but also form fewer colonies (hence, the lower efficiency of plating or EOP) under limiting conditions. The spot assay measures both EOP by the frequency of colony formation and growth rate by the colony size.

2.1. Materials

2.1.1. Yeast strains

To study the function of yeast proteins, mutations are introduced to the desired strain by crossing, one-step gene replacement (Sherman *et al.*, 1974), or "plasmid shuffling" (Boeke *et al.*, 1987). The method of plasmid shuffling uses the drug 5-fluoroorotic acid (FOA) to select against the plasmid carrying the *URA3* marker, as described in the next section. The plasmid carrying the mutant allele will be "shuffled in" with the residing *URA3* plasmid carrying the wild-type allele in the strain deleted for the chromosomal copy of the gene of interest. Thus, the former, incoming plasmid becomes the sole source of the protein under study.

2.1.2. Media

The YPD medium is made of 1% (w/v) yeast extract (Difco), 2% (w/v) bacto peptone (Difco), and 2% (w/v) glucose (Sherman, 1991). The SD medium contains 0.145% (w/v) yeast nitrogen base without amino acid and ammonium sulfate (Difco), 0.5% (w/v) ammonium sulfate, and 2% (w/v) glucose (Sherman, 1991). To make YPD and SD media, the sterile 40% glucose solution is added to 2% after autoclaving the rest. The pH of SD media should be ~6.0 without adjustment. Adjust this to pH 5.6, if necessary.

The SC medium is the same as SD except, in addition, it contains 0.009% (w/v) of each 20 amino acids, inositol and uracyl, 0.002% (w/v) adenine (hemi sulfate salt, Sigma A9126), and 0.0009% *para*-aminobenzoic acid (PABA). To make this, 0.2% premixed amino acid/base powder (AA powder, see below) is added to the yeast nitrogen base-ammonium sulfate solution prior to adding glucose. After autoclaving, add glucose to 2%. To make AA powder, mix 2 g each of all 20 amino acids, uracil and inositol, 0.5 g adenine, and 0.2 g PABA. An economical alternative to AA powder (AA* powder) contains 0.2 g adenine, 1 g arginine, 1.4 g aspartate, 0.4 g histidine, 1 g isoleucine, 2 g leucine, 1 g lysine, 0.4 g methionine, 1 g phenylalanine, 2 g threonine, 1 g tryptophan, 1 g tyrosine, 2.8 g valine, and 0.4 g uracil; 0.2% (w/v) of this can replace 0.2% of the original AA powder in SC. AA* powder can replace AA powder, except when SC-FOA or SC-HisxLeu is prepared for plasmid shuffling and Gcn⁻ assay (see below).

The SC dropout medium is made just as SC, but full AA powder is replaced with one lacking the component covered by the selectable marker. For instance, SC-Leu (SC minus Leu) lacks leucine for selection of Leu⁺ transformants.

A selective medium of SD supplemented with the required components is prepared by adding the following stock solutions to the SD medium, according to Table 6.1: 10 mM adenine (Ade), 40 mM tryptophan (Trp), 100 mM leucine (Leu), 20 mM uracil (Ura), 100 mM histidine (His), and 50 mM isoleucine and valine (Ilv). All of these solutions except Trp are sterilized by autoclaving. Tryptophan is heat labile in solution, so Trp should be filter sterilized and stored at 4° for not longer than 3 months. To grow *leu* mutants in SD, always add Ilv together with Leu.

To make a solid medium, 2% agar (Difco) is added to the solution lacking glucose or galactose (below) and autoclaved. After adding the sugar component and cooling to ~60°, each 25 ml is decanted into a sterile Petri dish (90 to 100 mm in diameter).

2.1.3. Solutions

The following are used for yeast transformation: 1 M lithium acetate, TE (10 mM Tris–HCl, pH7.5, 1 mM EDTA), 50% polyethylene glycol (PEG) (MW 3000 to 8000; e.g., MW 8000 Fisher Scientific), sterile 1 M sorbitol,

Table 6.1 Amount of amino acid or base solutions to be added to SD medium

Amino acid or base	Stock concentration	Milliliters added per liter medium	Milliliters added per 90-mm plate
Adenine (Ade)	10 mM	7.5	0.2
Tryptophan (Trp)	40 mM	10	0.1
Leucine (Leu)	100 mM	20	0.2
Uracil (Ura)	20 mM	10	0.25
Histidine (His)	100 mM	3	0.1
Isoleucine, valine (Ilv)	50 mM each	10	0.1

and 1 mg/ml ssDNA solution. ssDNA is made by dissolving 250 mg sermon sperm DNA (ssDNA, Sigma D1626) into 25 ml of TE and mixing for 2 to 4 h . After dissolving, use 18-gauge needles with syringe to pass the solution through 20 times. Then break ssDNA further with a sonic dismembrator (Model 500 Fisher Scientific or equivalent) for 30 sec. Check the homogeneity by pipetting. Treat the solution with an equal volume of phenol/chloroform/isoamylalcohol (25:24:1, pH 6.7, Fisher Scientific), and the aqueous phase is precipitated with ethanol. The purified DNA is suspended into the original volume of distilled water (final conc. is 1 mg/ml), aliquoted, and stored at $-20°$.

2.2. Procedures of yeast transformation

1. Prior to transformation, plasmid DNA is purified from an appropriate bacterial strain using a commercially available plasmid purification kit. The concentration of plasmid DNA typically obtained by Qiaprep Spin Miniprep Kit (Quiagen) or Wizard Plus SV Miniprep kit (Promega) is sufficient for this method, and the DNA solution can be used without dilution.
2. Grow the overnight culture of the strain to be transformed.
3. Dilute them 100-fold in 50 ml YPD in a flask (the starting A_{600} should be between 0.1 and 0.3).
4. Grow until $A_{600}= 0.5$ to 0.7 at an appropriate rate of rotation at 30°.
5. Transfer the culture to a 50-ml conical tube (Falcon) and spin it at 4.2k rpm ($3600g$) for 5 min at 4° using Beckman J6-MI or equivalent.
6. Wash the pelleted cells with 5 ml 1 M lithium acetate and spin again as in step 4.
7. Repeat step 6 to wash the cells again.
8. Resuspend the cells in 0.3 ml TE. The suspended cells are competent for transformation.

9. In a sterile microcentrifuge tube, add 50 μl ssDNA, 36 μl 1 M lithium acetate, 3 μl plasmid DNA, 30 μl competent cells, and 240 μl 50% PEG.

10. Rotate the DNA–cell mixture for 30 to 60 min at 30°. (PEG makes the mixture very viscous; thus it is very critical to make a uniform mixture here.)

11. Incubate the cells at 42° for 7 min, and then spin down the cells in a desktop centrifuge at 5k rpm for 1 min at room temperature.

12. Suspend cells in 50 μl 1 M sorbitol and spread them onto a solid selective medium using a sterile triangular glass rod.

13. Incubate the plate at 30° unless specified otherwise.

Note on step 3. Each transformation reaction requires 5 ml of yeast mid-logarithmic culture in YPD. Thus, this method makes competent cells for 10 reactions. Modify the scale of the culture according to the number of reactions needed.

Note on step 12. The incubation time to transform wild-type yeast is typically 2 to 3 days. Successful transformation should yield colonies of approximately equal sizes. If the introduced plasmid carries a dominant negative allele, the size of the transformants can become much smaller, and it may take a week to produce visible colonies.

2.3. Procedures of quantitative growth assay (spot assay)

1. Grow an overnight culture of yeast at the permissive temperature in the appropriate medium.

2. Collect the cells in 1.5-ml tubes by spinning with a desktop centrifuge (10k rpm, 1 min). Discard the supernatant by decantation.

3. Suspend the cells in 1 ml SD medium, and collect them again as in step 2.

4. Discard the supernatant by decantation and briefly spin them again. Completely remove the supernatant by pipetting.

5. Resuspend the washed cells in 1 ml SD medium.

6. Measure cell density at A_{600} after appropriate dilution. Suspend cells very well before taking cells for dilution and A_{600} measurement.

7. Dilute the original overnight culture to $A_{600} = 0.15$ into ~500 μl SD medium in sterile microcentrifuge tubes. For accurate pipetting, add 6 μl of the original culture to an appropriate volume of SD medium, such that final $A_{600} = 0.15$.

8. Starting with the diluted culture, make two 10-times serial dilutions (300 μl each) in sterile microcentrifuge tubes.

9. Mark the top side of the appropriate agar plates and set the plates on the template, indicating the positions of three consecutive spots from a dilution series in several rows. For clean results, diluted cultures should be spotted on a plate according to the template under it.

10. Spot 5 μl of the diluted cultures on the plate.
11. Allow the plates to dry for a while and incubate them at appropriate temperatures.

Note. The SD medium is recommended for dilution medium versus sterilized water or saline because the lack of glucose can change the cell physiology. Each time before starting a new dilution, vortex the culture tube a few seconds.

3. USE OF FOA TO ASSAY LETHAL MUTATIONS AND PERFORM PLASMID SHUFFLING

The classical genetic approach to analyze a recessive lethal allele is to form a heterozygous diploid carrying wild–type and mutant alleles. When sporulated, the resulting tetrad will form two viable and two nonviable progenies, if the mutant allele is lethal. In a second approach, FOA is used to evict the plasmid carrying the wild-type gene of interest and the *URA3* gene as the selectable marker (Boeke *et al.*, 1987). The Ura3p enzyme converts FOA into a cytotoxic compound (Boeke *et al.*, 1984). Thus, the growth on the FOA-containing medium produces clones without the *URA3* plasmid (i.e., selects against *URA3*). Taking advantage of this, the strain deleted for the chromosomal gene, which is complemented by the wild-type allele on the single-copy (sc) *URA3* plasmid, is transformed with the sc plasmid carrying the mutant allele and the selective marker other than *URA3*. If the mutant allele is lethal and does not complement the chromo-somal deletion, the resultant transformant will not produce FOA-resistant cells. If the mutant allele is not lethal, the mutant plasmid can be replaced with the resident *URA3* plasmid (plasmid shuffling). Plasmid shuffling is a convenient method to create an isogenic set of yeast strains carrying recessive mutations of a given gene.

3.1. Materials and procedures

3.1.1. Media
The SC-FOA medium is SC containing 1 g/liter FOA. To make solid medium (1.0 liter in total), autoclave a 500-ml solution of 1.45 g yeast nitrogen base without amino acid and ammonium sulfate, 5.0 g ammonium sulfate, 2.0 g complete AA powder, and 20 g agar. In the meantime, a 450 ml solution of 1.0 g FOA (US Biological c3051550) is filter sterilized using an appropriate filter unit (Falcon, 430769, pore size 0.22 μm). The two solutions are mixed together with 50 ml 40% glucose and then poured into Petri dishes. Note that it takes about 3 h to dissolve FOA; dissolving at

room temperature or at a temperature not higher than 50° is recommended so as not to damage the compound. FOA plates are stored in the dark at 4° for up to 6 months. Note also that FOA plates must contain uracil.

3.1.2. Procedures

1. Transform an appropriate yeast Ura$^+$ strain with the incoming plasmids marked by *LEU2* or *TRP1*.
2. Purify the transformants by streaking out on the solid selective medium and by isolating a single colony (single-colony isolation).
3. Make a patch with an isolated colony on a selective media plate together with patches of positive (the incoming plasmid carries a wild-type allele) and negative (the vector alone) controls.
4. The patches of the cells are grown for 1 to 2 days at 30°.
5. The patches are printed onto sterile velvet (Q Biogene 5000-006) that is set on a replica-plating base (Fisher Scientific 09-718-1).
6. Transfer the cells on the velvet to the FOA plate first by applying the plate onto the velvet and then to an appropriate selective medium to maintain the incoming plasmid.
7. The cells are grown for 2 to 5 days at 30°.
8. If FOA-resistant cells do not appear, the allele is judged to be lethal. FOA-resistant cells must appear from a control patch of transformants carrying the wild-type plasmid.
9. (Plasmid shuffling) If FOA-resistant cells appear, the FOA-resistant derivatives should be purified by streaking cells from the patch for single colonies on a new FOA plate. Three colonies should be picked for phenotype testing. It is also advisable to carry out plasmid shuffling on four or more of the six independent transformants for each mutant plasmid under study and ensure that all (or three out of four) give rise to FOA-resistant derivatives that exhibit the same phenotypes. It is not uncommon to isolate variants incapable of respiration on FOA (Pet$^-$ mutants), recognized by slow growth and lack of pigment on glucose medium and failure to grow on glycerol/ethanol medium. These should be discarded along with any other derivatives that show atypical phenotypes compared to the majority of five FOA-resistant clones.

Note on step 4. A common problem with this assay is the growth of FOA-resistant cells on the patch of the vector control. This is due to unknown chromosomal mutations, recombinational repair of the chromosomal deletion with the wild-type allele on the resident plasmid, or mutations in *URA3* carried by the resident plasmid. To avoid this, the amount of cells on the replica velvet needs to be carefully controlled. For this purpose, the cells can be printed onto two FOA plates (to choose one with the best pattern) or onto a blank SD plate first, and then onto an FOA plate (to remove cells with the blank plate).

Note on step 9. It is desirable to observe *URA3* plasmid loss with a high enough frequency that confluent growth of the replica-printed patch is observed as opposed to individual papillae.

4. Assay of Dominant Negative Mutants, Foreign Proteins, or Phenotypic Suppression by Overexpression

Plasmid shuffling as described above provides for a prime opportunity to analyze recessive mutations. However, this requires construction of the appropriate strain lacking the chromosomal wild-type allele and harboring the latter on a *URA3* episome. In construct, the characterization of dominant mutants does not require such genetic engineering. Dominant lethal or "dominant negative" mutant protein expressed from a plasmid perturbs the function of the endogenous wild-type protein. This effect may lead to conditional lethality, and hence can be measured by the phenotypic assays, as described in this chapter. Various truncated derivatives of eIF3 subunits are known to form partial, inhibitory eIF3 complexes, leading to dominant negative phenotypes (Evans *et al.,* 1995; Valasek *et al.,* 2003). The analysis of dominant negative mutants can be done with the transformants with the expression plasmids and therefore is simpler than that of recessive mutants. However, it should be noted that the interpretation of phenotypes produced by dominant negative mutants is complicated by the functional contribution of the wild-type protein.

Dominant negative mutants can be expressed from the natural promoter, unless they severely impair yeast growth or revert frequently by secondary mutations. Alternatively, these proteins are expressed from an inducible promoter, to suppress any toxic effects caused by dominant negative mutants before the mutant activity is measured. Induction of the mutant protein is achieved simply by moving the yeast construct from the non-inducing medium to the inducing medium, which contains a specific compound to activate transcription from the inducible promoter. Two examples of commonly used inducible promoters are the galactose-inducible *GAL1* promoter (p_{GAL1}) and the copper-inducible promoter (p_{CUP1}). p_{CUP1} was used to promote transcription of a dominant negative eIF4G mutant (Dominguez *et al.,* 1999), and elegant genetic experiments proved that this effect was due to formation of an inhibitory complex with eIF4A (Dominguez *et al.,* 2001). Yeast expression plasmids carrying these promoters are described under *Materials.* We describe the growth assay conditions for the use of these plasmids.

In addition to studying dominant negative mutants, wild-type yeast eIFs were overexpressed in a Ts⁻ mutant of the partner protein to study interactions between the mutated and overexpressed factors. If the conditional lethal phenotype of an eIF mutant is due to its reduced affinity with the partner protein, overexpression of the partner from a high-copy (hc) plasmid increases its cellular concentration, thereby restoring the level of the protein complex by mass action (high-copy suppressor analysis). Ts⁻ mutants altering the eIF4G2 N-terminal segment, the eIF4G2 C-terminal HEAT domain, and eIF3i (Table 6.2) were shown to be suppressed by overexpression of eIF4E, eIF4A, and eIF3g, respectively, by this mechanism (Asano et al., 1998; Neff and Sachs, 1999; Tarun and Sachs, 1997). This concept was also used to screen for unidentified partners of a mutated protein from a yeast genomic library constructed with an hc plasmid vector. eIF3g/Tif35p and eIF3j/Hcr1p were identified as hc suppressors of tif34 and rpg1 mutants altering eIF3i and eIF3a subunits, respectively, as a result of such screening (Valasek et al., 1999; Verlhac et al., 1997).

Proteins from humans or other eukaryotic species can also be expressed in yeast and tested for their effects on yeast translation by a variety of phenotypic assays described in this chapter. These proteins can be cloned under control of p_{GAL1} or p_{CUP1}, as described above, or a yeast constitutive promoter, such as p_{GPD1}, originally promoting transcription of an abundant glycerol synthesis enzyme (Schena et al., 1995). Yeast translation initiation factors are also abundant, and some of the plasmids expressing them under the natural promoter can be used as a cloning vehicle, as described below.

The assays described here might reveal that the expressed foreign proteins show no discernible phenotypic difference from wild-type yeast. In this case, these proteins can be further tested with yeast mutants defective in different initiation factors, such as those listed in Table 6.2. It is possible that the effect of the expressed protein on the yeast factor is simply not strong enough to show a phenotype due to a weak similarity between the yeast factor and its homologue, to which the expressed protein binds in the original species. If so, the expression of the foreign protein might exacerbate the phenotype caused by the conditional phenotype of the yeast factor mutant.

4.1. Materials

4.1.1. Yeast expression plasmids

Two examples of yeast–E. coli shuttle vectors commonly used to clone a gene under an inducible promoter are pEMBLyex4 (URA3 2μ) carrying the galactose-inducible GAL1 promoter (p_{GAL1}) (Cesareni and Murray, 1987) and pYELC5 (LEU2 2μ) carrying the copper-inducible CUP1 promoter (Macreadie, 1990). The dominant mutant (or wild-type as control) allele or foreign gene is to be cloned into one of these plasmids. A variety of galactose or copper-inducible fusion plasmids are also available. These include pYEX-4T (URA3 2μ) (AMRAD Biotech) (Dominguez et al., 1999)

Table 6.2 Yeast mutants altering essential translation initiation factors

Factor	Mutation	Defective in	Mutant strain	Isogenic wild-type	Empty marker	Reference
eIF1	$sui1-1$	AUG selection	Y220	Y218	$leu2\ ura3$	Cui et al., 1998
eIF1A	$tif11^{67-70}$	43S assembly	H3580	H2999	$ura3$	Fekete et al., 2005
	$tif11^{98-101}$	AUG selection	H3584	H3583	$ura3\ trp1$	Fekete et al., 2005
eIF2α	$SUI2^{S51A}$	eIF2 phosphorylation	H1817	H1816	$ura3$	Dever et al., 1992
eIF2β	$SUI3-2$	tRNA$_i^{Met}$ binding	KAY57	KAY56	$ura3$	Asano et al., 2000
eIF2γ	$gcd11^{K250R}$	GTP binding	Unnamed[a]	Unnamed[a]	$ura3$	Erickson and Hannig, 1996
eIF2Bε	$gcd6^{S576N}$	eIF2 GEF activity	GP3758	GP3751	$leu2$	Gomez and Pavitt, 2000
eIF3a	$rpg1-1$	eIF3 function	YLV314L	W303	$ade2\ ura3$ $his3$	Valasek et al., 1998
eIF3b	$prt1-1$	48S function	H1676	H2879	$leu2\ ura3$	Nielsen et al., 2004
eIF3i	$tif34-1$	eIF3b/g binding	KAY11	KAY8	$ura3$	Asano et al., 1998
eIF4A	$tif1^{A79V}$	RNA helicase	SS8-3A	SS8-3D	$ade2\ trp1$	Schmid and Linder, 1991
eIF4E	$cdc33-1$	m^{7}G-cap binding	YAS1888	YAS538	$ade2\ leu2$ $ura3\ trp1$	Tarun and Sachs, 1997
eIF4G2	$tif4632-430$	eIF4E binding	YAS2002	YAS1951	$ade2\ ura3$	Tarun and Sachs, 1997
	$tif4632-1$	eIF4A binding	YAS1998	YAS1951	$ade2\ ura3$	He et al., 2003
eIF5	$SUI5^{G58S}$	eIF2 GAP activity	KAY321	KAY314	$ura3\ trp1$	Singh et al., 2005
	$tif5-7A$	MFC assembly	KAY328	KAY314	$ura3\ trp1$	Singh et al., 2005

[a] Unnamed strains created by plasmid shuffling using transformants of DRD72 carrying wild-type or $gcd11$ plasmid.

and pEG(KT) (*URA3* 2μ) (Mitchell *et al.*, 1993) to express N-terminal GST fusion from p*CUP1* and p*GAL1*, respectively, and pAV1427 (p*GAL1* *URA3* 2μ) for N-terminal FLAG-epitope His6 tagging (Gomez *et al.*, 2002). pAV1427 is a derivative of pEMBLyex4 encoding N-terminal FLAG-His6-tagged Gcd6p. A desired gene can be cloned into this plasmid by replacing the *GCD6* ORF using the flanking *Mlu*I and *Bam*HI sites, of which the reading frame of the former is 5′-ACG (Thr) CGT (Arg)-3′.

Of expression vectors with a constitutive promoter, pG-1 carries p*GPD1* before *Bam*HI and *Sal*I sites, followed by a *PGK* transcription terminator (Schena and Yamamoto, 1988). YCpSUI3 (*LEU2 CEN*) and YEpSUI3 (*LEU 2μ*) contain the *Nde*I–*Hind*III eIF2β ORF fragment flanked by its natural promoter and terminator (Asano *et al.*, 1999). YCpTIF5 (*LEU2 CEN*) and YEpTIF5 (*LEU 2μ*) likewise contain the *Nde*I–*Sal*I eIF5 ORF fragment flanked by its natural promoter and terminator (Asano *et al.*, 1999). Any ORF DNA segment can be cloned into the unique *Nde*I and *Hind*III or *Sal*I sites of these plasmids; of these sites, the ATG triplet of the *Nde*I site (5′-CAT ATG-3′) is the start codon.

4.1.2. Yeast strains

For p*GAL1* constructs, it is advisable to use strains that contain *GAL2*, encoding the galactose permease. Yeast strains originating from the commonly used wild-type strain S288c may be *gal2*, as this mutation occurs in S288c. Galactose induction occurs in *gal2* mutants but is believed to occur both less efficiently and more slowly. Pep4p and Prb1p protease-deficient strains such as BJ1991, BJ5457, etc. (listed in Jones, 1991) are commonly used as hosts to express foreign proteins. Table 6.2 lists yeast translation factor mutants available to study the effect of expressed proteins including hc suppressor analysis.

4.1.3. Media

The SGal and SCGal media are the same as SD and SC, except that 4% galactose and 2% raffinose replace 2% glucose. To make these, the solutions of yeast nitrogen base and ammonium sulfate and of galactose and raffinose are separately prepared into 50% each of the final volume and mixed together after autoclaving. SD medium containing 0.5 mM CuSO$_4$ is used to induce the p*CUP1*-dependent transcription.

4.2. Procedures

The wild-type or dominant negative mutant alleles or foreign protein-encoding genes are cloned into the expression vectors, and the resulting plasmids are introduced to yeast strains with an appropriate reporter. The resulting transformants are purified by single-colony isolation on the selective medium, precultured in the same medium, and then cultured to a scale required for the respective assays. If an inducible promoter is used, the medium used in the purification and preculturing steps should be chosen to suppress expression from the inducible promoter. For instance, if the spot assay is used, the overnight culture should be grown in the liquid noninducing medium and then spotted onto the solid inducing medium and the solid noninducing medium as a control. The plates with the spots are incubated at 18°, 30°, 37°, or other temperatures to check for temperature sensitivity.

5. ASSAY OF STRINGENCY IN START CODON SELECTION

The eIF2 or eIF5 mutations increasing the eIF2 GTPase activation (Huang *et al.*, 1997; Singh *et al.*, 2005) or the eIF1 mutations promoting its spontaneous release (Cheung *et al.*, 2007; Maag *et al.*, 2005) relax the stringency of start codon selection and allow translation from UUG codons (Sui⁻ phenotype). To assay this, *HIS4,* which encodes a histidine synthesis enzyme, is often used as a reporter. The *his4-303* mutant requires histidine in the growth medium, due to alteration of the *HIS4* start codon to AUU. It was shown that its translation predominantly starts from the third codon of the *HIS4* open reading frame, which is UUG, in most Sui⁻ mutants due to relaxed stringency in start codon selection (Huang *et al.*, 1997). Thus, the simplest and most reliable assay of Sui⁻ mutants is to check the histidine requirement in a *his4-303* background by plate assays. In this assay, a His⁺ phenotype (growth in the absence of histidine) indicates a Sui⁻ phenotype. It is easy to assay for a Sui⁻ phenotype of plasmid-borne dominant negative mutants or foreign proteins by this method, as the expression plasmids can be introduced to a *his4-303* strain by transformation. However, to study recessive mutants altering yeast proteins, the mutations need to be introduced to a *his4-303* or another appropriate reporter strain by crossing or plasmid shuffling. Before making such a strain, the Sui⁻ phenotype can be determined more quantitatively by the second assay using *HIS4-lacZ* plasmids. In this assay, the strain bearing any mutation will be separately transformed with the *HIS4-lacZ* plasmid and its mutant *his4-lacZ* allele altering the first AUG codon. The ratio of expression from the *his4-lacZ* to the *HIS4-lacZ* allele (UUG/AUG ratio) is an indicator of the phenotype. If this ratio is significantly higher than the ratio obtained with the isogenic wild-type control, the mutant can be judged as Sui⁻.

It should be noted that observing a His⁺ phenotype in *his4-303* strains harboring a Sui⁻ mutation is dependent on transcriptional induction of *his4-303* by Gcn4p and, hence, translational induction of *GCN4* by eIF2α phosphorylation. Thus, the presence of any *gcn* mutation in the *his4-303* strain will confound the assay by suppressing the Sui⁻ phenotype. Moreover, it is possible that a Sui⁻ phenotype of a mutation can be dampened if it also reduces the efficiency of *GCN4* translational induction. This would be a complication for a mutation that reduces the rate of scanning as a means of increasing UUG selection, because the impairment of scanning interferes with induction of *GCN4* translation (see below). Measuring the ratio of expression from the UUG to AUG reporter normalizes for any Gcn⁻ or Gcd⁻ effect of a given eIF mutation on transcriptional activation of the *HIS4* promoter by Gcn4p.

Sui⁻ mutants altering eIF2 subunits or eIF5 (but not those altering eIF1) are, in general, dominant. Thus, the plasmid encoding one of these mutants can be introduced to a wild-type strain as a positive control, when the effect of a dominant negative mutant or foreign protein is tested.

5.1. Materials

5.1.1. Plasmids and yeast strains

The low-copy plasmid p367 carries *URA3* and the wild-type *HIS4-lacZ* (*HIS4^{AUG}-lacZ*) fusion (Cigan et al., 1988). p391 is a derivative of p367, with the first AUG codon of *HIS4-lacZ* ORF altered to UUG (*his4^{UUG}-lacZ*) (Cigan et al., 1988). Each of these plasmids is used to transform yeast strains carrying translation mutations or expressing foreign proteins. Two examples of Sui⁻ mutant plasmids that can be used as a positive control for Sui⁻ assays are p2192 and p2187 encoding yeast Sui⁻mutant eIF2β and eIF5, respectively, on an sc *LEU2* plasmid (Huang et al., 1997).

Yeast strain 76–8D (*MATa ura3–52 leu2–3,112 his4–303*) is used for a dominant Sui⁻ test (Cigan et al., 1988). The *his4–303* start codon is altered to AUU. 76–8D is also available from K. Asano (stock # KAY84) or A. Hinnebusch (stock # F252).

5.1.2. Buffers and solutions

Buffers and solutions used for β-galactosidase assay are Breaking Buffer [0.1 M Tris–HCl, pH 8.0, 20% glycerol (v/v), 1 mM 2-mercaptoethanol], Z-buffer (0.06 M Na$_2$HPO$_4$ · 7H$_2$O, 0.04 M NaH$_2$PO$_4$ · H$_2$O, 0.01 M KCl, 0.001 M MgSO$_4$ · 7H$_2$O, 0.05 M 2-mercaptoethanol), PMSF [40 mM phenylmethylsulfonyl fluoride (PMSF) in 90% ethanol, store at −20°], 4 mg/ml 2-nitrophenyl-β-D-galactopyranoside (ONPG) (solution in Z-buffer, prepared immediately before use), and 1 M Na$_2$CO$_3$. To make a 1-liter stock solution of Z-buffer, mix salt components first, adjust the pH to 7.0 with HCl, bring the volume to 1000 ml, and store at 4° or aliquot and then store at −20°.

5.2. Procedures of Sui⁻ test by histidine requirement

The appropriate strain or transformant carrying *his4-303* is streaked out onto SC and SC-His plates and compared for the growth in the absence or presence of histidine. Alternatively, the cells are grown in patches on the solid medium and replica plated onto SC and SC-His plates, just as described in the preceding section. The His$^+$ strain is judged to be Sui$^-$. If a weak His$^+$ phenotype is observed, the spot assay is used to better compare the growth between the selective media. Detection of a weak Sui$^-$ phenotype in a *his4-303* strain can be facilitated by addition of a small amount of histidine at a final concentration of 1 to 20 μM.

5.3. Procedures of Sui$^-$ test by β-galactosidase assay

5.3.1. Cell growth and disruption

1. Introduce p367 and p391 separately to an appropriate yeast strain by transformation, and streak out several independent transformants on a solid medium and incubate at 30° for 2 to 4 days.
2. Prepare a fresh overnight culture of transformants carrying p367 or p391 in SC-Ura medium. Use appropriate SC dropout medium if different reporter plasmids are chosen and they carry another selectable marker.
3. Add 1 ml of overnight culture to 50 ml of fresh medium and let grow until it reaches a cell density of 2 to 4 × 10^6 cells/ml (A_{600} ~ 1.0). It usually takes 8 h to grow in SC-Ura to this density.
4. Transfer the culture to a sterile centrifuge tube and spin at 3600×g (4.2 k rpm in a Beckman J6-MI).
5. Resuspend in 125 μl breaking buffer and transfer to a 1.5-ml microcentrifuge tube. If the assay is done in 2 days, this is the last step for the first day; freeze the cell suspension at −70°.
6. Add 6.25 μl of 40 mM PMSF. Then add glass beads (Glass beads, acid-washed, Sigma G8772) to the meniscus.
7. Vortex four times for 15 sec at 4°. Check for cell breakage by examination in the microscope (checking is optional).
8. Add 250 μl breaking buffer, then vortex twice for 1 min. This and the preceding vortexing steps are critical for obtaining dense, active extracts. During vortexing, the meniscus should be V-shaped due to strong rotation. Mix and transfer the supernatant (~350 μl) to an ice-cold, fresh microcentrifuge tube.
9. Centrifuge for 15 min in a microcentrifuge at 4° and transfer the supernatant to a new microcentrifuge tube.

Note on step 1. The assay needs to be done with several independent transformants to obtain average values. Every time they are assayed, replenish them by growing a patch on fresh medium. If necessary, transformants can be stored in 15% glycerol at −80°.

Note on step 4 and onward. These steps should be done on ice and all spins should be done at 4° in a cold room.

5.4. β-galactosidase assay

1. Incubate blank 450 μl Z-buffer at 28° for 2 min. Then add extracts to a final volume of 0.5 ml (50 μl of extract). Include about four controls with Z-buffer alone.
2. Add 100 μl ONPG, and gently vortex and incubate at 28°. Be sure to note this time as time 0.
3. When the color turns yellow stop the reaction by adding 250 μl 1 M Na_2CO_3. Note the time. Make sure that the final volume is 850 μl. Stop two blanks for the shortest time and two for the longest.
4. Read A_{420}: 1 A_{420} unit is defined as 0.0045 nmol/ml.
5. Determine the total protein concentration of the extracts by Bradford assay and bovine serum albumin (BSA) as a standard, e.g., with Bio-Rad Protein Assay (Cat. #500-0006).
6. Calculate specific β-galactosidase activity by nanomoles ONPG cleaved per minute per milligram total protein.

$$\beta\text{-gal. unit} = \frac{A_{420} \times 0.85/0.0045}{\text{time (min)} \times \text{vol. of extract (ml)} \times \text{total protien (mg/ml)}}$$

5.5. UUG/AUG ratio calculation

The UUG/AUG ratio is a measure of the frequency of translation from UUG codons compared to normal initiation frequency from AUG codons. This is calculated as the percentage of β-galactosidase activity from the transformant carrying p391 ($HIS4^{UUG}\text{-}lacZ$) to the activity from the transformant carrying p367 ($HIS4^{AUG}\text{-}lacZ$). The typical UUG/AUG ratio values differ from strain to strain, ranging between 2 and 10%. However, the standard deviation for measurement with each Sui$^+$ strain should be within 20% of the obtained value. Sui$^-$ mutants should show a statistically significant increase in this value compared to the isogenic wild-type control strain ($p < 0.05$).

6. ASSAY OF TRANSLATION INITIATION ACTIVITIES WITH *GCN4* AS REPORTER

Proper amino acid response of yeast to amino acid availability is mediated by controlled derepression of *GCN4* translation in a manner depending on uORFs 1 to 4 (see Fig. 6.1). Gcd$^-$ mutations decrease (1) the assembly of TC independent of eIF2a phosphorylation (e.g., eIF2B mutations), or (2) the rate of TC loading on scanning 40S subunits.

Hence, they reduce the proportion of scanning ribosomes that have re-acquired TC by the time they reach uORF4, leading to an increased fraction that reinitiates downstream at *GCN4* instead. Other mutations impairing different eIFs impair *GCN4* translation derepression under starvation conditions, hence scoring as Gcn⁻ (see below for mechanisms). Thus, *GCN4* can serve as a specific reporter for translation initiation activities. There are two methods to measure the level of *GCN4* expression. One is to use a *GCN4-lacZ* reporter and measure β-galactosidase activity from the reporter construct (Fig. 6.2A). The other indirect method uses the *GCN4*-dependent

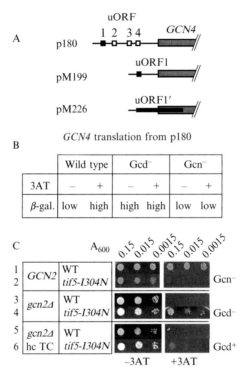

Figure 6.2 Assay of translation initiation activities using *GCN4* as the reporter. (A) Structure of the leader region of *GCN4-lacZ* fusion in the indicated plasmids. p180 contains the wild-type regulatory region. (B) Expected *GCN4-lacZ* expression levels for wild-type, general control derepressed (Gcd⁻) and general control nonderepressible (Gcn⁻) mutants bearing p180 in the presence (+) or absence (−) of 3AT in the medium. (C) Example of the 3AT test for Gcn⁻ (rows 1 and 2) and Gcd⁻ (rows 3 and 4) phenotypes and the suppression of the Gcd⁻ phenotype by hc TC (rows 5 and 6). The isogenic *GCN2⁺* strain bearing *TIF5* (row 1) or *tif5-I304N* (row 2) and the isogenic *gcn2Δ TIF5* or *tif5-I304N* strain transformant carrying a *URA3* vector (rows 3 and 4) or p1780-IMT (rows 5 and 6) are spotted onto SC-HisxLeu with (+) or without (−) 50 m*M* 3AT (rows 1 and 2) or SD with (+) or without (−) 30 m*M* 3AT (rows 3 to 6). The culture was diluted to the indicated A_{600} and 5 μl of diluted culture was spotted at each position. Cells in rows 1 and 2 and in rows 3 to 6 were incubated at 36° and 30°, respectively. (Data taken from Singh *et al.*, 2005.)

transcription of amino synthesis enzymes. Yeast overcomes the growth inhibition caused by amino acid starvation by inducing *GCN4* translation and the attendant expression of amino acid enzymes. For instance, the drug 3-amino-1,2,4-trizole (3AT) causes histidine starvation. Thus, yeast growth in the presence of 3AT is the result of Gcn4p activity and indicates strong *GCN4* expression. Among the variety of drugs used to score these phenotypes, we focus on 3AT, a histidine enzyme (His3p) inhibitor, as this is one of the most commonly used drugs to study expression from *HIS3* reporters, e.g., in two hybrid assays.

To assay the Gcd$^-$ phenotype as an indicator of reduced TC level or slow TC binding to the 40S subunit, we look for constitutive derepression of *GCN4* translation independent of Gcn2p kinase. In typical Gcd$^-$ mutants, the wild-type *GCN4-lacZ* plasmid p180 (see Fig. 6.2A) should produce β-galactosidase activity that is (1) higher than the activity obtained with the Gcd$^+$ strain, and (2) cannot be elevated by histidine starvation in the presence of Gcn2p (Fig. 6.2B, Gcd$^-$). Note, however, that Gcd$^-$ mutations that confer only partial derepression will show induction, albeit of a lesser magnitude than in *GCN2$^+$* cells starved with 3AT. It is also advisable to assay p227, the p180 derivative lacking all four uORFs, to ensure that the mutations of interest have no effect on its expression. This eliminates possible effects on transcription of *GCN4-lacZ*.

To use 3AT resistance as the measure of Gcn4p activity, the putative Gcd$^-$ mutation needs to be introduced to a *gcn2Δ* strain. If the mutation is Gcd$^-$, it will confer 3AT resistance (Fig. 6.2C). The Gcd$^-$ phenotype resulting from a low TC level or slow TC recruitment to the ribosome should be suppressed by overexpressing all eIF2 subunits and tRNA$_i^{Met}$ from the plasmid p1780-IMT (Asano *et al.*, 1999) (see Fig. 6.2C).

Gcd$^-$ mutants can be assayed in *GCN2$^+$* backgrounds, as they are resistant to amino acid analogs that compete with the cognate amino acids for incorporation into proteins. Wild-type cells are sensitive to these drugs because the drugs do not cause amino acid starvation. These drugs are briefly reviewed under the section on "Use of drugs other than 3AT to study yeast Gcd$^-$ and Gcn$^-$ phenotypes."

To assay the Gcn$^-$ phenotype, we seek failure to induce *GCN4* translation under 3AT-induced histidine starvation. Thus, Gcn$^-$ mutations should not increase β-galactosidase produced from p180 in the presence of 3AT (see Fig. 6.2B, Gcn$^-$). Gcn$^-$ mutants should be sensitive to 3AT (see Fig. 6.2C). Alternatively, Gcn$^-$ mutants can be scored with sulfometuron methyl (SM) (Cheung *et al.*, 2007).

There are three different possible mechanisms of Gcn$^-$ mutations: (1) failure to initiate at uORF1 (leaky scanning of uORF1), (2) failure to resume scanning after uORF1 translation, and (3) slow rate of scanning between uORFs 1 and 4. The first two Gcn$^-$ phenotypes arise from the fact that only ribosomes that translate uORF1 and resume scanning can skip uORF4 when TC loading is impaired. The third mechanism reflects the fact that a decrease in the rate of scanning will allow a greater fraction of scanning ribosomes to rebind TC before reaching uORF4 when the concentration of TC is reduced by eIF2α phosphorylation, thereby decreasing the proportion that can reach *GCN4*.

The following *GCN4-lacZ* fusion plasmids with a modified leader region are employed to investigate the molecular basis of a Gcn⁻ phenotype (see Fig. 6.2A). pM199 contains uORF1 only and is used to measure the efficiency of reinitiation of *GCN4* translation following uORF1 translation. pM226 is a derivative of pM199 with frameshift mutations in uORF1. The latter extends uORF1 to a site 130 nt downstream of the *GCN4* start codon, making the ribosome unable to reinitiate *GCN4* translation after translating the altered uORF1. Under these conditions, *GCN4* can be translated only by the ribosomes that had failed to initiate translation at uORF1 (e.g., by leaky scanning). Therefore, increased expression from this reporter would indicate increased frequency of leaky scanning of uORF1. Thus, both mechanisms (1) and (2) can be directly implicated from the increase in expression from pM226 and the decrease in expression from pM199, respectively. Indeed, both Gcn⁻ eIF5 and eIF5B mutants increased *GCN4* translation from pM226, indicative of leaky scanning of uORF1 (Shin *et al.*, 2002; Singh *et al.*, 2005). Interestingly, Gcn⁻ eIF1A and eIF3b mutants did not alter translation from pM199 or pM226 (Fekete *et al.*, 2005; Nielsen *et al.*, 2004). A mechanism consistent with this phenotype is the third mechanism to assume a slow migration of the ribosome on the *GCN4* leader, such that the ribosome would never reach *GCN4* ORF even under strong starvation by 3AT.

6.1. Materials

6.1.1. Plasmids

The *GCN4-lacZ* plasmids used are p180, p227 (Mueller and Hinnebusch, 1986), pM199, and pM226 (Grant and Hinnebusch, 1994). pHQ414 (*gcn2Δ::hisG:URA3:hisG*) (Qiu *et al.*, 1998) is used to disrupt the chromosomal *GCN2*. p1780-IMT is an hc plasmid encoding *IMT SUI2 SUI3 GCD11* and *URA3* (Asano *et al.*, 1999).

6.1.2. Yeast strains

GCN⁺ strains should be used to score Gcn⁻ phenotypes of a mutation of interest. H1894 (*MATa ura3–52 leu2-3,-112 trp1–63 gcn2Δ*) (Kawagishi-Kobayashi *et al.*, 1997) or H2557 (*MATα ura3–52 leu2-3 trp1–63 GAL2+ gcn2Δ*) (Marton *et al.*, 1993) is used as a typical *gcn2Δ* host strain to score Gcd⁻ phenotypes.

6.1.3. Media

Appropriate SC dropout or supplemented SD medium containing 3AT is used. 3AT is supplemented from stock solution, when the medium is cooled to ~60°. To make a 3AT stock solution, 3AT (Fluka 09540) is dissolved into distilled water to 1 M by stirring for 1 h at 60°, filter sterilized using an appropriate filter unit (e.g., Falcon 43079, pore size 0.22 μm), aliquoted, and stored at −20°. A test of the Gcn⁻ phenotype must use the complete AA powder instead of its economical alternative, AA*. SM (0.5 μg/ml) should be added to SC lacking leucine, isoleucine, and valine.

To enhance the sensitivity of the Gcn⁻ test using 3AT, 3AT is added to SC-HisxLeu, which is SC-His supplemented with 40 mM leucine. Excess leucine suppresses the general control response and, therefore, may make a weak Gcn⁻ mutant more sensitive to 3AT.

6.2. Gcd⁻ phenotype test by 3AT resistance

For recessive mutations, introduce $gcn2\Delta$ by transforming the mutant strain or its isogenic wild type with the EcoRI–XbaI fragment of pHQ414 to Ura⁺. Integration at the correct position is confirmed by 3AT sensitivity (for otherwise wild-type strain), Southern blotting, or PCR using primers hybridizing to pHQ414 and $GCN2$ flanking regions. If necessary, the $URA3$ allele inserted at the $gcn2\Delta$ locus is removed by homologous recombination (or "looping out") by growing $gcn2\Delta::URA3$ cells on an SC-FOA plate. For dominant mutations or foreign genes, introduce the expression plasmids to a $gcn2\Delta$ strain (such as H1894 or H2557) by transformation.

The desired strains or transformants are subjected to *Spot Assay* by spotting their culture onto appropriate SC-His (and other dropout as needed) or SD solid medium with or without 10 to 50 mM 3AT and incubating the plates at 30°. Histidine should not be present because it relaxes the general control response. Testing at a high temperature (such as 37°) is not recommended because this also relaxes the general control response. Weak Gcd⁻ mutants form visible colonies on 10 mM 3AT plates after 6 to 8 days.

6.3. Gcn⁻ phenotype test by 3AT sensitivity

The test of the Gcn⁻ phenotype requires that the relevant mutation is introduced to a HIS^+ GCN^+ strain. Desired strains or transformants are subjected to *Spot Assay* with SC-HisxLeu medium with or without 30 to 50 mM 3AT. Plates are incubated at 30° at an appropriate restrictive or semirestrictive temperature. For testing his^- GCN^+ derivatives, use SM (0.5 μg/ml) (see below).

6.4. Gcd⁻ or Gcn⁻ phenotype test by β-galactosidase assay

Transform mutants and isogenic wild-type control with p180. After single-colony isolation, grow yeast and assay β-galactosidase, essentially as described in the preceding section. To induce histidine starvation, 3AT is added to the culture to 10 mM after 2 h of initial growth at step 3 of "Cell Growth and Disruption." Again, be sure to remove histidine from the medium when 3AT is added.

6.5. Test of mechanisms causing Gcn⁻ phenotypes

Transform Gcn⁻ mutants and isogenic wild-type control with pM199 and pM226. After single-colony isolation, grow yeast in SC-Ura and assay β-galactosidase, as exactly described in the preceding section. Note that yeast must be grown at the temperature at which Gcn⁻ phenotypes are observed. In the case of most eIF5 Gcn⁻ mutants, Gcn⁻phenotypes were observed at 36°, but not at 30° or 33°. Thus, transformants with pM199 and pM226 were assayed at 36° (Singh *et al.*, 2005).

6.6. Use of drugs other than 3AT to study yeast Gcd⁻ and Gcn⁻ phenotypes

Gcd⁻ mutants can be assayed in $GCN2^+$ backgrounds, as they are resistant to analogs that compete with the cognate amino acids for incorporation into proteins (Niederberger *et al.*, 1986), due to constitutively elevated amino acid pools resulting from constitutively derepressed Gcn4p. Wild-type cells are sensitive to such analogs because the analogs do not inhibit biosynthesis of the cognate amino acids and, hence, do not derepress Gcn4p. The best ones are 1,2,4-triazolealanine (TRA), 5-fluorotryptophan (5-FT), and azetidine-2-carboxylic acid (AZC), which mimic His, Trp, and Pro, respectively. A combination of 0.5 mM TRA and 0.25 mM 5-FT (in SD, see the Note below) can be used to increase discrimination between wild-type and Gcd⁻ strains (Ramirez *et al.*, 1992). Pavitt *et al.* were the first to use AZC to score Gcd⁻ mutants altering eIF2B subunits (Richardson *et al.*, 2004).

Besides 3AT, Gcn⁻ mutants are sensitive to other antimetabolites (which may or may not be amino acid analogs) that interfere with amino acid biosynthesis, as these mutants cannot increase expression of the inhibited enzymes by inducing their transcription via Gcn4p. The most effective and specific inhibitors are 3AT and sulfometuron methyl (SM); SM inhibits Ilv2p, involved in synthesis of isoleucine and valine. This is a good choice for scoring Gcn⁻ phenotypes in a *his* background, because adding the required histidine overcomes the starvation effect imposed by 3AT.

Note. As 3AT cannot be used with *his* mutants, TRA cannot be used with *his* mutants and 5-FT cannot be used with *trp* auxotrophs. Finally, Pro may be substituted for NH_4 as a nitrogen source to increase uptake of TRA and 5-FT, and 5-FT must be used with minimal supplements to avoid competition with other amino acids for uptake.

7. POLYSOME PROFILING

The ultimate test of *in vivo* translation deficiency is to measure the abundance of polysome by sucrose gradient-velocity sedimentation, in which extracts prepared from growing yeast are layered onto a sucrose density gradient and resolved by ultracentrifugation in a swinging bucket rotor. Polysome runoff occurs when mutants inhibit initiation but have normal elongation rates, or at least have a stronger effect on initiation than on elongation. If translation *initiation* is defective, this reduces the number of ribosomes translating a given mRNA and hence reduces the abundance of polysomes, due to defective ribosome loading onto mRNA. If translation *elongation* is defective, this increases the number of ribosomes on a given mRNA and hence increases the abundance of polysomes, due to slow migration of translating ribosomes on the mRNA (Anand *et al.,* 2003; Ortiz and Kinzy, 2005).

Cycloheximide is added to the living cells just prior to harvesting to freeze the polysomes *in vivo* and preserve them during extract preparation. This is necessary because without the drug, elongation and termination continue in the extract, whereas new initiation events do not occur, resulting in polysome runoff. Elongation defects will also prevent polysome runoff in the absence of cycloheximide. Indeed, polysomes isolated from elongation factor mutants are more stable in the absence of cycloheximide, as described below.

Because not all of the ribosomes are engaged in translation, a substantial fraction of ribosomes exists as "vacant" 80S ribosome. Thus, vacant 80S ribosome increases when polysome decreases, and the polysome abundance is reliably measured as the polysome-to-monosome (P/M) ratio. In theory, the P/M ratio becomes smaller with initiation defects, while it becomes larger with elongation defects. As described below, vacant ribosomes can be demonstrated by specifically dissociating them into large and small subunits, taking advantage of their salt sensitivity.

Figure 6.3A shows an example of translation initiation defects observed with a slow-growing *tif5–7A* mutant altering eIF5. The *tif5–7A* mutation reduced the P/M ratio, hence the initiation frequency, at the permissive temperature of 30° (top panel), and this trend was enhanced by shifting the culture to the limiting temperature of 37° (third and fourth panels). To determine whether the difference in P/M ratio was accompanied by accumulation of vacant 80S ribosomes, vacant 80S ribosomes were dissociated

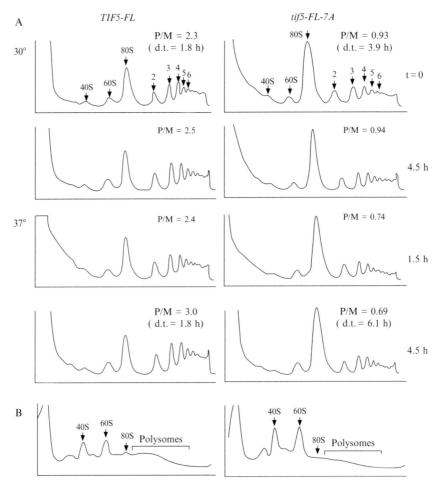

Figure 6.3 Polysome profiling of *TIF5* and *tif5–7A* strains. (A) Isogenic *TIF5* and *tif5–7A* strains were cultured in appropriate volumes of YPD at 30°. When grown to $A_{600}\sim1.0$, 300 ml of the culture was transferred to a flask equilibrated at 30°, supplemented with cycloheximide, and then collected for polysome profiling (top panel). Still another portion of the culture was transferred to a flask containing an appropriate amount of YPD equilibrated at 37°. After 1.5 and 4.5 h of temperature shift, 300 ml of the culture was withdrawn and subjected to polysome profiling (third and fourth panels). The original culture was supplemented with YPD prewarmed at 30° and subjected to polysome profiling after 4.5 h (second panel). At each step, culture was diluted with appropriate volumes of prewarmed YPD to maintain the culture A_{600} of about 1.0 when cells were collected next time. (B) Extracts prepared at time 0 were resolved with a sucrose gradient containing NaCl, as described in the text. (Adapted from Asano *et al.*, 2000.)

into subunits by salt treatment during the velocity sedimentation. Figure 6.3B indeed indicates that vacant 80S ribosomes, as visualized by 40S and 60S subunits, increased in the *tif5–7A* mutant.

7.1. Materials

7.1.2. Buffers and solutions

Breaking Buffer K includes 20 mM Tris–HCl, pH7.5, 50 mM KCl, 10 mM MgCl$_2$, 1 mM DTT, 200 μg/ml heparin, 50 μg/ml cycloheximide, manufacture-recommended amounts of EDTA-free Complete^{TM} protease inhibitors (Roche Applied Science), and 1 μg/ml each of pepstatin A, leupeptin, and aprotinin (Roche Applied Science). A 10 mg/ml cycloheximide (Sigma C-7698) solution in water is also used to supplement yeast growth media.

To set up a density gradient, 5% and 45% sucrose solutions (w/v) in 20 mM Tris–HCl, pH 7.5, 50 mM KCl, 10 mM MgCl$_2$, and 1 mM DTT are used. To make the sucrose solutions, consider that dissolving sucrose into water substantially increases the volume of the solution, and start by adding sucrose to a volume of water of about one-third of the final volume required. For a longer-term storage, the recommendation is to sterilize the solution by autoclaving. To do this, make the sucrose solution to 90% of the final volume and mark the meniscus on the bottle containing it. Autoclave the solution in the bottle and restore it to the original marked volume with sterile water. After autoclaving, add salt components from the stock solutions and then water to the final volume. The resulting sucrose solution can be stored at room temperature. Add DTT immediately before use.

7.2. Preparation of whole cell extracts

1. Grow 300 ml of yeast culture in a desired medium to $A_{600} = 1.5$ to 2.0 (mid-log phase). To aerate the cells well, grow them in a 1-liter flask with rotation at 250 rpm.
2. Five minutes prior to harvesting, add cycloheximide to 50 μg/ml and continue growing.
3. Collect the cells in ice-cold 500-ml centrifuge bottles (Nalgene). Top off with ice before closing the bottle.
4. Centrifuge immediately in a GSA-3 rotor or equivalent at 7 k rpm (6500×g) for 10 min at 4°.
5. After decanting, suspend the cells in 10 ml BBK and transfer to a conical 15-ml tube.
6. Spin in a J6-MI or equivalent centrifuge at 4.2k rpm for 7 min at 4°.
7. Remove the supernatant and estimate the cell volume. Expect ~1 ml of cells at this point. Resuspend the cells in 1.5 volumes of ice-cold BBK and add 1 volume of cold glass beads (425 to 600 μm, Sigma G8772).

8. Break the cells by vortexing at 4° with eight consecutive cycles of 45 to 60 sec, with 1 min intervals. Use a water/ice mixture bath or KCl/water/ice bath for quick chilling between cycles.
9. Spin in the J6-MI or equivalent at 1.0 k rpm for 5 min at 4° to pellet large debris and glass beads.
10. Transfer the supernatant to ice-cold microcentrifuge tubes and spin at 14 k rpm for 10 min. Repeat as needed until the supernatant is clear.
11. Transfer the cleared lysates to a fresh ice-cold microcentrifuge tube.
12. Make a 1/300 dilution and measure A_{260}. Gradients should be loaded with 25 A_{260} units, which corresponds to around 1 mg of total protein.

7.3. Procedures of sucrose gradient-velocity sedimentation

1. Using equal volumes of 5% and 45% sucrose, set up a sucrose gradient in a Beckman No. 331372 tube. Gradient preparation can be facilitated using a Gradient Master (Biocomp) according to the manufacturer's recommendations.
2. Equilibrate the temperature of the gradient at 4° before loading the sample.
3. Carefully layer 25 A_{260} units of cell extract with a pipettor. To balance the weight of two tubes for ultracentrifugation, an amount of sucrose gradient is removed very slowly from one tube prior to layering the cell extracts, such that the removed amount is equivalent to the difference in the volumes of cell extract to be loaded on the two tubes.
4. Centrifuge in an SW41Ti rotor or equivalent at 39k rpm, 4° for 2.5 h.
5. Place the tube onto a Tube Piercer (ISCO, Inc.), which connects to a Type 11 Optical unit of a UA-6 UV/Vis detector or equivalent.
6. Monitor and record A_{254} as the gradient sample flows into the UV unit.
7. Calculate the polysome to monosome (P/M) ratio, using an appropriate graphic software or photocopying the original chart, physically cutting the area under monosome and polysome peaks and weighing the pieces of paper (in milligrams).

Note on steps 5 and 6. The model 185 or equivalent density gradient fractionator (ISCO, Inc., Lincoln, NE) that is frequently used for gradient analyses was discontinued. We use a unit assembled from the components recommended by the same company. The sucrose gradient sample in the tube is pumped up to the optical unit by a peristaltic pump. We use a Tris pump from ISCO, Inc., which connects to a fractionator unit, Foxy Jr. Fraction Collector from ISCO, Inc. The speed of the pump is set at 1 ml/min. Use 60% sucrose dyed with an appropriate amount of bromophenol blue to pump up the sucrose gradient inside the tube into the optical unit. The sucrose that comes out from the tube is constantly monitored by the optical unit using a filter for A_{254}, which can be replaced with the desired absorption unit.

The optical unit signal is relayed into the UA-6 detector, which has a built-in recorder. The polysome profile is recorded on the recorder chart paper. The parameters are set on the recorder according to the experimental requirement. The sensitivity is kept at 1.0 A unit in full scale, speed at 150 cm/h, and noise filter at 1.5.

7.4. Modification to identify vacant ribosomes

To identify "vacant" 80S ribosomes, cell extracts are prepared exactly as above and layered on and resolved in the same 5% to 45% sucrose gradient except that it contains 0.7 M NaCl.

7.5. Modification to identify defects in elongation

To identify elongation defects by leaving out cycloheximide, 600 ml of yeast culture is grown and paired into two flasks at steps 1 and 2 of "Preparation of whole cell extracts." The 300-ml culture in one flask is treated with cycloheximide, collected, and washed and disrupted in BBK (with cycloheximide) as described above. The other 300-ml culture is not treated with cycloheximide, collected, and washed and disrupted in BBK lacking cycloheximide. It is important to know that even the polysomes isolated from elongation mutants eventually run off in the absence of cycloheximide. Thus, it is critical to prepare, layer, and spin the cell extracts with or without cycloheximide in parallel at a reasonable speed. The Kinzy group finds this method much more reliable in evaluating specific elongation defects.

ACKNOWLEDGMENTS

The methods described here are modified from standard procedures practiced by many yeast scientists. We thank Alan Hinnebusch for detailed reviews and numerous comments and his current and former colleagues for frank discussions about the methods described. Thanks are also due to Terri Kinzy for personal communications, Beth Montelone for helping us set up the initial yeast experiments and proofreading the manuscript, Ernie Hannig for strain information, and the former colleagues in the Asano laboratory for discussions. The research activities in the Asano laboratory are supported by NIH R01 GM64781.

REFERENCES

Anand, M., Chakraburtty, K., Marton, M., Hinnebusch, A. G., and Kinzy, T. G. (2003). Functional interactions between yeast translation eukaryotic elongation factor (eEF) 1A and eEF3. *J. Biol. Chem.* **278**, 6985–6991.

Asano, K., Phan, L., Anderson, J., and Hinnebusch, A. G. (1998). Complex formation by all five homologues of mammalian translation initiation factor 3 subunits from yeast *Saccharomyces cerevisiae*. *J. Biol. Chem.* **273**, 18573–18585.

Asano, K., Krishnamoorthy, T., Phan, L., Pavitt, G. D., and Hinnebusch, A. G. (1999). Conserved bipartite motifs in yeast eIF5 and eIF2Be, GTPase-activating and GDP-GTP exchange factors in translation initiation, mediate binding to their common substrate eIF2. *EMBO J.* **18,** 1673–1688.

Asano, K., Clayton, J., Shalev, A., and Hinnebusch, A. G. (2000). A multifactor complex of eukaryotic initiation factors eIF1, eIF2, eIF3, eIF5, and initiator tRNA^{Met} is an important translation initiation intermediate *in vivo. Genes Dev.* **14,** 2534–2546.

Boeke, J. D., LaCroute, F., and Fink, G. R. (1984). A positive selection for mutants lacking orotidine-5′-phosphate decarboxylase activity in yeast: 5-Fluoro-orotic acid resistance. *Mol. Gen. Genet.* **197,** 345–346.

Boeke, J. D., Trueheart, J., Natsoulis, G., and Fink, G. R. (1987). 5-Fluoroorotic acid as a selective agent in yeast molecular genes. *Methods Enzymol.* **154,** 164–175.

Castilho-Valavicius, B., Yoon, H., and Donahue, T. F. (1990). Genetic characterization of the *Saccharomyces cerevisiae* translational initiation suppressors *sui1, sui2* and *SUI3* and their effects on *HIS4* expression. *Genetics* **124,** 483–495.

Cesareni, G., and Murray, J. A. H. (1987). Plasmid vectors carrying the replication origin of filamentous single-stranded phages. *In* "Genetic Engineering: Principals and Methods" (J. K. Setlow and A. Hollaender, eds.), Vol. 9, pp. 135–154. Plenum Press, New York, NY.

Cheung, Y.-N., Maag, D., Mitchell, S. F., Fekete, C. A., Algire, M. A., Takacs, J. E., Shirokikh, N., Pestova, T., Lorsch, J. R., and Hinnebusch, A. (2007). Dissociation of eIF1 from the 40S ribosomal subunit is a key step in start codon selection *in vivo. Genes Dev.* **21(10),** 1217–1230.

Cigan, A. M., Pabich, E. K., and Donahue, T. F. (1988). Mutational analysis of the *HIS4* translational initiator region in *Saccharomyces cerevisiae. Mol. Cell. Biol.* **8,** 2964–2975.

Cui, Y., Dinman, J. D., Kinzy, T. G., and Peltz, S. W. (1998). The Mof2/Sui1 protein is a general monitor of translational accuracy. *Mol. Cell. Biol.* **18,** 1506–1516.

Dever, T. E. (2002). Gene-specific regulation by general translation factors. *Cell* **108,** 545–556.

Dever, T. E., Feng, L., Wek, R. C., Cigan, A. M., Donahue, T. D., and Hinnebusch, A. G. (1992). Phosphorylation of initiation factor 2a by protein kinase GCN2 mediates gene-specific translational control of *GCN4* in yeast. *Cell* **68,** 585–596.

Dever, T. E., Chen, J. J., Barber, G. N., Cigan, A. M., Feng, L., Donahue, T. F., London, I. M., Katze, M. G., and Hinnebusch, A. G. (1993). Mammalian eukaryotic initiation factor 2a kinases functionally substitute for GCN2 in the *GCN4* translational control mechanism of yeast. *Proc. Natl. Acad. Sci. USA* **90,** 4616–4620.

Dominguez, D., Altmann, M., Benz, J., Baumann, U., and Trachsel, H. (1999). Interaction of translation initiation factor eIF4G with eIF4A in the yeast *Saccharomyces cerevisiae. J. Biol. Chem.* **274,** 26720–26726.

Dominguez, D., Kislig, E., Altmann, M., and Trachsel, H. (2001). Structure and functional similarities between the cenral eukaryotic initiation factor (eIF) 4A-binding domain of mammalian eIF4G and the eIF4A-binding domain of yeast eIF4G. *Biochem. J.* **355,** 223–230.

Erickson, F. L., and Hannig, E. M. (1996). Ligand interactions with eukaryotic translation initiation factor 2: Role of the g-subunit. *EMBO J.* **15,** 6311–6320.

Evans, D. R. H., Rasmussen, C., Hanic-Joyce, P. J., Johnston, G. C., Singer, R. A., and Barnes, C. A. (1995). Mutational analysis of the Prt1 protein subunit of yeast translation initiation factor 3. *Mol. Cell. Biol.* **15,** 4525–4535.

Fekete, C. A., Applefield, D. J., Blakely, S. A., Shirokikh, N., Pestova, T., Lorsch, J. R., and Hinnebusch, A. G. (2005). The eIF1A C-terminal domain promotes initiation complex assembly, scanning and AUG selection *in vivo. EMBO J.* **24,** 3588–3601.

Gietz, R. D., and Sugino, A. (1988). New yeast-*Escherichia coli* shuttle vectors constructed with *in vitro* mutagenized yeast genes lacking six-base pair restriction sites. *Gene* **74,** 527–534.

Gomez, E., and Pavitt, G. D. (2000). Identification of domains and residues within the epsilon subunit of eukaryotic translation initiation factor 2B (eIF2b) required for guanine nucleotide exchange reveals a novel activation function promoted by eIF2B complex formation. *Mol. Cell. Biol.* **20,** 3965–3976.

Gomez, E., Mohammad, S. S., and Pavitt, G. P. (2002). Characterization of the minimal catalytic domain within eIF2B: The guanine-nucleotide exchange factor for translation initiation. *EMBO J.* **21,** 5292–5301.

Grant, C. M., and Hinnebusch, A. G. (1994). Effect of sequence context at stop codons on efficiency of reinitiation in GCN4 translational control. *Mol. Cell. Biol.* **14,** 606–618.

He, H., von der Haar, T., Singh, R. C., Ii, M., Li, B., McCarthy, J. E. G., Hinnebusch, A. G., and Asano, K. (2003). The yeast eIF4G HEAT domain interacts with eIF1 and eIF5 and is involved in stringent AUG selection. *Mol. Cell. Biol.* **23,** 5441–5445.

Hinnebusch, A. G. (1992). General and pathway-specific regulatory mechanisms controlling the synthesis of amino acid biosynthetic enzymes in *Saccharomyces cerevisiae*. *In* "The Molecular and Cellular Biology of the Yeast *Saccharomyces*: Gene Expression" (J. R. Broach, E. W. Jones, and J. R. Pringle, eds.), pp. 319–414. Cold Spring Harbor Laboratory Press, Cold Spring Harbor, NY.

Hinnebusch, A. G. (2005). Translational regulation of gcn4 and the general amino acid control of yeast. *Annu. Rev. Microbiol.* **59,** 407–450.

Hinnebusch, A. G., Dever, T. E., and Asano, K. (2006). Mechanism of translation initiation in the yeast *Saccharomyces cerevisiae*. *In* "Translational Control in Biology and Medicine" (M. B. Mathews, N. Sonenberg, and J. W. B. Hershey, eds.). Cold Spring Harbor Laboratory Press, Cold Spring Harbor, NY.

Huang, H., Yoon, H., Hannig, E. M., and Donahue, T. F. (1997). GTP hydrolysis controls stringent selection of the AUG start codon during translation initiation in *Saccharomyces cerevisiae*. *Genes Dev.* **11,** 2396–2413.

Ito, H., Fukada, Y., Murata, K., and Kimura, A. (1983). Transformation of intact yeast cells treated with alkali cations. *J. Bacteriol.* **153,** 163–168.

Jones, E. W. (1991). Tackling the protease problem in *Saccharomyces cerevisiae*. *Methods Enzymol.* **194,** 428–453.

Kawagishi-Kobayashi, M., Silverman, J. B., Ung, T. K., and Dever, T. E. (1997). Regulation of the protein kinase PKR by the vaccinia virus pseudosubstrate inhibitor K3L is dependent on residues conserved between the K3L protein and the PKR substrate eIF2a. *Mol. Cell. Biol.* **17,** 4146–4158.

Maag, D., Fekete, C. A., Gryczynski, Z., and Lorsch, J. R. (2005). A conformational change in the eukaryotic translation preinitiation complex and release of eIF1 signal recognition of the start codon. *Mol. Cell* **17,** 265–275.

Macreadie, I. G. (1990). Yeast vectors for cloning and copper-inducible expression of foreign genes. *Nucl. Acids Res.* **18,** 1078.

Marton, M. J., Crouch, D., and Hinnebusch, A. G. (1993). GCN1, a translational activator of *GCN4* in *S. cerevisiae*, is required for phosphorylation of eukaryotic translation initiation factor 2 by protein kinase GCN2. *Mol. Cell. Biol.* **13,** 3541–3556.

Mitchell, D. A., Marshall, T. K., and Deschenes, R. J. (1993). Vectors for the inducible overexpression of glutathione S-transferase fusion proteins in yeast. *Yeast* **9,** 715–722.

Mueller, P. P., and Hinnebusch, A. G. (1986). Multiple upstream AUG codons mediate translational control of *GCN4*. *Cell* **45,** 201–207.

Natarajan, K., Meyer, M. R., Jackson, B. M., Slade, D., Roberts, C., Hinnebusch, A. G., and Marton, M. J. (2001). Transcriptional profiling shows that Gcn4p is a master

CHAPTER SEVEN

LOCALIZATION AND CHARACTERIZATION OF PROTEIN–PROTEIN INTERACTION SITES

Chingakham Ranjit Singh *and* Katsura Asano

Contents

Abstract

This chapter aims to describe methods to identify and characterize protein–protein interactions that were developed during our studies on translation initiation factor complexes. Methods include the two-hybrid assay, the GST

Molecular, Cellular, and Developmental Biology Program, Division of Biology, Kansas State University, Manhattan, Kansas

Methods in Enzymology, Volume 429
ISSN 0076-6879, DOI: 10.1016/S0076-6879(07)29007-X

pull-down assay, and the coimmunoprecipitation (co-IP) assay. The two-hybrid assay provides for a convenient start to find the minimal interaction domains, which generally produce well-behaved recombinant proteins suited for various *in vitro* interaction assays. Emphasis is placed on demonstrating physiological relevance of identified interactions. The effective strategy is to find mutations that reduce the interaction by genetic or site-directed mutational approaches and obtain correlations between their effects *in vitro* (GST pull down) and effects *in vivo* (co-IP).

1. INTRODUCTION

Most protein complexes assemble and function in a highly coordinated fashion to regulate a variety of cellular activities. In eukaryotic translation initiation, sequential binding of the eukaryotic initiation factor 2(eIF2)/GTP/Met-tRNA$_i^{Met}$ ternary complex (TC) and eIF4F/mRNA complex to the 40S ribosomal subunit mediates formation of the 43S and 48S preinitiation complexes, respectively (for review of the pathway, see Hinnebusch *et al.*, 2006; Pestova *et al.*, 2006). The multisubunit factor eIF3 and numerous other factors promote these assembly processes by binding directly or indirectly to the 40S subunit. In the yeast *Saccharomyces cerevisiae*, formation of the 43S complex occurs in concert with eIF1A binding to the 40S subunit and formation of a multifactor complex (MFC) containing eIF1, eIF2, eIF3, eIF5, and Met-tRNA$_i^{Met}$, with the latter positioned at the P-site. The 48S complex then scans for the first AUG codon in the mRNA with the help of mRNA unwinding by the eIF4A helicase and its cofactor eIF4B. Correct tRNA$_i^{Met}$ anticodon pairing to the AUG promotes hydrolysis of the GTP bound to eIF2, which is mediated by the GTPase activation function of the eIF5-N-terminal domain (NTD), and the ribosomal conformational change. The release of P$_i$ that is produced by GTP hydrolysis is tightly coupled to AUG recognition by the 48S complex and directly signals dissociation of eIF1, eIF2-GDP, and eIF5. Subsequent joining of the 60S subunit and release of the remaining eIFs to produce the 80S initiation complex are stimulated by a GTPase switch eIF5B.

The MFC was originally identified by systematic, pairwise two-hybrid interaction assays between all the eIFs encoded by the yeast genome (Asano and Hinnebusch, 2001). This analysis identified interactions between eIF1 and eIF3c, eIF5 and eIF2β, and eIF3c and eIF5 in addition to those between subunits of eIF2 and eIF3. The interaction sites were identified by determining the minimal binding domains. Conserved amino acid residues found in the minimal domains were mutated, and the mutant proteins were analyzed *in vitro* by interaction assays using recombinant glutathione-S-transferase (GST) fusion proteins (GST pull-down assay) and *in vivo* by coimmunoprecipitation

(Co-IP) assays (Asano *et al.*, 1998, 1999, 2000; Phan *et al.*, 1998; Valásek *et al.*, 2004). Specifically, we focused on the eIF5-CTD (C-terminal domain), because this was found to bind simultaneously to eIF2β and eIF3c, hence potentially mediating a gigantic eIF2/eIF5/eIF3 complex composed of at least nine polypeptides (Asano *et al.*, 1999). The *tif5-7A* mutant altering the eIF5-CTD residues conserved with the eIF2Bε-CTD showed a Ts$^-$ phenotype (Asano *et al.*, 1999) and disrupted eIF2 binding to eIF3 *in vivo* as shown by coimmunoprecipitation. The latter experiment demonstrated that formation of the MFC is dependent on eIF5-CTD acting as an assembly core (Asano *et al.*, 2000; Singh *et al.*, 2004). In this chapter, we describe the methods that have been employed and modified to study the MFC, anticipating that they will be applicable to the study of other multiprotein complexes.

2. THE USE OF TWO-HYBRID ASSAY TO IDENTIFY PROTEIN–PROTEIN INTERACTION SITES

The two-hybrid assay is a powerful genetic technique that uses the transcriptional activity from a reporter gene to measure protein–protein interaction (Bartel *et al.*, 1993). The assay requires construction of "two hybrids," of which one is a DNA-binding domain fused to a protein of interest, and the other is a transcription activation domain fused to another protein in question. These two hybrid proteins are expressed in a yeast strain carrying one or more reporter genes regulated by the transcription factor that supplied the DNA-binding domain. If the two proteins interact, they form a functional activator due to the close proximity between the activation domain and the DNA-binding domain; as a result, the reporter gene will be expressed. The level of expression can be assayed to measure the strength of the protein–protein interaction. Frequently used reporter genes described in this section are Gal4p activator-driven *HIS3* or *lacZ* genes. Positive two-hybrid constructs produce reconstructed Gal4p activators, which then turn on transcription of *HIS3* or *lacZ*. Expression from these reporters is assayed as described below.

The open reading frame (ORF) segment encoding each of the proteins identified in a multiprotein complex is cloned into two-hybrid vectors, now commercially available from various companies. Every pairwise combination of activation and DNA-binding domain fusion plasmids is introduced to yeast carrying appropriate reporters and tested for expression of the reporter(s). To produce all possible combinations of two-hybrid yeast constructs, we use the ability of a haploid yeast strain carrying one plasmid to mate with another of the opposite mating type carrying a second plasmid. In this way, it is easy to produce a diploid strain carrying both plasmids.

Once interacting protein pairs are found, the minimal interacting domains are identified by creating and testing truncated versions of the two-hybrid constructs by polymerase chain reaction (PCR)-based recombinant techniques.

2.1. Materials

2.1.1. Plasmids and yeast strains

DNA fragments encoding each of the known or predicted subunits of a protein complex and their deletion or mutant derivatives are cloned into the two-hybrid expression vectors pGBT9 (*TRP1*) and pGAD424 (*LEU2*) (Bartel *et al.*, 1993) or their commercially available equivalents (Clontech, Stratagene, etc.), fusing the yeast proteins to the Gal4 (or appropriate transcription factor) DNA-binding domain (DBD) and activation domains (AD), respectively. We recommend introducing an *Nde*I site at the 5′ end of the ORF (in which the ATG triplet of the recognition site sequence 5′-CAT ATG-3′ corresponds to the start codon of the ORF) following the site used to clone the DNA fragment, if the *Nde*I site is not present in the ORF. Then the *Nde*I site is used to move the DNA segment to GST fusion or yeast expression vectors (see below).

Y190 (α *leu2 trp1GAL-lacZ GAL-HIS3*) and Y187 (a *leu2 trp1GAL-lacZ*) (Harper *et al.*, 1993), which contain chromosomal copies of the *lacZ* and *HIS3* genes under the regulation of the upstream activation sequence (UAS) recognized by Gal4p, were transformed with the resulting pGBT9- and pGAD424-derived plasmids, respectively, as described (Schiestl *et al.*, 1993).

2.1.2. Buffers and solutions

Z buffer (Miller, 1972) contains 16.1 g/liter of Na_2HPO_4, 5.5 g/liter of NaH_2PO_4, 0.75 g/liter of KCl, and 0.246 g/liter of $MgSO_4$. Adjust the pH to 7.0. At the time of use, add 2-mercaptoethanol at 2.7 ml/liter. Prepare a 2.5 *M* stock solution of 3-amino-1,2,4-triazole (3AT; Fluka 09540) by dissolving 21 g of 3AT in 100 ml of water, and sterilize by filtration. Prepare a concentrated stock by dissolving X-Gal in *N,N*-dimethylformamide at a concentration of 20 mg/ml. Store at $-20°$.

2.1.3. Yeast growth media

Synthetic complete medium and plates lacking leucine and tryptophan (SC-Leu-Trp) as well as lacking leucine, tryptophan, and histidine (SC-Leu-Trp-His) (Sherman, 1991): 1.45 g of Difco yeast nitrogen base without ammonium sulfate and amino acids, 5 g ammonium sulfate, and 2 g of appropriate amino acid drop-out powder are dissolved into water to 950 ml (for plates add 20 g of agar). Autoclave and cool to $< 65°$, and then add 50 ml of 40% D-glucose (Lee *et al.*, this volume).

2.2. Procedure of identifying interacting partners

The transformants of Y190 (Trp⁺) and Y187 (Leu⁺) are mated together as described in Fig. 7.1 to produce Trp⁺ Leu⁺ diploid strains bearing all pairwise combinations of the pGBT9 and pGAD424 constructs (method modified from Bendixen *et al.*, 1994). The Y190/Y187 diploids contain two reporter genes, *GAL-HIS3* and *GAL-lacZ*, which allowed us to judge the strength of the two-hybrid interactions by two independent assays: the growth of cells on medium containing 3AT, a sensitive indicator of *HIS3* expression (Wolfner *et al.*, 1975), and the degree of production of blue products from 5-bromo-4-chloro-3-indoyl-β-D-galactopyranoside (X-Gal), a chromogenic substrate for β-galactosidase (Breeden and Nasmyth, 1987), respectively, as described below.

2.2.1. Assay for GAL-HIS3 expression levels

The collection of Y190/Y187 yeast hybrids containing all combinations of pGAD424 and pGBT9 constructs was grown in patches on SC-Leu-Trp, incubated overnight at 30°, and replica plated to the same medium also lacking histidine (SC-Leu-Trp-His) and supplemented with 3AT at 5, 10, 15, 20, 25, and 30 m*M*. The strength of each interaction was scored qualitatively by the density of cell growth observed at 30° for up to 6 days after plating. Stronger two-hybrid interactions allowed cells to produce more *HIS3* product from the *GAL* promoter and, hence, to grow on medium containing higher concentrations of 3AT.

2.2.2. Assay for GAL-lacZ expression levels

1. Permeabilize yeast colonies or patches printed to a 45-mm nitrocellulose filter (BA85, Schleicher & Schuell) by placing the filter on a sheet of aluminum foil floating on the surface of liquid nitrogen for 5 to 10 sec.
2. Place the filter with the cell-side up on Whatman filter paper presoaked with 3 ml of Z buffer containing 1 mg/ml X-Gal.
3. Incubate the filters at 30° for 3 to 4 h for the appearance of the blue product of X-Gal cleavage by β-galactosidase.

Note 1. It is important to test expression from both reporters by these assays to confirm the authenticity of the interactions observed, with the following fact in mind: interactions that can be detected by the β-galactosidase assay with X-Gal are in general strong and correspond to the extent of *GAL-HIS3* expression allowing cells to grow on medium containing >30 m*M* 3AT.

Note 2. In the case of interaction between yeast eIF2β and eIF5, this was detected only with the combination of AD-eIF2β and DBD-eIF5 and not with that of AD-eIF5 and DBD-eIF2β (Asano and Hinnebusch, 2001;

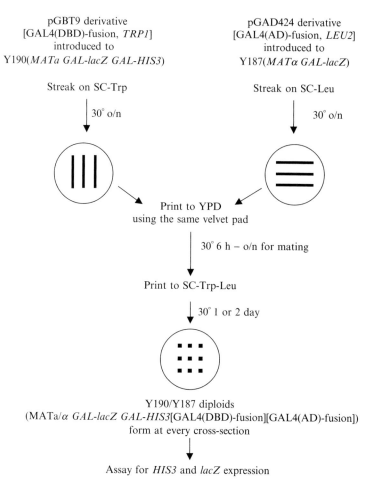

Figure 7.1 Construction of diploid yeast strains containing all pairwise combinations of Gal4p DBD and AD fusions. Y190 (left) or Y187 (right) transformants bearing pGBT9 or pGAD424 derivatives were inoculated in parallel lines on SC–Trp and SC–Leu plates, respectively. After incubating overnight at 30°, both plates were replica plated at 90° angles to a single velvet pad (Q Biogene) and then plated to rich medium (YPD). After incubating overnight at 30°, during which mating occurs at the cross sections between the transformants of opposite mating type, the mating plate was replica plated to SC–Trp–Leu. Only the diploids grew on this medium and were subsequently tested for growth on SC–Trp–Leu–His plates containing 3AT and for β-galactosidase activity with X–Gal as substrate (see text for further details).

Asano *et al.*, 1999). Thus, we recommend testing both combinations of a potentially interacting pair of proteins fused to either DBD or AD.

2.2.3. Troubleshooting

A common problem with two-hybrid assays is that some DBD fusion constructs alone autoactivate *GAL* promoters to turn on expression from *GAL-HIS3* or *GAL-lacZ*. Even if this happens, however, a positive two-hybrid interaction can still increase the reporter expression beyond the basal level promoted by the DBD fusion alone. The difference beyond the basal level can be detected using 3AT plate assays by carefully controlling the amount of the cells transferred to the replica velvet by printing to blank plates prior to printing to test plates.

If no positive interaction is observed between any pair of protein complex partners, it is possible to construct GST fusion plasmids and test interactions by GST pull-down assays, as described in the next section. Alternatively, it is possible to construct an AD fusion plasmid DNA library by shearing the subunit ORF DNA segments by sonication to an average size of ~600 kb and cloning them into pGAD424. The plasmid DNA library is used to screen for any pGBT9-derived fusion plasmids as the bait. AD or DBD-fused full-length proteins are sometimes unstable in yeast, while fusions with shorter segments can be much more stable. If this is the case, a protein segment of the partner protein can be identified by this method (Asano and Hinnebusch, 2001; Asano *et al.*, 1998; Bartel *et al.*, 1996).

2.3. Procedure of identifying interacting domains by two-hybrid assay

Plasmids encoding truncated protein fragments fused to the Gal4p AD or DBD (see below) are constructed by synthesizing DNA fragments containing the corresponding coding regions by PCR using oligonucleotides that introduce restriction enzyme sites at both ends and inserting them into pGBT9 or pGAD424 (Bartel *et al.*, 1993). The Y190 transformants carrying the resulting plasmids were mated with the Y187 transformants carrying the other two-hybrid plasmid encoding the partner protein of a desired length, as shown in Fig. 7.1. The resulting diploid strains are tested for expression of *GAL-HIS3* or *GAL-lacZ*, as described above.

2.3.1. Design of deletion constructs and evaluation of the results

The endpoints of either N-terminal or C-terminal deletions can be designed at regular intervals, which can vary depending on the size of the protein. As noted under "Troubleshooting" above, some of the DBD-fused constructs are likely to autoactivate *GAL* promoters. Thus, it is recommended that deletion constructs be made starting with an AD-fusion plasmid, if possible. Some of the deletion constructs may also lose protein stability and produce a negative signal for this reason. Misinterpretation caused by these unexpected results can be avoided by testing a reasonable number of N-terminal and C-terminal deletion constructs created at regular intervals. Alternatively, deletion constructs can be designed based on computer-predicted domains. In this case, it is recommended for the same reason that truncated versions be tested with several different endpoints per predicted domain at each terminus. The N-terminal end of the shortest N-terminal deletion construct that shows a positive interaction defines the N-terminus of the minimal domain. The C-terminus of the minimal domain is defined similarly. It is recommended to create a two-hybrid plasmid encoding the minimal domain, which is truncated to both termini as defined above, and confirm a positive interaction with its partner.

3. GST PULL-DOWN ASSAY

The GST pull-down assay is one of the most useful *in vitro* techniques to study the interaction of two or more proteins. This is based on affinity chromatography and makes use of a purified, GST-fused protein (the bait) that is immobilized to a selected glutathione-linked resin to capture and "pull down" a binding partner (the prey) by a small g force generated by a microcentrifuge.

In the GST pull-down assay, the prey proteins can be prepared from a variety of sources, such as crude or purified, recombinant, or endogenous proteins. The source can even be whole cell extracts (WCE) of the original organism, and unknown protein partners can be identified by this method in combination with mass spectrometry technology. The caveat to this technique is detection of nonspecific interactions. Thus, it is always necessary to include a negative control with binding by GST alone. In principle, nonspecific interactions can result from either experimental parameters, or the unfolding or incorrect folding of proteins used in the assay. For the former, some protein interactions are sensitive to changes in salt concentration, decreasing specific interactions and increasing nonspecific interactions. In our experience, the quality of employed proteins, in particular of the prey, matters more. The GST fusion, in general, produces a high-quality bait, because GST is a well-folded, soluble protein, and probably enhances the solubility and proper folding of the fused protein. Thus, GST fusion proteins are generally expected to retain the original propensity for interaction.

As a source of the prey protein, we find it convenient to use ^{35}S-labeled proteins expressed from a plasmid in reticulocyte lysates. Commercially available kits, such as the Promega TnT T7 or SP6 RNA polymerase-coupled system, are useful for this purpose. In this system, the gene, either wild-type or mutant, encoding a suspected partner protein is cloned under the T7 or SP6 promoter. The purified plasmid is used as a template for transcription by the bacteriophage RNA polymerases. Then the transcript is translated in the same test tube in a "cap-independent" manner by the reticulocyte ribosomes and translation factors. The merit of this system is that it generates a fresh and hopefully well-folded protein as the prey for the assay. The interactions between most yeast eIFs have been assayed with this system as the source of the prey. The disadvantage of this system is the impurity of the prey, coming from protein and RNA components of reticulocyte lysates. Thus, it is crucial at some point to confirm the interaction using a purified protein as the prey and including RNase to rule out bridging by RNA components. We also found that ^{35}S-labeled mammalian eIF3 subunits expressed in reticulocyte lysate cannot bind GST-fused bait efficiently (Shalev et al., 2001). This was likely because the expressed subunits are readily incorporated into reticulocyte translation complexes, as demonstrated for mammalian eIF2β as an early example (Pathak et al., 1988). If this prevents interaction, the prey proteins must be expressed in wheat germ extracts or purified from yeast or bacteria.

The methods of conducting the GST pull-down assay and evaluating the results carefully are described later. The experimenter should keep in mind the general cautions as discussed above. We also describe a modification of the method to study complex formation by three interacting proteins.

3.1. Materials

3.1.1. Plasmids

All recombinant proteins are expressed in BL21 $[F^- ompT\ hsdS_B(r_B^- m_B^-)]$ or BL21(DE3) carrying corresponding expression plasmids. GST fusion proteins are expressed from an isopropyl-β-D-thiogalactoside (IPTG)-inducible promoter present on the plasmids derived from a pGEX vector series (GE Healthcare Life Sciences). Using pGEX-4T-1 as the vector, we created pGEX-TIF5 (Phan et al., 1998) and pGEX-TIF35 (Asano et al., 1998), both containing the unique NdeI site (5'-CATATG-3') overlapping with yeast eIF5 and eIF3g start codons (underlined), respectively, which are in-frame with the GST coding sequence located 5' to them. In addition, these plasmids have BamHI*/SalI and PstI*/BglII/HindIII/SphI, respectively, following the fusion ORF (* indicates the site not unique to the plasmids). Thus, NdeI and one of these sites are used to conveniently prepare a variety of GST fusion plasmids. The DE3 prophage integrated into the BL21(DE3) genome expresses T7 RNA polymerase dependent on IPTG. Thus, this strain is a useful host for expressing recombinant proteins from the T7 promoter. Useful cloning vehicles for this purpose are the pET- (Novagen) and pT7- (Tabor and Richardson, 1987) series of vectors. The FLAG peptide (DYKDDDDK) or hexahistidine (His6) tags are introduced to either terminus of the expressed proteins by PCR using oligonucleotides that include the coding sequence of the tags or their complementary sequences. These plasmids can also be used as a template for ^{35}S-labeled protein synthesis using the T7 RNA polymerase-coupled reticulocyte lysate system.

3.1.2. Media

LB-Ampicillin: One liter of LB-ampicillin is prepared with 10 g tryptone, 10 g NaCl, 5 g yeast extract, and 1 liter distilled H_2O. The pH is adjusted to 7.2 with HCl, and autoclaved. For LB plates, add 15 g agar/liter before autoclaving. One milliliter of ampicillin is added after autoclaving from a stock concentration of 100 mg/ml; a 25 mM M IPTG solution in sterile water (filter-sterilized) is used as a stock solution.

3.1.3. Reagents, buffers, and solutions

Glutathione Sepharose 4B resin (Pharmacia 17–0756–01), phosphate-buffered saline (PBS) (0.14 M NaCl, 2.7 mM KCl, 10.1 mM Na_2HPO_4, pH 7.3), and Binding buffer [20 mM HEPES/KOH, pH 7.5, 75 mM KCl, 0.1 mM EDTA, 2.5 mM $MgCl_2$, 1% skim milk, 1 mM dithiothreitol (DTT), 0.05% Triton X-100] are used. The Washing buffer is the same as the Binding buffer except that it lacks skim milk. Staining solution (0.125% Coomassie Brilliant Blue R-250, 50% methanol, 10% acetic acid) and de-staining solution (40% methanol, 10% acetic acid) are used as well as Ponceau S stain (0.2% Ponceau S, 1% acetic acid).

The stock solution of RNase A (from bovine pancreas, Sigma R4875) is made at 10 mg/ml. After dissolving into 10 mM sodium acetate (pH 5.2), it was heated for 15 min at 90° and allowed to cool at room temperature. The final pH is adjusted to 7.4 by adding ∼0.1 ml of 1 M Tris–HCl (pH 7.4) and stored at −20°.

3.2. Procedure of GST pull-down assay

3.2.1. Preparation of bacterial lysate containing GST fusions

1. Grow the cells at 30° in 250 ml of LB-Amp in 1-liter flasks with good aeration.
2. Add 0.1 mM IPTG to induce GST fusions when the A_{600} is ∼0.8. The induction time (at 30°) depends on the nature of fusions, but typically it is 1 h.
3. Harvest the cells by centrifugation at 8000 rpm (10,000×g) for 10 min in a Sorvall GSA rotor or equivalent, and discard the supernatant.
4. Suspend the cell pellets in 12.5 ml of PBS, and transfer into a 50-ml centrifuge tube.

5. Sonicate the cell suspension for 6 min (2 min on and 30 sec off) with a Fisher Scientific sonicator with the centrifuge tube held in ice.

6. Add 625 μl of 20% Triton X-100, and rotate the tubes for 30 min at 4° in a cold room.

7. Centrifuge the tubes to clear the cell lysate (12k rpm, 10 min, 4°) in a microcentrifuge.

8. Transfer only the supernatant containing the GST fusion proteins in a fresh 15-ml tube.

9. To determine the yield of GST fusions in the cell lysate, incubate 0.2 ml of lysates with 5 μl of glutathione resin for 30 min at room temperature on a rocker.

10. Wash the resins three times with 0.2 ml ice-cold PBS.

11. Add 10 μl of 2× sodium dodecyl sulfate polyacrylamide gel electrophoresis (SDS–PAGE) loading buffer and heat the tubes at 95° for 2 min to elute the bound GST fusion proteins.

12. Load the supernatant on an SDS–PAGE gel. Include 5 μg and other amounts of bovine serum albumin (BSA) protein in separate lanes to quantify the amount of GST fusion protein in the lysate. Electrophorese until the front dye reaches the bottom of the gel.

13. Stain the SDS–PAGE gel with Coomassie Blue for 1 h to overnight on a rocker at room temperature.

14. Destain the gel using destaining solution until the gel background becomes clear and dry it on a gel dryer.

15. Store the cell lysates containing GST fusion protein at −80° in aliquots. This helps to minimize degradation of the GST fusion proteins by repeated thawing.

3.2.2. Preparation of the resin adsorbed to GST fusion proteins

1. Use 5 μl of glutathione resin (bed volume) for one reaction of a GST pull-down assay. To maintain a 5 μl bed volume, take out 6.66 μl of the original glutathione resin–buffer mixture in a fresh microcentrifuge tube. If the same GST fusion protein is used in multiple reactions, combine and handle GST fusions together through step 8.

2. Repeat step 1 for all of the GST fusions used in the set of experiments.

3. Centrifuge the tubes at 5000 rpm for 2 min at room temperature.

4. Discard the supernatant and wash the resin with 10 times bed volume of ice-cold PBS.

5. Add cell lysates, such that every 5 μl bed volume of resin binds an equal molar amount of GST fusion proteins.

6. Incubate for 30 min on a rocker at room temperature.

7. Centrifuge the tubes and remove the supernatant slowly so as not to disturb the resin at the bottom of the tube.

8. Wash three times with 10-times bed volume of ice-cold PBS. Equilibrate the resin with 10 times bed volume of ice-cold binding buffer. If the resin is taken for multiple reactions in a single microcentrifuge tube, split the buffer–resin mix into fresh tubes (55 μl each) where the binding reaction will take place. Spin the tubes at normal parameters described above and discard the supernatant. Washed resin in the tube should be placed on ice prior to setting up binding reactions.

3.2.3. Preparation of the prey proteins

If ^{35}S-labeled proteins are expressed in the TNT/T7 system (Promega) as the prey, set up the reaction using a recommended amount of ^{35}S-labeled L-methionine (1.175 mCi/mmol, Easy Tag^{TM} Perkin Elmer NEG709A001MC) in a volume of $5 \times (n + 1)$ μl, where n is the number of reactions using the particular construct. The resultant ^{35}S-labeled protein should be ready and fresh when the resins adsorbed to GST fusion proteins are prepared. Treat the remaining ^{35}S-labeled proteins with an exactly equal amount of SDS–PAGE loading buffer and keep on ice, to use later as inputs. This treatment will minimize the possible degradation of the ^{35}S-labeled proteins for a few hours during the binding reaction.

If purified proteins are used as the prey, they are purified appropriately. We find it convenient to use FLAG- or His6-tagged constructs to purify the relevant recombinant proteins by affinity chromatography. The method of purification of FLAG-tagged proteins from both yeast and bacteria has previously been described for the example of FLAG-eIF5 (Asano et al., 2002). To purify His6-tagged proteins, a variety of kits are available, including the His Bind Purification Kit from Novagen (cat. no. 70239–3).

Alternatively, extracts/lysates of bacteria or yeast expressing the prey protein can be directly applied to the GST pull-down assay, if antibodies specific to the prey protein are available. If the prey is expressed in an organism distantly related to the organism under study (e.g., bacterial lysates in the case of studying yeast proteins), the use of the former's extracts is not a problem because there is little chance that the proteins in the extracts specifically affect the interaction of interest. Rather, the proteins in the extracts can serve as blocking reagents (like dry milk in the binding buffer of the assay). The cell lysates/extracts from the same organism are a good source of endogenous protein complexes, such as eIF2 and eIF3 in the case of translation initiation complexes (Asano et al., 1999). However, the control reaction must be set up with the GST-bait mutant altering the binding sites for the endogenous complexes.

3.2.4. Binding reaction for binary interaction

1. Add 100 μl of fresh binding buffer containing 5 μl of the ^{35}S-labeled protein expressed in the TNT/T7 system (Promega) or an appropriate

amount of purified recombinant protein into each tube containing 5 μl resin attaching GST fusions.

2. Incubate the reaction in a cold room for 90 min on a rotator.

3. Wash the resin three times with 100 μl of ice-cold Washing buffer. Centrifuge each time for 2 min at 5k rpm at 4° and discard the supernatant.

4. Elute the protein from the washed resin by suspending in 10 μl of SDS–PAGE loading buffer and heat at 95° for 2 min.

5. Run the eluates along with a 10% input amount of the expressed protein on a 12% SDS–PAGE gel (the percentage of the gel can be adjusted according to the molecular weight of the proteins in the assay).

6. If the protein of interest is radiolabeled with ^{35}S, stain the gel with Coomassie Blue to visualize the GST fusion proteins. After staining, treat it with destaining solution and dry in a gel dryer.

7. Expose the dried gel on a phosphorimaging screen for an appropriate time and analyze the protein bands. Quantitate the amounts of bound ^{35}S-labeled proteins using STORM or TYPHOON (Molecular Dynamics).

8. If the prey is a nonradioactive protein, transfer the proteins on the gel to a nitrocellulose membrane (e.g., Hybond ECL, Amersham Pharmacia) for immunoblotting.

9. Treat the nitrocellulose membrane for 1 or 2 min with Ponceau S stain to visualize the amount of GST fusion proteins used in the assay. Excess stain can be removed simply by washing with PBS until the background is clear.

10. Detect the level of bound proteins by specific antibodies of the protein of interest using appropriate antibodies, e.g., with the ECL™ chemiluminescent system (Amersham Pharmacia) following the manufacturer's recommendation.

Note. It is generally not recommended that the equilibrium dissociation constants (K_d) be measured with the GST pull-down assay, because the prey protein can readily dissociate during the washing step of the assay, even if the reaction reaches equilibrium before this. It is also not clear if the reaction reaches the equilibrium at all at the beginning. Furthermore, the assay can erroneously produce too high values: high local concentrations of protein on the beads may lead to excessively high retention. Binding of the GST fusion protein to the beads may alter the affinity of the interaction. Finally, it is difficult to know the specific activity of the GST fusion protein, even though the GST may increase the solubility of the fused protein, as mentioned above. Thus, this assay is best suited to detect the interaction and examine the effect of mutations on the interaction.

To evaluate the nature of the reaction under study and increase the quality of the produced data, however, we recommend taking a titration

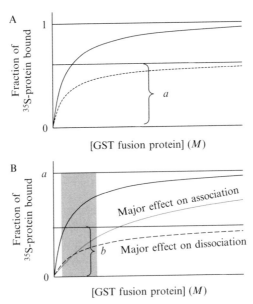

Figure 7.2 Titration curves for well-behaved GST pull-down assays. (A) Because the GST pull-down assay involves an extensive washing step, a certain fraction (*a*) of complex dissociates and lowers the plateau level. The *a* value would be lower if the dissociating rate of the complex is higher. (B) Predicted effect of mutations altering GST fusions. When the GST fusion protein carries a mutation altering the mutual binding site, this may decrease the dissociation rate more than the association rate (dashed curve). In this case, the plateau level decreases further to *b*. Alternatively, the mutation may decrease the association rate without altering the dissociation rate (dotted curve). If this is the case, the plateau level stays at *a*. Considering both options, the range of the optimal concentration of the GST fusion protein for mutational studies can be determined in the area shown in gray.

curve of the reaction by binding a trace amount of the prey protein to different amounts of the GST fusion. Under favorable conditions where the reaction reaches equilibrium, the plot of the fraction of ^{35}S-labeled prey protein (f) bound to a known concentration of GST fusion protein (G) follows a hyperbolic profile under the equation $G = K_d/[(1/f) - 1]$, as shown by the continuous line in Fig. 7.2A. This holds true when G is much greater than the concentration of the ^{35}S-labeled prey protein, and when the binding reaction reaches a plateau at 100% of the input amount. If the reaction under study follows this profile, it is a well-behaved reaction.

If the reaction follows a hyperbolic profile but reaches a plateau at a fraction (value a; $0 < a < 1$) of the input amount under the equation $G = K_d/[(a/f) - 1]$ (dotted line in Fig. 7.2A), the reaction may reach an equilibrium but a substantial amount of the prey protein would be dissociating from GST fusion proteins. If a is too low (less than 0.05), consider decreasing the number of washing steps to two times (step 3 above) or modifying the composition of Binding and Washing buffers (usually decreasing salt concentrations helps). Keep this titration curve in mind to determine the amount of GST fusion proteins used to test the effect of mutations introduced to either prey or bait (gray zone in Fig. 7.2B).

3.3. Modification to verify direct interactions

As mentioned above, it is possible that interaction between the GST fusion protein and ^{35}S-labeled prey protein is bridged by RNA or protein components of the reticulocyte lysate. To exclude bridging by protein, the prey protein should be purified and tested at some stage of the study. To exclude bridging by RNA, RNase can be added at any step in the binding reaction. Any RNase, single-stranded or double-stranded specific, can be used for this purpose, but we use RNase A, which cleaves 3′ end of unpaired C and U residue, as this is one of the most commonly used RNases. We recommend the following procedure.

1. Perform the binding reaction and wash the resin once, as described in steps 1 to 3.
2. Add 100 μl Washing buffer, and then RNase A to 10 μg/ml.
3. Incubate at room temperature for 20 min.
4. Wash the resin twice more and analyze the complex as in steps 3 to 10.

3.4. Modification to assay the effect of a third protein on protein–protein interaction

The GST pull-down assay is modified by adding a third protein in the reaction. This is done to test the ability of the third protein to compete with, bridge, or enhance the interaction between the GST-fused bait and the prey. It is recommended that the prey concentration be set at least 10 times lower than the bait concentration, which is usually 0.2 to 1 μM, to maintain a hyperbolic binding profile of the prey–bait interaction that is expected under favorable conditions. If ^{35}S-labeled protein expressed in the TnT/T7 system is used, the method described above is expected to meet this requirement.

To assay competition by the third protein, add the third protein in different amounts to a fixed amount of the GST-fused bait prior to adding the bait. The amount of the prey bound to the bait should become lower with the increasing amount of the third protein. To assay bridging or enhancement, add the third protein in excess of the bait prior to adding the prey. The third protein should bind rapidly to the bait due to high concentrations of both, allowing a substantial amount of the prey to be pulled down due to bridging by the third protein. As a control, a reaction should be set up with no third protein added. In this reaction, little (in the case of enhancement) or no (in the case of bridging) prey should be pulled down by the GST-fused bait alone.

To evaluate the bridging/enhancement assay, add different excess amounts of the third protein to fixed concentrations of the GST bait and the prey protein. The concentration of the binary complex made of the bait and the third proteins can be determined by the amount of the third protein precipitated in the reaction, assuming an appropriate stoichiometry (usually 1:1) of the complex. Then plot the fraction of the prey precipitated against the binary complex concentration to take a titration curve.

1. Purify the protein to be used as the third binding partner and check its concentration against a known amount of protein by running on an SDS–PAGE gel.
2. Distribute the resin bound to the GST fusion protein into an appropriate number of tubes. Each tube should contain 5 μl resin and 200 μl of Binding buffer.
3. Assemble the reaction first with different amounts of the purified third protein in each tube starting from zero to an increasing concentration.
4. Add 5 μl of TnT expressed ^{35}S-radiolabeled protein in each tube and incubate for 90 min in cold room on a rotator.
5. Follow the washing steps as described above to run on an SDS–PAGE, followed by phosphoroimaging analysis.
6. After staining the gel with Coomassie Blue, quantitate the amount of the third protein bound on the GST fusion protein. If the third protein is not visible, spare a part of the precipitated fraction and detect it by immunoblotting, for instance, with anti-His antibodies (Sigma, H1029), if the third protein is His6 tagged.
7. Quantify the amount of the bound ^{35}S-labeled protein using phosphoroimaging.
8. Draw a graph of the amount of ^{35}S-labeled protein against the amount of purified protein bound on the GST fusion proteins.

4. SITE-DIRECTED MUTAGENESIS TO STUDY PROTEIN–PROTEIN INTERACTIONS

The ultimate proof for any protein–protein interaction is to detect the relevant interaction *in vivo* in a wild-type strain (by Co-IP assay, as described in the next section), and observing a decrease in this interaction in the isogenic mutant altering the binding site of the protein under study. Thus, it is very critical to find mutations that reduce the interaction of interest.

If the interaction is essential for cell growth or any other activity whose alteration produces a phenotype, a straightforward way to find such mutations is to use genetics; the yeast shuttle plasmid to express the protein under study is modified, such that DNA encoding the minimal binding domain is flanked by unique restriction enzyme sites. The binding domain-encoding DNA is mutagenized by error-prone PCR and the resulting plasmid library is used to screen for mutants showing a desired phenotype. However, this method is laborious and needs to be done carefully so as not to spend too much time on false mutants.

To find appropriate mutations using an alternative approach, start by altering a cluster of amino acids present in the minimal binding domain, which is highly conserved throughout evolution. These residues can be altered to alanines, such that they disrupt the local structure of the binding site. If the structure of the protein or a homolog is known, charged surface (solvent-exposed) residues can be altered to polar residues with an equivalent size, e.g., basic residues to glutamine and acidic residues to serine. However, alanine substitutions are recommended for the initial, exploratory stage of the study, as they are expected to produce a stronger effect on the interaction of interest, if they alter the authentic binding site.

The created mutations are first introduced to appropriate plasmids to study their effects on protein interactions in a two-hybrid or GST pull-down assay. The GST pull-down assay is recommended at this stage because the introduced mutations can destabilize the protein in yeast and produce a false negative signal in the two-hybrid assay. On the other hand, the GST pull-down assay can use equal amounts of wild-type and mutant proteins to evaluate the results appropriately. Once the mutations that impair the interaction *in vitro* are found, they are introduced to appropriate yeast strains for *in vivo* assays.

4.1. Procedures of site-directed mutagenesis

Oligodeoxyribonucleotide-directed site-directed mutagenesis is efficiently performed by modifying PCR reactions. If a unique restriction enzyme site is present less than 100 bp from the desired mutation site, an oligo covering both the restriction site and the mutation site is designed and used to generate a restriction DNA fragment containing the mutation site by PCR. The mutant fragment is cloned into appropriate plasmids. If there is no such restriction site, two complementary oligos covering the mutation site are designed and each is used to synthesize a mutant DNA segment with the mutation site at either the 5′- or 3′-end of the fragment. The two mutant DNA fragments are combined as template for a second PCR to generate a larger DNA fragment with the mutation site in its middle. Finally, this is cloned into appropriate plasmids. Mutant plasmid construction using these methods is described in detail in the supplementary file of Yamamoto *et al.* (2005).

5. Co-IP Assay

In Co-IP, an endogenous protein complex found in the WCE is purified or "pulled down" in one step by small-scale immunoaffinity chromatography and the composition of the coprecipitated proteins is analyzed by immunoblotting. While technically similar to GST pull-down, it is important to note that the Co-IP assay method should be used to analyze protein–protein interactions that occur *in vivo*, while the GST pull-down assay method should be used to study interactions that occur *in vitro* (even if the whole cell extract is used as the source of the prey).

To perform Co-IP, any antibody, polyclonal or monoclonal, can be used after attaching to protein A- or protein G-conjugated resin, and the experiment can be controlled by using the same resin attached to commercially available IgG as a negative control. For a better controlled experiment, we recommend introducing an epitope tag to the protein of interest, and immunoprecipitation of the tagged protein with commercially available monoclonal antibodies raised against the epitope tag. The tags commonly used are FLAG (DYKDDDDK) and HA (for the influenza hemagglutinin peptide; YPYDVPDYA) epitopes. We describe below the Co-IP assays specifically using these epitopes. As a negative control, use cell extracts prepared from an isogenic strain encoding the untagged protein. To evaluate the amount of protein precipitated and the quality of the experiment, always analyze the pellet fractions together with input and supernatant fractions.

5.1. Materials

5.1.1. Yeast strains

A FLAG or HA epitope-coding sequence is conveniently introduced to either terminus of the ORF of any protein by designing an oligo containing the epitope-coding sequence or its complementary sequence and using it for PCR to make a DNA fragment encoding the tagged protein. The DNA segment is then cloned into an appropriate yeast expression vector. For the Co-IP assay, it is desirable to construct a yeast strain expressing the tagged protein as the sole source. This can be done by the method of "plasmid shuffling," which uses the drug 5-fluoroorotic acid (FOA) to select against the plasmid carrying the URA3 marker (Boeke *et al.*, 1984). The plasmid carrying the tagged allele will be "shuffled in" to replace the resident URA3 plasmid carrying the wild-type, unmodified allele in a strain deleted for the chromosomal copy of the gene of interest (Boeke *et al.*, 1987).

5.1.2. Media

Selection of yeast media for coimmunoprecipitation depends upon the experimental design and strain backgrounds. Generally, YPD and SC drop-out media are used (Sherman, 1991). YPD medium consists of 1% (w/v) yeast extract (Difco) and 2% (w/v) bacto peptone (Difco). SC medium has a 0.2% (w/v) mixture of amino acid powder (0.2 g adenine, 1 g arginine, 1.4 g aspartate, 0.4 g histidine, 1 g isoleucine, 2 g leucine, 1 g lysine, 0.4 g methionine, 1 g phenylalanine, 2 g threonine, 1 g tryptophan, 1 g tyrosine, 2.8 g valine, and 0.4 g uracil), 0.145% (w/v) yeast nitrogen base without amino acid and ammonium sulfate (Difco), and 0.5% (w/v) ammonium sulfate. Glucose is added to 2% from a sterile stock of 40% glucose in both YPD and SC media. The SC drop-out medium is the same as SC, but lacks one or more amino acids/bases in the amino acid powder that are covered by the selectable marker, e.g., SC-Leu (SC minus Leu) deficient in leucine is used for selection of Leu^+ transformants.

5.1.3. Buffers and solutions

Buffer A consists of 100 mM KCl, 20 mM Tris–HCl, pH 7.5, 0.1 mM EDTA, 5 mM $MgCl_2$, 7 mM 2-mercaptoethanol, 5 mM NaF, 1 mM phenylmethylsulfonyl fluoride, complete protease inhibitors (Roche Applied Science), and 1 μg/ml each of pepstatin A, leupeptin, and aprotinin. Buffer A + T consists of Buffer A and 0.1% Triton X-100. Buffer A + 2T contains Buffer A and 0.2% Triton X-100; 0.1 M glycine–HCl, pH 3.5, is used to renature anti-FLAG affinity resin.

5.2. Procedure

5.2.1. Cell culture and whole cell extract (WCE) preparation

1. Grow the selected yeast cells overnight in 15-ml culture tubes and make a subculture in 100 ml of suitable liquid medium with a starting A_{600} of 0.2 to 0.4 until the culture reaches the mid-exponential phase ($A_{600} \sim 1$).
2. Harvest the cells by centrifugation at 4000 rpm for 10 min at 4° in a Beckman J6-MI or equivalent.
3. Wash the cell pellets with 5 ml of ice-cold sterilized double-distilled water.
4. Centrifuge at the same condition as above and remove the supernatant.
5. Resuspend the cell pellets in 1.5 cell volume of Buffer A and 1 cell volume of glass beads.
6. Vortex to disrupt the cells for 30 sec followed by 30 sec on ice and continued vortexing for six cycles.
7. Transfer the cell suspension without the glass beads to a fresh microcentrifuge tube with the help of a pipette and microcentrifuge for 10 min at 13,000 rpm at 4°.

8. Transfer the clear middle layer to an ice-cold microcentrifuge tube and measure the protein concentration by the Bradford assay.

   Note on Step 2. When formaldehyde is used to cross-link protein complexes, the cell culture is treated with 1% formaldehyde (e.g., from 37% formaldehyde, Fisher F79-500) for 10 min on ice (see below).
   Note on Step 7. To maximize the recovery of WCE, puncture the bottom of the tube with a heated needle *with the cap of the tube open.* Then set the tube over a second fresh microcentrifuge tube. Place these in a 50-ml conical tube (this can be reused for this purpose). Spin at 1k rpm for 1 min at room temperature in a Beckman J6-MI or equivalent to recover WCE in the second tube.

5.3. Preparation of immunoaffinity resin adsorbed to antibodies

5.3.1. Anti-FLAG affinity resin preparation

1. Take 100 μl of Mouse Anti-FLAG M2 affinity-resin (Sigma A2220) in a microcentrifuge tube for six reactions.
2. Centrifuge the tube at 5k rpm for 2 min at 4° and remove the storage solution.
3. Wash the resin with 600 μl of 0.1 M glycine–HCl, pH 3.5.
4. Continue washing three times with 1 ml ice-cold PBS. Equilibrate the resin by adding 500 μl Buffer A containing 0.1% Triton X-100 and divide among the six tubes.
5. Centrifuge and remove the supernatant; the resin is now ready for the binding reaction.

5.3.2. Anti-HA affinity resin preparation

1. Place 0.15 g protein A Sepharose CL-4B (Amersham Pharmacia, 17-0780-01) in a tube and swell the resin with 150 μl of sterilized distilled water by rotating for 5 min at room temperature.
2. Centrifuge the tube and discard the supernatant and repeat the swelling process again with 1 ml of sterilized distilled water by rotating the tube for 5 min at room temperature.
3. Centrifuge and discard the supernatant.
4. Add 100 μl of Buffer A containing 0.1% Triton X-100 and 10 μl SA (10 mg/ml) along with 18 μl of anti-HA monoclonal antibody (COVANCE, PRB-101C) kept in −80°. Rotate the tube for 1.5 h in 4°.
5. Centrifuge the tube at 4° and discard the supernatant. The tube should be kept on ice from this step on.
6. Wash the resin twice with 1 ml of ice-cold Buffer A containing 0.1% Triton X-100 and divide among the six tubes. Centrifuge and remove

the supernatant; the resin (~10 μl resin) is now ready for the next binding reaction with the whole cell extract.

5.4. Immunoprecipitation reaction and detection of precipitated proteins

1. Add 200 μg WCE (as estimated by the Bradford assay) to a tube containing an appropriate immunoaffinity resin. Fill up to 200 μl with Buffer A and add Triton X-100 to 0.1%.
2. Incubate the resin–WCE mixture at 4° for 90 min with constant rocking.
3. Microcentrifuge the tube to remove the supernatant. Keep the supernatant to analyze a portion (usually 10 μl or 5%) by immunoblotting.
4. Wash the resin three times with 200 μl of Buffer A containing 0.1% of Triton X-100. Each time, collect the resin by briefly spinning the tube.
5. Elute the bound proteins with 10 μl of 2× Laemmli buffer (Laemmli, 1970) and heat for 2 min at 95°.
6. Resolve the immunoprecipitated (pellet) fraction with SDS–PAGE gel of appropriate acrylamide concentration, together with 20% of input (original WCE) and 5% of supernatant fractions.
7. Detect precipitated proteins by immunoblotting with appropriate antibodies.

5.4.1. Troubleshooting

A common problem with this assay occurs when the epitope is sterically sequestered. This can be solved by using a small amount of detergent in the immunoprecipitation reaction. Specifically, the Triton X-100 concentration may be increased to 0.2% and SDS may be additionally added to 0.1 to 0.2%.

Like the GST pull-down assay, this assay is biased toward dissociation. To detect weaker interactions, washing conditions can be modified as described in the section describing the GST pull-down assay. Alternatively, the immune complexes can be cross-linked using formaldehyde, as noted above in step 2 of WCE preparation.

ACKNOWLEDGMENTS

All of the assays described here are modified from biochemical/genetic techniques practiced by many yeast scientists and, in particular, those developed and modified by members of the Hinnebusch group. We are greatly indebted to Alan Hinnebusch and the present and former members of his laboratory for frank discussion about these methods. Special thanks are due to Beth Montelone for skillful proofreading of the manuscript. We also thank Yasufumi Yamamoto, Tsuyoshi Udagawa, and Bumjun Lee for discussions. The research activities in the Asano laboratory are supported by NIH R01 GM64781.

REFERENCES

Asano, K., and Hinnebusch, A. G. (2001). Protein interactions important in eukaryotic translation initiation. *In* "Two Hybrid Systems, Methods and Protocols" (P. N. MacDonald, ed.), Vol. 177, pp. 179–198. Humana Press, Inc., Totowa, NJ.

Asano, K., Phan, L., Anderson, J., and Hinnebusch, A. G. (1998). Complex formation by all five homologues of mammalian translation initiation factor 3 subunits from yeast *Saccharomyces cerevisiae. J. Biol. Chem.* **273**, 18573–18585.

Asano, K., Krishnamoorthy, T., Phan, L., Pavitt, G. D., and Hinnebusch, A. G. (1999). Conserved bipartite motifs in yeast eIF5 and eIF2Be, GTPase-activating and GDP-GTP exchange factors in translation initiation, mediate binding to their common substrate eIF2. *EMBO J.* **18**, 1673–1688.

Asano, K., Clayton, J., Shalev, A., and Hinnebusch, A. G. (2000). A multifactor complex of eukaryotic initiation factors eIF1, eIF2, eIF3, eIF5, and initiator tRNA^{Met} is an important translation initiation intermediate *in vivo. Genes Dev.* **14**, 2534–2546.

Asano, K., Lon, P., Krishnamoorthy, T., Pavitt, G. D., Gomez, E., Hannig, E. M., Nika, J., Donahue, T. F., Huang, H.-K., and Hinnebusch, A. G. (2002). Analysis and reconstitution of translation initiation *in vitro. Methods Enzymol.* **351**, 221–247.

Bartel, P. L., Chien, C. T., Stemglanz, R., and Fields, S. (1993). Using the two-hybrid system to detect protein-protein interactions. *In* "Cellular Interactions in Development: A Practical Approach" (D. A. Hartley, ed.), pp. 153–179. Oxford University Press, Oxford.

Bartel, P. L., Roecklein, J. A., SenGupta, D., and Fields, S. (1996). A protein linkage map of *Escherichia coli* bacteriophage T7. *Nature Genet.* **12**, 72–77.

Bendixen, C., Gangloff, S., and Rothestein, R. (1994). A yeast mating-selection scheme for detection of protein-protein interactions. *Nucleic Acids Res.* **22**, 1778–1779.

Boeke, J. D., LaCroute, F., and Fink, G. R. (1984). A positive selection for mutants lacking orotidine-5′-phosphate decarboxylase activity in yeast: 5-Fluoro-orotic acid resistance. *Mol. Gen. Genet.* **197**, 345–346.

Boeke, J. D., Trueheart, J., Natsoulis, G., and Fink, G. R. (1987). 5-Fluoroorotic acid as a selective agent in yeast molecular genes. *Methods Enzymol.* **154**, 164–175.

Breeden, L., and Nasmyth, K. (1987). Cell cycle control of the yeast HO gene: Cis- and trans-acting regulators. *Cell* **48**, 389–397.

Harper, J. W., Adami, G. R., Wei, N., Keyomarsi, K., and Elledge, S. J. (1993). The p21 Cdk-interacting protein Cip1 is a potent inhibitor of G1 Cyclin-dependent kinases. *Cell* **75**, 805–816.

Hinnebusch, A. G., Dever, T. E., and Asano, K. (2006). Mechanism of translation initiation in the yeast *Saccharomyces cerevisiae. In* "Translational Control in Biology and Medicine" (M. B. Mathews, N. Sonenberg, and J. W. B. Hershey, eds.). Cold Spring Harbor Laboratory Press, Cold Spring Harbor, NY.

Laemmli, U. (1970). Cleavage of structural proteins during the assembly of the head of bacteriophage T4. *Nature* **227**, 680–685.

Miller, J. H. (1972). "Experiments in Molecular Genetics." Cold Spring Harbor Laboratory Press, Cold Spring Harbor, NY.

Pathak, V. K., Nielsen, P. J., Trachsel, H., and Hershey, J. W. B. (1988). Structure of the b subunit of translational initiation factor eIF-2. *Cell* **54**, 633–639.

Pestova, T. V., Lorsch, J. R., and Hellen, C. U. T. (2006). The mechanism of translation initiation in eukaryotes. *In* "Translational Control in Biology and Medicine" (M. B. Mathews, N. Sonenberg, and J. W. B. Hershey, eds.). Cold Spring Harbor Laboratory Press, Cold Spring Harbor, NY.

Phan, L., Zhang, X., Asano, K., Anderson, J., Vornlocher, H. P., Greenberg, J. R., Qin, J., and Hinnebusch, A. G. (1998). Identification of a translation initiation factor 3 (eIF3) core complex, conserved in yeast and mammals, that interacts with eIF5. *Mol. Cell. Biol.* **18**, 4935–4946.

Schiestl, R. H., Manivasakam, P., Woods, R. A., and Gietz, R. D. (1993). Introducing DNA into yeast by transformation. *Methods* **5**, 79–85.

Shalev, A., Valasek, L., Pise-Masison, C. A., Radonovich, M., Phan, L., Clayton, J., He, H., Brady, J. N., Hinnebusch, A. G., and Asano, K. (2001). *Saccharomyces cerevisiae* protein Pci8p and human protein eIF3e/Int-6 interact with eukaryotic initiation factor 3 core complex by binding to cognate eIF3b subunits. *J. Biol. Chem.* **276**, 34948–34957.

Sherman, F. (1991). Getting started with yeast. *Methods Enzymol.* **191**, 3–21.

Singh, C. R., Yamamoto, Y., and Asano, K. (2004). Physical association of eukaryotic initiation factor 5 (eIF5) carboxyl terminal domain with the lysine-rich eIF2b segment strongly enhances its binding to eIF3. *J. Biol. Chem.* **279**, 49644–49655.

Tabor, S., and Richardson, C. C. (1987). DNA sequence analysis with a modified bacteriophage T7 DNA polymerase. *Proc. Natl. Acad. Sci. USA* **84**, 4767–4771.

Valásek, L., Nielsen, K. H., Zhang, F., Fekete, C. A., and Hinnebusch, A. G. (2004). Interaction of eIF3 subunit NIP1/c with eIF1 and eIF5 promote preinitiation complex assembly and regulate start codon selection. *Mol. Cell. Biol.* **24**, 9437–9455.

Wolfner, M., Yep, D., Messenguy, F., and Fink, G. R. (1975). Integration of amino acid biosynthesis into the cell cycle of *Saccharomyces cerevisiae*. *J. Mol. Biol.* **96**, 273–290.

Yamamoto, Y., Singh, C. R., Marintchev, A., Hall, N. S., Hannig, E. M., Wagner, G., and Asano, K. (2005). The eukaryotic initiation factor (eIF) 5 HEAT domain mediates multifactor assembly and scanning with distinct interfaces to eIF1, eIF2, eIF3 and eIF4G. *Proc. Natl. Acad. Sci. USA* **102**, 16164–16169.

CHAPTER EIGHT

In Vivo Stabilization of Preinitiation Complexes by Formaldehyde Cross-Linking

Leos Valášek,* Bela Szamecz,* Alan G. Hinnebusch,[†] *and* Klaus H. Nielsen[‡]

Contents

Abstract

Translation initiation starts with the formation of the 43S preinitiation complex (PIC) consisting of several soluble factors, including the ternary complex (TC; eIF2-GTP-Met-tRNA$_i^{Met}$), which associate with the small ribosomal subunit. In the next step, mRNA is recruited to form the 48S PIC and the entire machinery starts scanning the 5′ untranslated region of the mRNA until the AUG start codon is encountered. The most widely used method to separate 40S and 60S ribosomal subunits from soluble factors, monosomes and polysomes,

* Institute of Microbiology, AS CR, Prague, Czech Republic
† National Institute of Child Health and Human Development, National Institutes of Health, Bethesda, Maryland
‡ Department of Molecular Biology, University of Århus, Århus C, Denmark

Methods in Enzymology, Volume 429
ISSN 0076-6879, DOI: 10.1016/S0076-6879(07)29008-1

is sucrose density centrifugation (SDC). Since PICs are intrinsically unstable complexes that cannot withstand the forces imposed by SDC, a stabilization agent must be employed to detect the association of factors with the 40S subunit after SDC. This was initially achieved by adding heparin (a highly sulfated glycosaminoglycan) directly to the breaking buffer of cells treated with cycloheximide (a translation elongation inhibitor). However, the mechanism of stabilization is not understood and, moreover, there are indications that the use of heparin may lead to artifactual factor associations that do not reflect the factor occupancy of the 43S/48S PICs in the cell at the time of lysis. Therefore, we developed an alternative method for PIC stabilization using formaldehyde (HCHO) to cross-link factors associated with 40S ribosomal subunits *in vivo* before the disruption of the yeast cells. Results obtained using HCHO stabilization strongly indicate that the factors detected on the 43S/48S PIC after SDC approximate a real-time *in vivo* "snapshot" of the 43S/48S PIC composition. In this chapter, we will present the protocol for HCHO cross-linking in detail and demonstrate the difference between heparin and HCHO stabilization procedures. In addition, different conditions for displaying the polysome profile or PIC analysis by SDC, used to address different questions, will be outlined.

1. INTRODUCTION

Initiation of translation is a complicated process that involves many soluble proteins called eukaryotic initiation factors (eIF) that interact with the 40S ribosomal subunit to facilitate the recruitment of the ternary complex (TC) consisting of eIF2, Met-tRNA$_i^{Met}$ and GTP, and mRNA to form the 43S and 48S preinitiation complexes (PICs), respectively. The 48 PIC searches for the correct AUG start codon on the mRNA in a process called scanning, after which subunit joining occurs and the ribosome enters the elongation cycle. For a general review of initiation, see Hershey and Merrick (2000).

We will first introduce several eIFs that will be mentioned throughout the text and appear in the figures. In budding yeast *Saccharomyces cerevisiae*, the most complex initiation factor, eIF3, is composed of five essential subunits, TIF32, NIP1, PRT1, TIF34, and TIF35, and one nonessential subunit, HCR1. Although its mammalian counterpart is composed of 13 subunits, yeast eIF3 has, nevertheless, been shown to possess the critical activities of mammalian eIF3, including recruitment of TC and mRNA to the 40S ribosomes (Phan *et al.*, 1998, 2001). Genetic and biochemical data have also implicated yeast eIF3 in scanning and AUG recognition (Nielsen *et al.*, 2004; Valášek *et al.*, 2004), suggesting that its activity is required throughout the initiation process. Along with eIF3, eIF1 and eIF1A have also been implicated in TC recruitment (Algire *et al.*, 2002; Majumdar *et al.*, 2003) in addition to their critical roles in scanning and AUG recognition

(Pestova and Kolupaeva, 2002; Pestova et al., 1998). Thus, it appears that these three factors cooperate extensively during several steps of the initiation pathway. Finally, eIF5 is the GTPase-activating protein (GAP) that stimulates the GTPase activity of eIF2 in the TC when the initiator tRNA$_i^{Met}$ base pairs with the AUG start codon (Hershey and Merrick, 2000). In yeast, eIF3 is found to associate with eIFs 1, 2, and 5 in the multifactor complex (MFC) that can be found free of ribosomes *in vivo* and was demonstrated to stimulate efficiency of the initiation process (Asano et al., 2000).

Comprehensive characterization of the translation initiation machinery and defects caused by its mutant variants requires a technique that will enable a researcher to visualize, *in vivo*, which factors are bound to the 40S ribosomal subunit in wild-type (WT) versus mutant cells. A commonly used method to separate the 40S ribosomal subunit from smaller soluble factors and heavier ribosomal species is to subject whole cell extracts (WCE) to sucrose density centrifugation (SDC). However, PICs when left untreated cannot withstand the forces during SDC and leave behind empty 40S ribosomal subunits with no bound initiation factors (Fig. 8.1, right panel).

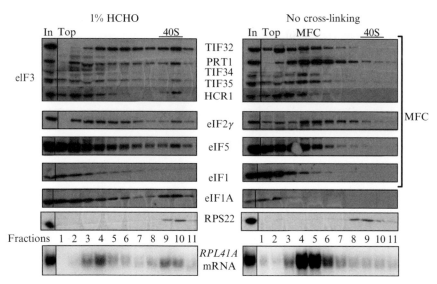

Figure 8.1 Stabilization of PIC using 1% HCHO (left panel) where a fraction of factors and mRNA is associated with the 40S ribosomal subunit in contrast to the right panel, in which no cross-linking agent was used and all factors and mRNA sediment at the top of the gradient. Immunodetection is performed by chemiluminescence (ECL^{TM}, Amersham Pharmacia Biotech) using horseradish peroxidase-conjugated secondary antibodies (Amersham Pharmacia Biotech). The probe against *RPL41A* mRNA was generated by using the Rediprime II random prime labeling system (Amersham) and [α-^{32}P]dCTP (Redivue 6000, Ci/mmol; Amersham) according to the vendor's protocol. (Reproduced with permission from Nielsen et al., 2006.)

Thus, the only way to inspect the steady-state composition of PICs *in vivo* is to use a stabilization agent prior to SDC. Heparin (a highly sulfated glycosaminoglycan) was demonstrated to stabilize PICs when added to the breaking buffer prior to yeast cells lysis (Asano *et al.*, 2000, 2001). Heparin has been shown to inhibit initiation of protein synthesis and to interact with initiation factors (Hradec and Dusek, 1978; Waldman *et al.*, 1975); however, it is not known how heparin stabilizes PICs. Importantly, our results, presented below, indicated that heparin may lead to accumulations of PICs that are not a reflection of the actual PICs in the cell at the time of lysis. The development of a new strategy to stabilize PIC was therefore undertaken using the known properties of formaldehyde (HCHO), used in chromatin immunoprecipitation assays to stabilize protein–DNA interactions in the living cell (Ren and Dynlacht, 2004) and for fixing *in vitro* translational extracts as described below (Phan *et al.*, 1998). HCHO has long been known for its ability to reversibly cross-link DNA–protein, RNA–protein, and protein–protein complexes (Jackson, 1978; Solomon and Varshavsky, 1985).

In this chapter we will describe our newly developed protocol for HCHO cross-linking as a way to stabilize PICs, in addition to WCE preparation and SDC, followed by a protocol for Western and Northern blot analysis to visualize the factors and RNAs associated with the 40S ribosomal subunit. We will also describe a special modification of the latter protocol, designated resedimentation of the 40S fractions, which can be used for further clarification of the results. Typical examples will be presented to document the differences between the use of heparin and HCHO cross-linking stabilization of the PICs. Finally, we will outline various protocols for SDC and discuss their possible applications including HCHO cross-linking for detection of 48S PICs in polysomes, known as halfmers.

2. RATIONALE BEHIND THE CHOICE OF HCHO AS A STABILIZATION AGENT

The most common and established way of monitoring the rate of translation is determination of the polysome content. This analysis is carried out by adding the translation elongation inhibitor cycloheximide to the culture just prior to harvesting the cells in order to preserve the polysomes. If no cycloheximide is added, the ribosomes will run off the mRNAs during extract preparation, resulting in the disappearance of polysomes. In our search for an agent to stabilize PICs we reasoned that if HCHO cross-links elongating ribosomes similar to cycloheximide, it could become a good candidate for further testing. Indeed, 1% HCHO under the conditions we employed was found to produce a polysome profile in a WT strain nearly identical to cycloheximide (see Fig. 8.5A and B later in this chapter); thus we rationalized that our cross-linking conditions were adequate to stabilize polysomes and PICs without introducing artifactual associations (see below).

In Fig. 8.1, a WT strain was grown to ∼1 OD_{600} after which WCE, stabilized with 1% HCHO cross-linking, was prepared, subjected to SDC (7.5% to 30% gradient), and fractions were collected. To visualize the PICs, Western and Northern blot analyses were carried out on the first 11 fractions with fractions 9 to 11 containing the 40S ribosomal subunits (PICs), using specific antibodies against chosen initiation factors and probes against $tRNA_i^{Met}$ and specific mRNAs, respectively (Nielsen et al., 2006). We used a probe against the mRNA encoding RPL41A, due to the short length of this mRNA (340 nucleotides), which ensures that free mRNP complexes sediment more slowly than 40S subunits. Another short mRNA that can be used is MFA2 (Jivotovskaya et al., 2006); however, we found this mRNA to be more susceptible to degradation than RPL41A mRNA. A possible drawback of using RPL41A is its short, 22-nucleotide-long 5′ untranslated region (UTR) (Yu and Warner, 2001), since its requirement for eIFs with respect to 40S binding could be atypical. The presence of 40S ribosomal subunits can be detected by immunoblotting using anti-RPS22 antibody that recognizes the small ribosomal protein 22. The Western and Northern blot analyses of fractions show that all factors and mRNA are present in the fractions where the 40S ribosomal subunits sediment (except eIF1) (see Fig. 8.1, left panel). Conversely, with WCE, in which no cross-linking agent was used, all factors and mRNA are found primarily in the upper fractions (see Fig. 8.1, right panel). Under the latter conditions, a multifactor complex (labeled MFC in Fig. 8.1) composed of eIF3, TC, eIF5, and eIF1 is observed free of the 40S ribosomal subunit as originally reported (Asano et al., 2000).

Since subjecting a WCE to SDC would normally be considered an in vitro experiment, it is important to explain the difference between what we call in vivo and in vitro experiments. An in vivo experiment, in our interpretation, means that PICs assembled in the living cell are stabilized either with HCHO treatment of the cells or by adding heparin to the breaking buffer used to prepare WCEs. These are then subjected to SDC, followed by Western and Northern blot analysis to visualize the factors and mRNA that cosediment in the fractions containing the 40S ribosomal subunits. In contrast, an in vitro experiment means that translation-competent WCE is made from an appropriate strain to which radiolabeled mRNA, tRNA, or purified factors (either WT or mutant) are added and where the translational activity can be measured using, for example, a reporter mRNA encoding luciferase (Asano et al., 2002). The latter translation-competent extracts can also be subjected to SDC to investigate the factor/Met-tRNA/mRNA association with the 40S ribosomal subunit and HCHO is routinely used to fix these extracts before loading them on sucrose gradients (Phan et al., 1998).

2.1. *In vivo* protocols

2.1.1. HCHO cross-linking

1. Cells are grown to OD_{600} ~1 in 200 ml of medium (either rich YEPD medium or defined SC medium, depending on the type of experiment) in a 1-liter Erlenmeyer flask. If the cell culture has to be heat shocked, cells are collected by centrifugation (10 min at $3000 \times g$), the supernatant is removed, and the pellet is resuspended in a prewarmed medium and incubated at the desired temperature.

2. Cells are then transferred to precooled centrifuge bottles (e.g., 500-ml Nalgene bottle; the exact size based on the culture volume in the experiment) that contain 25% of the total culture volume of crushed ice (50 g ice/200 ml of culture) to quickly cool the cells by inverting the centrifuge bottle five times.

3. HCHO (Mallinckrodt, cat. no. 5016-02) from a 37% stock solution is added to a final concentration of 1% relative to the original volume of the culture (5.4 ml 37% HCHO/200 ml culture) to the cooled cells by inverting the centrifuge bottle 10 times and leaving the bottle on wet ice for 1 h.

4. HCHO cross-linking is stopped by the addition of glycine to a final concentration of 0.1 M, from a 2.5 M stock solution (10 ml glycine/200 ml original culture).

3. WHOLE CELL EXTRACT PREPARATION AND WCE FRACTIONATION

1. After cross-linking with HCHO and addition of glycine, the cells are collected by centrifugation (5 min at 7000 rpm in a Sorvall RC5B rotor).

2. The cells are washed by resuspending the cells in 10 ml of lysis buffer (20 mM Tris–HCl, pH 7.5, 50 mM KCl, 10 mM $MgCl_2$), and the cell mixture is transferred to a precooled 15-ml tube and centrifuged as above.

3. The pellet is resuspended in 1.3 times v/v lysis buffer supplemented with EDTA-free protease inhibitor tablet (Roche, cat. no. 11 873 580 001), 5 mM NaF, 1 mM dithiothreitol (DTT), 1 mM phenylmethylsulfonyl fluoride (PMSF), and 1 μg/ml of the following protease inhibitors, pepstatin A, aprotinin, and leupeptin. For Northern analysis, 0.2 mg/ml of diethyl pyrocarbonate (DEPC) is added in addition.

4. Approximately the same volume of glass beads (200 to 500 μm, Thomas Scientific, cat. no. 5663R50) as lysis buffer is added and the cells are lysed by vortexing rigorously for 30 sec, followed by 1 min on ice, eight times. (A pellet volume of 0.6 ml should be resuspended in approximately 0.78 ml lysis buffer and a volume ~0.78 ml of glass beads measured using an Eppendorf tube should be added.)

Table 8.1 Gradients for SDC visualizing different ribosomal species

Ribosomal species to be resolved	Sucrose gradient	Time of centrifugation (h)	OD_{260} units resolved (12 ml gradients)	Speed (rpm)
Polysome profile (Nielsen et al., 2004)	4.5%–45%	2.5	10–15	39,000
40S subunit (Nielsen et al., 2004)	7.5%–30%	5	15–25	41,000
40S–80S subunits (Asano et al., 2000)	15%–40%	4.5	15–20	39,000

5. The disrupted cells are pelleted (5 min at $3000 \times g$) and the supernatant is transferred to a precooled 1.7-ml Eppendorf tube. The WCE is clarified by two successive centrifugations at 11,000 rpm in an Eppendorf micro-centrifuge for 2 and 10 min, respectively, each time transferring the supernatant carefully to a new precooled Eppendorf tube, taking care to avoid the lipid layer.

6. OD_{260} is measured for the WCE (triplicate measurements are done to ensure the best comparison between the WCE obtained from WT and the mutant strains) and the same number of OD_{260} unit in a total volume of 200 μl of lysis buffer is loaded on the gradient (see Table 8.1 for the percentage of sucrose, amount to load, as well as time and speed for the different types of experiments).

7. The WCEs are separated by SDC using a Beckman SW41 rotor and 0.7-ml fractions are collected while scanning continuously at A_{254} using an ISCO gradient fraction collector. If the experiment is designed to measure polysome profiles only, the scanning at A_{254} is performed without fractionation.

4. ANALYSIS OF FRACTIONATED PREINITIATION COMPLEXES

4.1. Western blot analysis

1. Depending on whether the analysis requires both Western and Northern blot analysis, either the whole fraction or only a 0.2-ml aliquot of the fraction, respectively, is used for Western blot analysis.

2. A 6× sodium dodecyl sulfate (SDS)–polyacrylamide gel electrophoresis loading buffer is added [375 mM Tris–HC, pH 6.8, 12% SDS, 30% sucrose, 0.06% bromophenol blue (sodium salt), and 1.47% 2-mercaptoethanol] to a final dilution of 1:6 and the samples are boiled for 10 min, a treatment sufficient to reverse cross-linking induced by HCHO, prior to SDS–polyacrylamide gel electrophoresis and Western blot analysis (Jivotovskaya *et al.*, 2006).

3. This protocol may not be sensitive enough for some experiments. If it does not produce sufficiently strong signals after Western blotting, another, more sensitive approach is to precipitate the fraction with 1.7 volumes of 100% ethanol chilled on dry ice and store at −20° overnight. The samples are then sedimented at 13,000 rpm for 30 min in an Eppendorf microcentrifuge, washed with cold 100% ethanol, and the pellet is resuspended in 20 to 100 μl 1× SDS–polyacrylamide gel electrophoresis loading buffer depending on the original volume. The samples are then boiled and processed as described above (Nielsen *et al.*, 2004).

4.2. Northern blot analysis

1. All RNA work is carried out taking special precautions to avoid contamination with RNAs. Gloves and Eppendorf tubes were RNase free and deionized H_2O was DEP treated.

2. RNA is extracted from the remaining 0.5 ml of the fractions. Total RNA from the sucrose gradient fractions is precipitated overnight at −20° by the addition of 1.0 ml of ice cold 100% ethanol, 50 μl 3 M NaAc, and 80 μg of sheared denatured herring sperm DNA (the addition of this DNA presumably facilitates the transfer of small RNA species from the gel to the membrane; Maraia, 1991).

3. Precipitated RNA is pelleted by centrifugation at 13,000 rpm for 30 min in an Eppendorf centrifuge at 4°. RNA is resuspended in 300 μl of RNA lysis buffer (20 mM Tris–HCl, pH 7.4, 100 mM NaCl, 2.5 mM EDTA, 1% SDS).

4. The RNA is extracted two times with hot (70°) phenol equilibrated in RNA lysis buffer lacking SDS for 15 min using an Eppendorf Thermomixer (operated at 1000 rpm), which is sufficient to reverse the cross-linking. In between extractions, the Eppendorf tubes are centrifuged in a Beckman J-6 centrifuge for 10 min at 4200 rpm to obtain a phase separation perpendicular to the axis of rotation.

5. The RNA is precipitated by the addition of 600 μl of cold ethanol (at this stage the RNA can be stored at −80°) and pelleted by centrifugation at 13,000 rpm for 30 min in an Eppendorf centrifuge at 4°.

6. The RNA pellets are washed with 300 μl of cold 70% ethanol, briefly dried under vacuum in a speed vac (∼5 min with heating on), and resuspended in 70 μl RNA formamide loading buffer (10 ml deionized

formamide, 50 μl 20% SDS, and xylene cyanol and bromophenol blue dyes aliquoted and stored at −20°). After heating at 70° for 15 min, one-half is resolved on a 10% polyacrylamide-Tris-borate-EDTA-urea gel (Bio-Rad Laboratories).

7. General protocols are then used for Northern blot transfer and development (Nielsen *et al.*, 2004, 2006; Sambrook *et al.*, 1989).

5. SPECIAL CONSIDERATIONS AND THE RESEDIMENTATION PROTOCOL

For unknown reasons, eIF1 is weakly associated with the 40S ribosomal subunit when 1% HCHO cross-linking is used to stabilize PICs instead of heparin. Instead, a trailing of eIF1 evenly distributed throughout the gradient is observed. In addition, in certain cases we encountered difficulties in detecting clear peaks with other eIFs or mRNAs. As described in detail below, several approaches can be employed to overcome this obstacle. For instance, 2% HCHO cross-linking can be used instead of 1% HCHO, which improves the amount of eIFs cosedimenting with the 40S ribosome (Jivotovskaya *et al.*, 2006); however, under these conditions the extracts are not suitable for polysome profiles.

A genetic trick to improve the odds of observing the mRNA peak in the 40S-containing fractions is to delete *RPL11B*, one of two genes encoding the 60S protein RPL11. The absence of *RPL11B* in the WT strain reduced the steady-state level of 60S subunits and produced halfmers on the 80S and polyribosome peaks, as expected from a reduced rate of 40S to 60S joining (Rotenberg *et al.*, 1988). Consistent with this, deletion of *RPL11B* in the otherwise WT strain increased the mRNA level in the 40S region by ∼1.7-fold due to an accumulation of free 48S complexes (Jivotovskaya *et al.*, 2006). Unfortunately, in several cases this method was not sufficient to observe a clear 40S-associated mRNA peak.

Another approach that was specifically designed to eliminate trailing is resedimentation (a second SDC) of the fractions containing the 40S species (Jivotovskaya *et al.*, 2006). This technique greatly simplifies the interpretation of whether a factor or RNA species is associated with the 40S ribosomal subunit (see Fig. 8.4 and Typical examples).

1. For resedimentation experiments, we follow the conventional protocol as described above with the exception that 2-fold greater OD_{260} units of WCE are resolved on the first gradient.

2. Following the first sedimentation, the 40S fractions are pooled, diluted with 10 volumes of lysis buffer (but lacking all inhibitors). In this way, the sucrose is diluted so that the mixture can be concentrated.

interactions could underlie the cooperative interaction of MFC components with their independent binding sites on the ribosome (Valášek *et al.*, 2003). Currently, two interactions have been demonstrated between eIF2 and eIF3, one direct interaction between the TIF32 subunit in eIF3 and eIF2β, and one indirect interaction between the NIP1 subunit of eIF3 and eIF2β that is bridged by eIF5 (encoded by *TIF5*). It was hypothesized that mutating these two interactions, *tif5–7A* hc*TIF32*-Δ6, would lead to a reduction in the level of TC associated with the 40S ribosomal subunit; however, only HCHO cross-linking gave the predicted result (Nielsen *et al.*, 2004), while with heparin treatment no difference was observed (L. Valášek, K. Nielsen, and A. Hinnebusch, unpublished data).

The cause of the difference between heparin and HCHO stabilization is not understood; however, it was recently reported that single-stranded oligonucleotides could suppress the requirement for mammalian eIF3j in eIF3 binding to the 40S ribosomal subunit (Kolupaeva *et al.*, 2005). Thus, it is possible that the polyanionic heparin might function similarly to single-stranded oligonucleotides in stabilizing the eIF3–40S association. Nevertheless, the accumulated data strongly suggest that heparin stabilization does not reflect the steady-state composition of the PIC and can produce artifacts.

5.2. Two percent HCHO cross-linking

When investigating a temperature sensitive (Ts$^-$) mutant (*prt1-rmp1*), it was found that at the permissive temperature, 25°, the mutant and its isogenic WT strain grew with nearly identical doubling times and their polysome profiles were indistinguishable. This strongly suggested that the translation rate of the mutant was comparable to WT. However, when performing Western blotting analysis of *prt1-rmp1* cells grown at 25°, factor binding to the 40S ribosomal subunit seemed to be dramatically impaired. We rationalized that 1% HCHO cross-linking was not sufficient for a complete cross-linking of PICs in the *prt1-rmp1* mutant. This implied that some dissociation of PICs can occur during SDC for certain mutants that contain less stable PICs. Thus, we decided to increase the concentration of HCHO to 2% and showed that under these conditions, the composition of PICs in the *prt1-rmp1* mutant and WT strain displayed little difference, as initially expected (Nielsen *et al.*, 2006). Importantly, at nonpermissive temperature, 2% HCHO cross-linking of the *prt1-rmp1* mutant displayed a severe reduction in 40S association of all essential subunits of eIF3 and some other eIFs (Fig. 8.3, right panel). In contrast, no difference in mRNA and TC could be observed. In addition, an apparent improvement in the amount of all factors, including eIF1, associated with 40S ribosomes was observed in the WT strain when compared to the same strain treated only with 1% HCHO (Fig. 8.3, left panel versus Fig. 8.2, upper left panel). It should be noted that

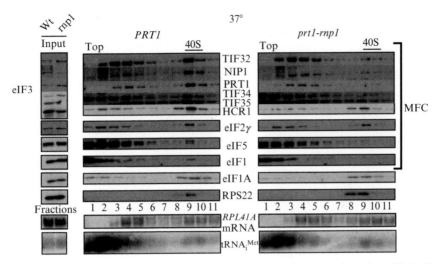

Figure 8.3 Western and Northern blot analysis of gradient fractions from SDC of WCE prepared from 2% HCHO cross-linking of a WT strain (*PRT1*) (left panel) and a mutant *prt1-rnp1* strain (right panel) after heat shock for 30 min, which displays a dramatic reduction in the amount of the eIF3 complex associated with the 40S ribosomal subunit (right panel). (Reproduced with permission from Nielsen *et al.*, 2006.)

the latter two experiments were not conducted in parallel; however, they illustrate fairly well the standard outcomes of these two variants. To conclude, the use of 2% HCHO may always be a better choice when examining PIC composition since the only concern is that it cannot be used for polysome profiles. With 2% HCHO, the cross-linking becomes more extensive and presumably the heavy polysomes get cross-linked to each other and therefore sediment as a bulk during the SDC. We are not aware of any artifactual associations in PICs that could be caused by 2% HCHO cross-linking, but we cannot, of course, completely rule them out.

5.3. Resedimentation experiments

Association of factors and especially mRNA with the 40S ribosomal subunit, visualized as a peak, may sometimes be misrepresented by factors and mRNA trailing from the immediately preceding fractions. This phenomenon complicates the data interpretation as it becomes difficult to distinguish the true amount of factors and mRNA associated with the 40S ribosomes from unbound portions of these components. A conspicuous example of mRNA trailing is displayed in mutants showing prominent polysome runoff with a resulting large accumulation of mRNA the top fractions. To circumvent this problem, it is possible to perform an extra SDC of the 40S fractions from the first SDC to minimize the trailing effect, as the trailing mRNA should be present primarily in the upper fractions after the second SDC. As can be seen in Fig. 8.4A, the *RPL41A* mRNA does not form an obvious peak in the 40S fractions of a mutant yeast strain (*tif32-td prt1-td*) in which the entire eIF3 complex has been depleted by the use of degron-tagging. (The *-td* stands for degron, a special tool used to make a conditional null-allele; Dohmen *et al.*, 1994; Labib *et al.*, 2000.) This was attributed to the accumulation of unbound mRNA in the upper fractions because of the polysome runoff that occurs in this mutant. Nevertheless, we still observed a strong and surprisingly evenly distributed signal running into the 40S-containing fractions—the trailing phenomenon. Deletion of *RPL11B* increased the amount of PICs and, thus, a stronger 40S peak was observed in the WT *rpl11bΔ* strain (Fig. 8.4B, left panel). However, in the mutant *tif32-td prt1-td rpl11bΔ* strain we still observed trailing without the appearance of any obvious 40S peak (see Fig. 8.4B, right panel). When the 40S fractions from this mutant were pooled and subjected to a second round of SDC (resedimentation), the trailing effect was clearly diminished and an obvious loss of mRNA association with 40S ribosomes was observed in the mutant strain when compared to the WT strain (Fig. 8.4C). We believe the resedimentation protocol by itself, without the need for deletion of *RPL11B*, in most cases, should be sufficient to distinguish 40S-bound mRNAs from trailing mRNAs.

Finally, it should also be mentioned that resedimentation did not completely eliminate trailing in all cases, indicating that factors and mRNAs are stripped from the PICs even during the second SDC. This again suggests that cross-linking of the PICs is not complete when 1% HCHO is used. Thus, the use of 2% HCHO cross-linking is recommended in order to ensure more complete stabilization of PICs.

5.4. Polysome profile analysis and halfmers

Varying the percentage of sucrose in the density gradients can be used to achieve optimal resolution of different ribosomal species (Table 8.1), including the complete profile of polysomes, 80S, and free subunits, the 40S to 80S region or just free 40S. Translation initiation is considered to be the rate-limiting step and a reduction in the polysome-to-monosome (P/M) ratio in a mutant strain serves as a hallmark of its defect. It should be noted that cycloheximide is a proven tool to preserve polysome profiles and unless Western or Northern analysis of eIF/mRNA association with ribosomes is to be conducted, there is no need to replace it with HCHO. As demonstrated in Fig. 8.5, only a small difference in the entire polysome profile was observed when cells were treated with HCHO (A) or cycloheximide (B). Slightly better resolution is obtained with cycloheximide, which should be used as the primary tool for determination of the P/M.

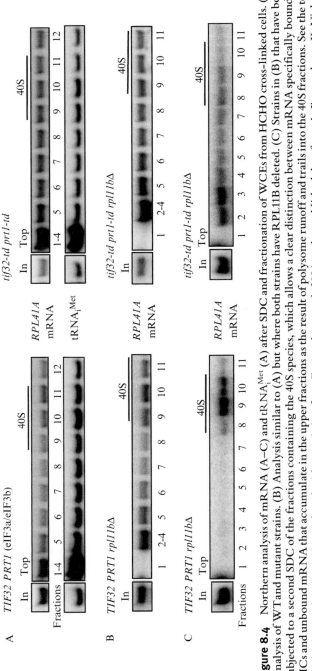

Figure 8.4 Northern analysis of mRNA (A–C) and tRNA$_i^{Met}$ (A) after SDC and fractionation of WCEs from HCHO cross-linked cells. (A) Analysis of WT and mutant strains. (B) Analysis similar to (A) but where both strains have RPL11B deleted. (C) Strains in (B) that have been subjected to a second SDC of the fractions containing the 40S species, which allows a clear distinction between mRNA specifically bound to PICs and unbound mRNA that accumulate in the upper fractions as the result of polysome runoff and trails into the 40S fractions. See the text for further explanation. (Reproduced with permission from Jivotovskaya et al., 2006, and unpublished data from A. Jivotovskaya, K. Nielsen, and A. Hinnebusch.)

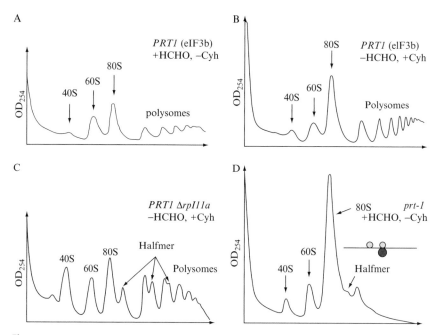

Figure 8.5   Polysome profiles examined in WCEs. (A) Wild-type strain that has been cross-linked with 1% HCHO prior to cell lysis (Nielsen *et al.*, 2004). (B) WT strain without HCHO treatment but with cycloheximide added (Nielsen *et al.*, 2004). (C) Cycloheximide added to an otherwise WT strain deleted for *RPL11A*, which exhibits halfmers due to a reduced level of 60S subunits (Rotenberg *et al.*, 1988). (D) HCHO cross-linking of mutant strain (*prt1–1*) showing a halfmer shoulder on the 80S peak (Nielsen *et al.*, 2004). (Reproduced with permission from Nielsen *et al.*, 2004, and unpublished data from K. Nielsen and A. Hinnebusch.)

Under normal growth conditions, a typical polysome profile will show peaks corresponding to free 40S and 60S ribosomal species and mRNAs containing different numbers of 80S elongating ribosomes appearing as monosomes, disomes, etc. However, each translationally active mRNA also contains one or more 48S PICs in the 5′ UTR that will eventually form 80S ribosomes when joined with the 60S subunit at the start codon. The 48S PICs should lead to the appearance of a small peak or shoulder (halfmer) on the heavy side of the monosomal and polysomal major peaks. These halfmers are normally not observed, however, possibly because the resolution of the available techniques is insufficient or possibly because the rate at which the 43S PIC scans the 5′ UTR and joins with the 60S subunit at the start codon is much faster than the rate of 43S PIC binding to the 5′ end of the mRNA. Halfmers can be observed using cycloheximide treatment when the cells contain mutations in the genes encoding either factors involved in the 60S ribosome biogenesis or large ribosomal proteins, such as

RPL11 (Fig. 8.5C). Alternatively, halfmers should be observed with mutant forms of those eIFs that promote postassembly processes such as scanning and AUG recognition. Indeed, this was the case of the *prt1–1* mutant where a weak but distinct halfmer was detected on the heavy side of the 80S peak when the cells were HCHO cross-linked (Fig. 8.5D). In this mutant, most of the 80S peak is composed of inactive 80S couples that are not bound to mRNA and therefore the halfmer peak is not expected to be prominent, even though the 80S peak is large. Interestingly, this halfmer was not detected if cycloheximide was used, not even in combination with heparin (see below). These results indicate that the composition of the *prt1–1* half-mer (see Fig. 8.5D) might be different from the halfmers occurring in the *rpl11Δ* mutant (see Fig. 8.5C). A possible explanation might be a difference in the rate-limiting step between these two mutant strains. The *rpl11Δ* strain has a defect in subunit joining because of the reduced 60S levels and, therefore, is expected to accumulate 48S PICs with Met-tRNA$_i^{Met}$ in the ribosomal P-site base paired with the AUG start codon after GTP hydrolysis on eIF2. These complexes are probably very stable since they do not require stabilization with heparin or by HCHO cross-linking before SDC. On the other hand, the *prt1–1* mutant is thought to have a defect in scanning such that 48S PICs in the process of scanning the 5' UTR would accumulate, and these halfmers would presumably require HCHO cross-linking for stabili-zation. The fact that inclusion of heparin did not stabilize formation of the *prt1–1* halfmer further indicates that the preservation of PICs by heparin is mechanistically different from that of HCHO cross-linking.

Two buffer solutions are used for preparation of the three different kinds of sucrose gradients. Buffer A contains 45% sucrose in 20 mM Tris–HCl, pH 7.5, 50 mM KCl, and 10 mM MgCl$_2$. Buffer B is identical except that it contains no sucrose. A 60% sucrose solution (autoclaved) and a 10× stock solution containing the buffer and salts (sterile filtrated) are used to prepare buffers A and B. They were then used to prepare a set of sucrose solutions to make the selected sucrose density gradient, e.g., 7.5% and 30% solutions for the 7.5% to 30% gradient according to Table 8.2. Fresh DDT was added to a final concentration of 1 mM from a 1 M stock solution immediately before making the gradients, which is conducted using a Gradient Master instrument (BioComp.).

6. FINAL REMARKS

Dissection of the roles played by the different initiation factors in translation initiation is an important task to achieve. The ability to deter-mine which eIFs or mRNAs remain associated with the 40S ribosomal subunit in 43S or 48S PICs in a particular mutant strain is imperative for

Table 8.2 Volume of buffer A and B for preparing the indicated sucrose solutions

Volume solutions for	30%			7.5%			4.5%		
# of gradients	2	4	6	2	4	6	2	4	6
Buffer A	10 ml	18 ml	26 ml	3 ml	5 ml	7 ml	2 ml	3 ml	4 ml
Buffer B	5 ml	9 ml	13 ml	15 ml	25 ml	35 ml	18 ml	27 ml	36 ml
1M DDT	15 μl	27 μl	39 μl	18 μl	30 μl	42 μl	20 μl	30 μl	40 μl

reaching this goal. Since SDC requires stabilization of the PICs, it is critical to employ the stabilization method that provides the best possible approximation of the composition of native PICs. Based on the examples discussed here, we believe that HCHO cross-linking is superior to heparin treatment in this regard, as heparin seems to misrepresent the 40S binding defect of eIFs in certain mutants such as *hcr1Δ* and *tif5-7A* hc*TIF32-Δ6* mutant strains, as described above. If the nature of this artifact is resolved, both methods may successfully complement each other. This would be desirable as one of the disadvantages of HCHO cross-linking not discussed above is a greater variation of obtained results requiring more independent experiments (three to five) than experiments performed with heparin. To summarize, cycloheximide should be used when the polysome-to-monosome ratio has to be calculated, whereas 2% HCHO cross-linking followed by resedimentation, if necessary, should be used to determine the PIC composition.

ACKNOWLEDGMENTS

We would like to thank Jon Lorsch for inviting us to write this chapter for *Methods in Enzymology*. L. V. was supported by the Wellcome Trust's Grant 076456/Z/05/Z, Howard Hughes Medical Institute Grant 55005626, Global Research Initiative Program Grant 1 R01 TW007271–01 from FIC NIH, Fellowship of Jan E. Purkyne from Academy of Sciences of the Czech Republic, and Institute Research Concept AV0Z50200510. K.H.N. was supported by the Danish National Research Foundation.

REFERENCES

Algire, M. A., Maag, D., Savio, P., Acker, M. G., Tarun, S. Z., Jr., Sachs, A. B., Asano, K., Nielsen, K. H., Olsen, D. S., Phan, L., Hinnebusch, A. G., and Lorsch, J. R. (2002). Development and characterization of a reconstituted yeast translation initiation system. *RNA* **8,** 382–397.

Asano, K., Clayton, J., Shalev, A., and Hinnebusch, A. G. (2000). A multifactor complex of eukaryotic initiation factors eIF1, eIF2, eIF3, eIF5, and initiator tRNA^{Met} is an important translation initiation intermediate *in vivo. Genes Dev.* **14,** 2534–2546.

Asano, K., Shalev, A., Phan, L., Nielsen, K., Clayton, J., Valášek, L., Donahue, T. F., and Hinnebusch, A. G. (2001). Multiple roles for the carboxyl terminal domain of eIF5 in translation initiation complex assembly and GTPase activation. *EMBO J.* **20,** 2326–2337.

Asano, K., Phan, L., Krishnamoorthy, T., Pavitt, G. D., Gomez, E., Hannig, E. M., Nika, J., Donahue, T. F., Huang, H. K., and Hinnebusch, A. G. (2002). Analysis and reconstitution of translation initiation *in vitro. Methods Enzymol.* **351,** 221–247.

Dohmen, R. J., Wu, P., and Varshavsky, A. (1994). Heat-inducible degron: A method for constructing temperature-sensitive mutants. *Science* **263,** 1273–1276.

Fraser, C. S., Lee, J. Y., Mayeur, G. L., Bushell, M., Doudna, J. A., and Hershey, J. W. (2004). The j-subunit of human translation initiation factor eIF3 is required for the stable binding of eIF3 and its subcomplexes to 40 S ribosomal subunits *in vitro. J. Biol. Chem.* **279,** 8946–8956.

Hershey, J. W. B., and Merrick, W. C. (2000). Pathway and mechanism of initiation of protein synthesis. *In* "Translational Control of Gene Expression" (N. Sonenberg, J. W. B. Hershey, and M. B. Mathews, eds.), pp. 33–88. Cold Spring Harbor Laboratory Press, Cold Spring Harbor, NY.

Hradec, J., and Dusek, Z. (1978). All factors required for protein synthesis are retained on heparin bound to Sepharose. *Biochem. J.* **172,** 1–7.

Jackson, V. (1978). Studies on histone organization in the nucleosome using formaldehyde as a reversible cross-linking agent. *Cell* **15,** 945–954.

Jivotovskaya, A. V., Valasek, L., Hinnebusch, A. G., and Nielsen, K. H. (2006). Eukaryotic translation initiation factor 3 (eIF3) and eIF2 can promote mRNA binding to 40S subunits independently of eIF4G in yeast. *Mol. Cell. Biol.* **26,** 1355–1372.

Kolupaeva, V. G., Unbehaun, A., Lomakin, I. B., Hellen, C. U., and Pestova, T. V. (2005). Binding of eukaryotic initiation factor 3 to ribosomal 40S subunits and its role in ribosomal dissociation and anti-association. *RNA* **11,** 470–486.

Labib, K., Tercero, J. A., and Diffley, J. F. (2000). Uninterrupted MCM2–7 function required for DNA replication fork progression. *Science* **288,** 1643–1647.

Majumdar, R., Bandyopadhyay, A., and Maitra, U. (2003). Mammalian translation initiation factor eIF1 functions with eIF1A and eIF3 in the formation of a stable 40S preinitiation complex. *J. Biol. Chem.* **278,** 6580–6587.

Maraia, R. (1991). The subset of mouse B1 (Alu-equivalent) sequences expressed as small processed cytoplasmic transcripts. *Nucl. Acids Res.* **19,** 5695–5702.

Nielsen, K. H., Szamecz, B., Valasek, L., Jivotovskaya, A., Shin, B., and Hinnebusch, A. G. (2004). Functions of eIF3 downstream of 48S assembly impact AUG recognition and GCN4 translational control. *EMBO J.* **23,** 1166–1177.

Nielsen, K. H., Valasek, L., Sykes, C., Jivotovskaya, A., and Hinnebusch, A. G. (2006). Interaction of the RNP1 motif in PRT1 with HCR1 promotes 40S binding of eukaryotic initiation factor 3 in yeast. *Mol. Cell. Biol.* **26,** 2984–2998.

Pestova, T. V., and Kolupaeva, V. G. (2002). The roles of individual eukaryotic translation initiation factors in ribosomal scanning and initiation codon selection. *Genes Dev.* **16,** 2906–2922.

Pestova, T. V., Borukhov, S. I., and Hellen, C. U. T. (1998). Eukaryotic ribosomes require initiation factors 1 and 1A to locate initiation codons. *Nature* **394,** 854–859.

Phan, L., Zhang, X., Asano, K., Anderson, J., Vornlocher, H. P., Greenberg, J. R., Qin, J., and Hinnebusch, A. G. (1998). Identification of a translation initiation factor 3 (eIF3) core complex, conserved in yeast and mammals, that interacts with eIF5. *Mol. Cell. Biol.* **18,** 4935–4946.

Phan, L., Schoenfeld, L. W., Valášek, L., Nielsen, K. H., and Hinnebusch, A. G. (2001). A subcomplex of three eIF3 subunits binds eIF1 and eIF5 and stimulates ribosome binding of mRNA and tRNA$_i^{Met}$. *EMBO J.* **20,** 2954–2965.

Ren, B., and Dynlacht, B. D. (2004). Use of chromatin immunoprecipitation assays in genome-wide location analysis of mammalian transcription factors. *Methods Enzymol.* **376,** 304–315.

Rotenberg, M. O., Moritz, M., and Woolford, J. L., Jr. (1988). Depletion of *Saccharomyces cerevisiae* ribosomal protein L16 causes a decrease in 60S ribosomal subunits and formation of half-mer polyribosomes. *Genes Dev.* **2,** 160–172.

Sambrook, J., Fritsch, E. F., and Maniatis, T. (1989). "Molecular Cloning, a Laboratory Manual." Cold Spring Harbor Laboratory Press, Cold Spring Harbor, NY.

Solomon, M. J., and Varshavsky, A. (1985). Formaldehyde-mediated DNA-protein cross-linking: A probe for *in vivo* chromatin structures. *Proc. Natl. Acad. Sci. USA* **82,** 6470–6474.

Valášek, L., Hašek, J., Nielsen, K. H., and Hinnebusch, A. G. (2001a). Dual function of eIF3j/Hcr1p in processing 20 S pre-rRNA and translation initiation. *J. Biol. Chem.* **276,** 43351–43360.

Valášek, L., Phan, L., Schoenfeld, L. W., Valásková, V., and Hinnebusch, A. G. (2001b). Related eIF3 subunits TIF32 and HCR1 interact with an RNA recognition motif in PRT1 required for eIF3 integrity and ribosome binding. *EMBO J.* **20,** 891–904.

Valášek, L., Mathew, A., Shin, B. S., Nielsen, K. H., Szamecz, B., and Hinnebusch, A. G. (2003). The yeast eIF3 subunits TIF32/a and NIP1/c and eIF5 make critical connections with the 40S ribosome *in vivo. Genes Dev.* **17,** 786–799.

Valášek, L., Nielsen, K. H., Zhang, F., Fekete, C. A., and Hinnebusch, A. G. (2004). Interactions of eukaryotic translation initiation factor 3 (eIF3) subunit NIP1/c with eIF1 and eIF5 promote preinitiation complex assembly and regulate start codon selection. *Mol. Cell. Biol.* **24,** 9437–9455.

Waldman, A. A., Marx, G., and Goldstein, J. (1975). Isolation of rabbit reticulocyte initiation factors by means of heparin bound to sepharose. *Proc. Natl. Acad. Sci. USA* **72,** 2352–2356.

Yu, X., and Warner, J. R. (2001). Expression of a micro-protein. *J. Biol. Chem.* **276,** 33821–33825.

CHAPTER NINE

MOLECULAR GENETIC STRUCTURE–FUNCTION ANALYSIS OF TRANSLATION INITIATION FACTOR EIF5B

Byung-Sik Shin *and* Thomas E. Dever

Contents

Abstract

Recently, significant progress has been made in obtaining three-dimensional (3-D) structures of the factors that promote translation initiation, elongation, and termination. These structures, when interpreted in light of previous biochemical characterizations of the factors, provide significant insight into the function of the factors and the molecular mechanism of specific steps in the translation process. In addition, genetic analyses in yeast have helped elucidate the *in vivo* roles of the factors in various steps of the translation pathway. We have combined these two approaches and use molecular genetic studies to define the structure–function properties of translation initiation factors in the yeast *Saccharomyces cerevisiae*. In this chapter, we describe our multistep approach in which we first characterize a site-directed mutant of the factor of interest using *in vivo* and *in vitro* assays of protein synthesis. Next, we subject the mutant gene to random mutagenesis and screen for second-site mutations that restore the factor's function *in vivo*.

Laboratory of Gene Regulation and Development, National Institute of Child Health and Human Development, National Institutes of Health, Bethesda, Maryland

Methods in Enzymology, Volume 429

ISSN 0076-6879, DOI: 10.1016/S0076-6879(07)29009-3

Following biochemical and *in vivo* characterization of the suppressor mutant, we interpret the results in light of the 3-D structure of the factor to define the structure–function properties of the factor and to provide new molecular insights into the mechanism of translation.

1. Introduction

Structure–function studies of translation factors have provided a wealth of information on both the functions of specific factors and the general mechanism of protein synthesis. To date, most structure–function studies of translation factors have typically focused on identifying RNA-binding domains and domains that mediate protein–protein interactions with other factors. At the same time, amino acid sequence conservation and site-directed mutagenesis experiments have helped define critical determinants for factor function. Finally, the recent elucidation of the structures of individual domains or of intact translation factors has provided new insight into the structure–function properties of the factors. However, a limitation of these structural analyses is that they present a static image and further study is typically limited to biochemical assays that may not reflect the full function or require all of the domains of the factor.

To further advance our understanding of the structure–function properties of translation initiation factors, and more precisely define the function of the factor eIF5B, we have exploited molecular genetic analyses in the yeast *Saccharomyces cerevisiae* to characterize mutant forms of eIF5B and then to screen *in vivo* for second-site suppressor mutations in eIF5B that restore (at least partially) the factor's function (Fig. 9.1; Shin *et al.*, 2002, 2007).

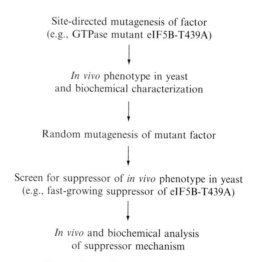

Figure 9.1   Flow scheme for intragenic suppressor analysis.

Biochemical characterizations to elucidate the functional consequences of the second-site suppressor mutations, combined with mapping of the mutations on the structure of eIF5B (Roll-Mecak *et al.*, 2000), have provided significant new insight into both the structure–function properties of eIF5B and the role of eIF5B in translation initiation. Similar strategies have proven powerful in the study of the eIF2α kinase PKR (Dey *et al.*, 2005) and the initiation factor eIF2γ (P. V. Alone and T. E. Dever, unpublished), indicating that this approach may be useful for studying those translation factors and regulators that have identifiable homologs in yeast.

The GTPase eIF5B, an ortholog of the bacterial translation factor IF2 (Choi *et al.*, 1998), catalyzes ribosomal subunit joining, the final step of translation initiation in eukaryotes (Pestova *et al.*, 2000). In the first step of translation initiation, the eIF2–GTP–Met-tRNA$_i^{Met}$ ternary complex binds the small ribosomal subunit forming a 43S preinitiation complex that also contains the factors eIF1, eIF1A, and eIF3 (reviewed in Hinnebusch, 2000). This complex binds an mRNA and scans to identify a start codon. The GTPase activity of eIF2 is functional on the scanning ribosome and eIF2–GTP is in equilibrium with eIF2–GDP + P$_i$. Upon AUG codon recognition, release of eIF1 is coupled to release of P$_i$ from eIF2–GDP (Algire *et al.*, 2005; Maag *et al.*, 2005). The eIF2–GDP and several of the other factors dissociate from the complex leaving eIF1A bound to the subunit and Met-tRNA$_i^{Met}$ in the ribosomal P site. Recent evidence indicates that eIF1A helps recruit eIF5B to the 40S subunit, and that subsequent GTP hydrolysis by eIF5B is required to release both eIF5B and eIF1A from the 80S ribosomal complex, which then enters the elongation phase of protein synthesis (Fringer *et al.*, 2007).

In yeast, the nonessential *FUN12* gene encodes eIF5B. Cells lacking the *FUN12* gene exhibit a severe slow-growth phenotype and polyribosome profile analyses revealed a runoff of polysomes and accumulation of inactive 80S couples (associated 40S and 60S subunits lacking mRNA), consistent with a defect in translation initiation (Choi *et al.*, 1998). *In vitro* translation extracts prepared from cells lacking eIF5B were defective for translation of an exogenous luciferase reporter mRNA; however, translational activity was restored by addition of recombinant eIF5B, indicating that eIF5B directly stimulated translation initiation (Choi *et al.*, 1998; Lee *et al.*, 1999). While purified eIF5B fails to bind isolated 40S and 60S ribosomal subunits, it binds to 80S ribosomes in a GTP-dependent manner (Pestova *et al.*, 2000; Shin *et al.*, 2002). In addition, eIF5B hydrolyzes GTP in an 80S ribosome-dependent reaction (Pestova *et al.*, 2000; Shin *et al.*, 2002). Finally, as described elsewhere in this volume, eIF5B catalyzes ribosomal subunit joining in reconstituted translation initiation systems from yeast and mammalian cells (Algire *et al.*, 2002; Lee *et al.*, 2002; Pestova *et al.*, 2000).

To decipher the role of GTP hydrolysis by eIF5B, we mutated conserved residues in the GTP-binding (G) domain and characterized the functions of the mutant factor both *in vivo* and *in vitro*. Having found that GTPase defective forms of eIF5B cause a severe slow-growth phenotype in yeast, yet retain subunit-joining activity *in vitro* (Shin *et al.*, 2002), we chose a genetic approach to identify intragenic mutations in eIF5B that restore the *in vivo* function of GTPase-defective forms of the factor. A plasmid encoding a GTPase-defective mutant of eIF5B was subjected to random mutagenesis to generate a library of eIF5B mutant plasmids. This library was introduced into an *fun12Δ* yeast strain lacking eIF5B and plasmids containing mutations that restored eIF5B function were identified as conferring a fast-growing phenotype. This genetic approach enabled us to identify physiologically important residues in eIF5B and further biochemical studies of the suppressor mutants revealed that GTP hydrolysis was critical for release of eIF5B from the 80S ribosome following subunit joining. Analysis of the intragenic suppressor mutations in light of the 3-D structure of aIF5B (Roll-Mecak *et al.*, 2000) enabled us to map critical ribosome-binding determinants and structurally important elements in the eIF5B (Shin *et al.*, 2002, 2007).

In this article we will describe the strategy (see Fig. 9.1) and methods we use to screen for intragenic suppressor mutations that restore eIF5B function *in vivo* and the biochemical assays that we use to characterize the eIF5B mutants. The *in vivo* methods are directly applicable to structure–function studies of other proteins, and several of the *in vitro* assays will be useful to characterize mutants in other translation factors.

2. METHODS

2.1. Genetic selection of intragenic suppressors

2.1.1. Site-directed and random mutagenesis of the FUN12 gene encoding yeast eIF5B

As the N-terminal region (residues 1 to 396) of yeast eIF5B is not required for viability *in vivo* or for biochemical activity *in vitro*, an N-terminally truncated form of eIF5B (eIF5B$^{397-1002}$) was used for mutational and suppressor screening experiments. Site-directed mutagenesis of conserved eIF5B G domain residues Thr-439 in Switch I (Shin *et al.*, 2002) and Gly-479 in Switch II (Shin *et al.*, 2007) was performed using the Quick-Change XL Site-Directed Mutagenesis Kit (Stratagene). This kit and related products from other vendors yield the desired mutation at greater than 90% efficiency. Following generation and characterization of eIF5B point mutants that confer a significant growth defect on minimal complete medium or under restrictive conditions (e.g., elevated temperature, amino acid starvation), we screened for intragenic suppressor mutations. A plasmid containing the mutant *fun12* allele was subjected to random mutagenesis by passage through the *Escherichia coli* mutator strain XL1-Red (Stratagene) according to the manufacturer's instructions. XL1-Red cells were transformed with ~100 ng of plasmid DNA, plated on LB medium containing 100 μg/ml ampicillin and incubated 2 days at 37°. Transformants (>500 colonies) were collected by adding 1 ml LB broth to the plates, and then colonies were pooled using a spreader and inoculated into 10 ml LB broth. Following growth at 37° for 16 h, the plasmid DNA was isolated. In previous studies we found that the plasmid DNA prepared from XL1-Red cells is not efficient for yeast transformation, so we introduced another step and passed the mutated plasmids through DH5α. Since the plasmid DNA has random mutations, it is best to pool colonies (>10^4) from the LB + ampicillin transformation plates rather than amplifying the library in broth culture.

2.1.2. Screening for intragenic suppressor mutants

To identify plasmids encoding functional forms of eIF5B, the pool of randomly mutated plasmids was used to transform the *S. cerevisiae* strain J111 (*MAT*α *ura3-52 leu2-3 leu2-112 fun12Δ*). As controls, the same strain was transformed with plasmids encoding the original eIF5B mutant (e.g., eIF5B-T439A), wild-type eIF5B, or an empty vector. Transformants growing faster than the control expressing the original eIF5B mutant were selected for further analysis. In our experience, we typically need to screen more than 10^4 transformants to identify intragenic suppressor mutants of eIF5B. For example, ~5 × 10^4 transformants were screened to obtain three intragenic suppressors of the eIF5B-T439A Switch I mutant (Shin *et al.*, 2002) and ~3 × 10^4 transformants were screened to identify two suppressors of the eIF5B-G479A Switch II mutant (Shin *et al.*, 2007).

Following the identification of fast-growing yeast transformants, it is necessary to confirm that the improved growth is linked to the plasmid and not due to a spontaneous mutation in the yeast genome. The plasmids were recovered from the fast growing yeast transformants using standard protocols that rely on the ability of the plasmids to propagate in *E. coli* (Strathern and Higgins, 1991). The isolated plasmids were then reintroduced into strain J111, and plasmids that again conferred a faster growth phenotype were identified. Finally, the suppressor mutation was first mapped by subcloning the smallest fragment of the suppressor allele that conferred the improved growth phenotype, and then identified by DNA sequencing. To ensure that the identified mutation confers the suppressor phenotype, we used site-directed mutagenesis to introduce the mutation into a wild-type eIF5B plasmid and then retested the growth phenotype in yeast.

While the nonessential nature of the *FUN12* gene greatly facilitated the genetic approach for screening for suppressor mutations, related strategies can be employed to study proteins encoded by essential genes. For essential genes, the approach is modified using a plasmid-shuffling protocol (Sikorski and Boeke, 1991). First, the viability of a strain carrying a chromosomal deletion of the essential gene of interest is maintained by a wild-type version of the gene on a plasmid marked with the counter-selectable yeast *URA3* gene. A second plasmid marked with a compatible gene of interest (typically *LEU2* or *TRP1*) and carrying a defective (mutant) version of the gene of interest is subjected to random mutagenesis. This library is then introduced into the yeast strain and independent transformants are selected and replica printed to medium containing the drug 5-fluoroorotic acid (5-FOA) to select against cells carrying the *URA3* plasmid (Boeke *et al.*, 1987). In this way, only cells containing versions of the mutant plasmid that restore the factor's function are capable of growing on the 5-FOA medium.

Following the genetic screens to identify intragenic suppressor mutations in eIF5B, we used a series of *in vivo* and *in vitro* analyses to characterize the activities of the eIF5B mutants. The first assay analyzes global protein synthesis in yeast cells by examining polysome profiles in sucrose density gradients. The second assay examines the translation of a reporter mRNA in crude extracts from an eIF5B deletion strain that is supplemented with various amounts of wild-type eIF5B, the original mutant form of the factor, or the suppressor mutant. Additional *in vitro* assays described below were designed to characterize the eIF5B GTPase activity and the interaction of the factor with the ribosome.

2.2. Polysome profile analysis

2.2.1. Preparation of sucrose density gradients

We prepare linear 7 to 47% sucrose gradients using the Gradient Master (BioComp Instruments, Inc., Canada), and we subject the gradients to centrifugation using a Beckman SW41 rotor. First, using RNase-free distilled water (DW) prepare 50 ml each of 7% and 47 % sucrose solutions in 20 mM Tris–Cl (pH 7.5), 50 mM KCl, 10 mM MgCl$_2$, and 2 mM dithiothreitol (DTT). Next, using the Gradient Master according to the manufacturer's instructions, prepare six gradients by dispensing approximately 8 ml into the thin-walled polyallomer ultracentrifuge tubes for a Beckman SW41 rotor. Store the gradients at 4° and be careful to avoid disturbances that might disrupt the gradients.

2.2.2. Preparation of yeast cell extracts

Cells are grown in a 1-liter flask containing 200 ml of rich YPD medium or of synthetic complete (SC) medium containing all amino acids and lacking only the nutrient(s) required to maintain selection of any plasmid(s) in the strain (Sherman, 1991). Cultures are typically inoculated at an $A_{600} = 0.1$ and incubated at 30° while shaking at 250 rpm. When the density of the culture reaches an $A_{600} = 1.0$, add cycloheximide (stock: 10 mg/ml in DW) to a final concentration of 50 μg/ml, and continue incubating for an additional 5 to 10 min to halt protein biosynthesis. The cycloheximide is added to block translation elongation and freeze polysomes; otherwise, ribosomes would continue elongating and run off mRNAs during extract preparation. During the incubation with cycloheximide, prepare centrifuge bottles to collect the yeast cells. Fill the 500-ml centrifuge bottles for a Sorvall G3 rotor about two-thirds full with crushed ice and store the bottles on ice. Pour the contents of the yeast culture into the ice-filled bottles and gently shake to rapidly cool the cells. Collect the cells by centrifugation at 5000 rpm for 5 min at 4° in a Sorvall GS3 rotor. After pelleting the cells, suspend the pellets in 2 ml Breaking Buffer [20 mM Tris–Cl (pH 7.5), 50 mM KCl, 10 mM MgCl$_2$, 2 mM DTT, 1× Complete Protease Inhibitor cocktail (EDTA-free, Roche), 0.5 mM 4-(2-aminoethyl)benzenesulfonyl fluoride (AEBSF), 5 μg/ml pepstatin, and 50 μg/ml cycloheximide]. Transfer the cell suspensions to 15-ml centrifuge tubes (Falcon), and collect the cells by centrifugation at 3000 rpm for 5 min in a Beckman JS-4.2 rotor. Resuspend the pellets in 0.5 ml Breaking Buffer, and add acid-washed glass beads (Sigma) to ~60% final volume. Working in the cold room (~4°), mix the tubes vigorously using a vortex for 1 min and then incubate the tubes on ice for 1 min. Repeat this cycle of mixing and cooling for a total of five times. Pellet any unbroken cells and the glass beads by spinning the tubes at 3000 rpm for 5 min, and then transfer the supernatants containing the cell lysates to 1.5 ml-microcentrifuge tubes. Clarify the lysates by centrifugation at 12,000 rpm for 10 min (4°), and measure the absorbance (A_{260}) of a 1:1000 dilution of the final supernatant. Typically, the A_{260} = ~100.

2.2.3. Polysome analysis

Carefully layer ~100 μl of yeast extract (A_{260} = 100) on the top of each sucrose gradient, and use Breaking Buffer (without protease inhibitors) to balance the volume in the centrifuge tubes. Subject the gradients to ultracentrifugation at 39,000 rpm for 2.5 h (4°) in a Beckman SW41 rotor. We fractionate the gradients using an ISCO tube piercing system and Model T11 gradient fractionation system with an attached UA-6 UV/VIS Detector (with a 254-nm filter). Before fractionation, flush the detector with DW, set the sensitivity to 1.0, and adjust the baseline of the chart recorder. Following centrifugation, transfer individual gradients to the ISCO tube piercing system, pierce the bottom of the tube, and withdraw the gradient using an in-line peristaltic pump at the slowest speed. Once the unloading has initiated, increase the pump speed to maximum and set the chart recorder speed at 150 (cm/h).

2.2.4. Comments

1. We have found it convenient to prepare the sucrose gradients a day before generating the cell extracts. We store the gradients at 4° and carefully try to avoid any disturbances that may disrupt the gradients.

2. If low amounts of cell extracts are loaded on the sucrose gradients, the sensitivity of UV/VIS detector can be increased to 0.5. Alternatively, if a huge ribosome peak (80S) is anticipated, the entire peak can be visualized when the sensitivity is set to 2.0.

3. To analyze the composition of individual fractions from the gradients, we connect the outlet of the detector to an ISCO Foxy Jr. Fraction Collector. Typically we use Northern and/or Western analyses to examine the distribution of specific mRNAs and translation initiation factors, respectively, in the gradients. See the chapter by Valášek et al. (this volume) for additional methods using sucrose density gradients to analyze initiation factor–ribosome complexes.

2.3. Preparation of yeast cell extracts for *in vitro* translation assays

Building on the work of Peter Sarnow and co-workers (Iizuka et al., 1994), Alan Sachs and colleagues have described a robust *in vitro* translation system from yeast that recapitulates the natural cap and poly(A) requirements for translation observed *in vivo* (Tarun and Sachs, 1995). We utilize their assay system in our studies exactly as has been described (Choi et al., 1998), except that we have introduced some modifications to the protocol for preparing the yeast cell extracts as described in this section.

Grow the strain of interest in YPD medium to $A_{600} = 1$ to 1.5. Typically a 2-liter culture is sufficient to yield a good quality extract for a wild-type strain; however, for slow-growing strains a larger volume of culture ($\sim$6 liters) may be required. Harvest the cells by centrifugation at 4000 rpm for 10 min in a Sorvall GS-3 rotor. Wash the cell pellet twice with 200 ml cold Buffer A [30 mM HEPES–KOH (pH 7.4), 100 mM potassium acetate, 2 mM magnesium acetate] and once with 30 ml of Buffer A containing 8.5% mannitol and 5 mM 2-mercaptoethanol (prepared by adding 35 μl of 2-mercaptoethanol to 100 ml of Buffer A + 8.5% mannitol). Next, resuspend the pellet in 6 ml of Buffer A containing 8.5% mannitol and 0.5 mM AEBSF and transfer to a 50-ml Falcon tube. Add acid-washed glass beads ($\sim$60% v/v) to the cell suspension and break the cells by hand shaking in the cold (4°) for 1 min: two cycles/sec over a 50 cm path. Then cool the cells on ice for 1 min. Repeat the cycles of shaking and cooling for a total of five times. Following breakage, transfer the cell suspension (avoiding the glass beads) to 50-ml centrifuge tubes, clarify the lysate by centrifugation at 15,000 rpm for 20 min in a Sorvall SS34 rotor, and then carefully remove the supernatant fraction avoiding the lipid layer on the top.

While preparing the cell extract, preequilibrate a Sephadex G-25 (superfine grade, autoclaved with DW before use) column (2.5 cm diameter × 8 cm tall) with 200 ml of Buffer A containing 2 mM DTT. Load the clarified supernatant (~4 ml) on the column, and start collecting ~0.2 ml fractions after 15 ml of elution volume (we collect in a 96-well plate using an ISCO Foxy Jr. Fraction Collector). The peak fractions in the 96-well plate are detected by their cloudy appearance (more easily seen against a dark background). Next, measure the A_{260} values of the peak fractions by preparing 1:500 dilutions in water. For extracts from wild-type strains, pool all fractions for which the A_{260} value of the dilution is > 0.18 (for extracts from slow-growing *fun12* mutant strains, we pool all fractions for which the A_{260} value of the dilution is > 0.14). Finally, flash-freeze ~50-μl aliquots of the pooled fraction in liquid nitrogen and store the frozen *in vitro* translation extracts at $-80°$.

2.3.1. Comments

1. All extract preparation steps should be carried out in a cold room if possible.
2. RNAs are critical components of the translation extract. Try to avoid RNase contamination, and use DEPC-treated DW for all steps.
3. Manual breaking of the cells is recommended. In our experience, extracts prepared from cells broken using a French Press occasionally lack activity.
4. Do not use cells if the A_{600} exceeds 1.5. The translational activity is significantly decreased in extracts prepared from dense cultures.

2.4. Expression and purification of eIF5B from yeast

Yeast eIF5B$^{397-1002}$ can be expressed and purified from either bacterial or yeast cells. Whereas purification of recombinant eIF5B from *E. coli* typically requires multiple chromatographic steps [GST- or nickel-affinity chromatography followed by ion-exchange chromatography (Algire *et al.*, 2002)], overexpressed GST-eIF5B$^{397-1002}$ can be purified to near homogeneity (>95% pure) from crude yeast extracts using Glutathione Sepharose 4B affinity chromatography.

The DNA encoding yeast eIF5B$^{397-1002}$ was subcloned to the yeast expression vector pEG-KT (Mitchell *et al.*, 1993) that is designed to express GST fusion proteins under the control of the yeast *GAL1* (galactose-inducible) promoter. The resulting pEG-KT-eIF5B plasmid was introduced into the wild-type yeast strain H1511 (*MATα ura3-52 leu2-3 leu2-112 trp1-Δ 63*). The resulting transformant was grown in 2 liters of S-raffinose medium [0.145% yeast nitrogen base (without amino acids and without ammonium sulfate), 0.5% ammonium sulfate, and 2% raffinose] plus required supplements. In S-raffinose medium the *GAL1* promoter is neither repressed nor induced and so GST-eIF5B expression is maintained at a low level. Once the A_{600} reaches 0.4 to 1.0, induce expression of GST-eIF5B by adding

100 ml of a 40% galactose stock solution to achieve a final galactose concentration of 2%, and incubate the culture with shaking at 30° for 14 h.

Harvest the cells by centrifugation and suspend the cell pellet in 50% pellet volume of Cell Breaking Buffer: $1\times$ phosphate buffered saline (PBS) solution containing $1\times$ Complete Protease Inhibitor cocktail (EDTA-free, Roche), 0.5 mM AEBSF, and 5 μg/ml of pepstatin. Next, working in the cold room, freeze the cell suspension in liquid nitrogen. Using a 10-ml serological pipette, dispense individual drops of the cell suspension into 500 ml of liquid nitrogen in a 1-liter Dewar. The cells will freeze as beads in the liquid nitrogen, and the beads are then collected and broken using a coffee grinder (such as the Miracle MC 200, Miracle Exclusives Inc., Hicksville, NY). To avoid thawing of the cell beads during the grinding, prechill the grinder cup by filling it with dry ice pellets or liquid nitrogen. In addition, add dry ice pellets or liquid nitrogen to the cell beads immediately prior to grinding. After grinding for ~2 min, a very fine cell powder with the consistency of ground coffee is obtained. Dissolve this cell powder completely in 100 ml of Cell Breaking Buffer. Note that dry ice crystals remaining in the cell powder following cell breakage will generate many bubbles when dissolved in the Cell Breaking Buffer. Therefore, it is recommended that the dry ice be allowed to completely sublime prior to dissolving the cell powder in buffer (and to allow all bubbles to dissipate prior to capping tubes containing the dissolved cell powder).

After dissolving the cell powder, clarify the extract by centrifugation at 15,000 rpm for 30 min in a Sorvall SS34 rotor. Next, add 1 ml of a 50% slurry of Glutathione Sepharose 4B (GE Healthcare) to the extract (about 80 ml), and gently mix the suspension at 4° for 2 h. The following steps describing washing of the resin and elution of the eIF5B can be performed either in batch using 1.5-ml microcentrifuge tubes, as described below, or chromatographically using a disposable column. Wash the resin extensively using more than 20-times the resin volume of $1\times$ PBS buffer. To elute the eIF5B, add 500 μl of $1\times$ PBS buffer containing 20 units of thrombin to the washed resin in a 1.5-ml microcentrifuge tube. Gently rock the mixture (we typically use either a Nutator or Labquake Shaker) at room temperature for 2 h, and then continue mixing at 4° overnight. The progress of the thrombin digestion can be monitored by testing for the presence of protein released into the supernatant.

Once the digestion is complete, recover the released eIF5B by first pelleting the resin (bound with uncleaved GST–eIF5B and with cleaved GST) by centrifugation at 2000 rpm for 2 min in a microcentrifuge. Remove the supernatant containing released eIF5B to another tube, then wash the resin by adding 500 μl of $1\times$ PBS (without thrombin) and mix gently. Next, pellet the resin, and combine the wash supernatant with the supernatant obtained just prior to the wash. The pooled supernatants are then dialyzed against eIF5B Storage Buffer [20 mM Tris–Cl (pH 7.5),

100 mM NaCl, 2 mM DTT, and 10% glycerol]. Finally, the eIF5B solution (typically ~40 μM) can be concentrated using a Microcon YM-30 (Millipore, Bedford, MA) if a higher concentration of the factor is required.

2.5. Ribosome purification

2.5.1. Purification of crude 80S ribosomes from yeast

We have found that the choice of yeast strain can have a substantial impact on the quality of the purified ribosomes. We use the strain F353 (*MATα trp1 leu2-Δ1 his3-Δ200 pep4::HIS3 prb1-Δ1.6 GAL⁺*) lacking the vacuolar proteinases A (PEP4) and B (PRB1), which are required both for protein degradation and for activation of other vacuolar proteases. The following protocol has been developed in conjunction with Jon Lorsch and members of his laboratory. Inoculate 2 liters of YPD medium with F353 and grow at 30° until the $A_{600} = 1.0$ to 1.5 (overgrowing the cells yields inactive ribosomes). Harvest the cells by centrifugation and suspend the cell pellet using 50% of the pellet volume of Ribosome Buffer [30 mM HEPES–KOH (pH 7.4), 100 mM potassium acetate, 12.5 mM magnesium acetate, 2 mM DTT, 1× Complete Protease Inhibitor cocktail (EDTA-free, Roche), 0.5 mM AEBSF, 5 μg/ml of pepstatin, and 1 mg/ml of heparin]. Freeze the cell suspension in liquid nitrogen and break the cell beads using a chilled coffee grinder as described above in the section on eIF5B purification. After grinding the cells, dissolve the cell powder completely in 100 ml of Ribosome Buffer (remember to allow any dry ice in the cell powder to sublime prior to capping any of the extracts).

Remove any unbroken cells and cell debris by centrifugation at 15,000 rpm for 20 min in a Sorvall SS34 rotor, and then transfer the clarified supernatant to 25-ml polycarbonate tubes designed for use in the Beckman Type70 Ti rotor (21 ml of lysate per tube). Next, insert a 5-ml serological pipette into the tube and slowly add 3 ml of ice-cold sucrose cushion [30 mM HEPES–KOH (pH 7.4), 100 mM potassium acetate, 12.5 mM magnesium acetate, 2 mM DTT, and 1M sucrose] to the bottom of the tube. Pellet the 80S ribosomes through the sucrose cushion by centrifugation at 40,000 rpm for 3 h in a Type70 Ti rotor. Discard the supernatant and dissolve the ribosome pellet in 5-ml of Ribosome Buffer, and then stir for 1 h on ice. Clarify the ribosome solution by centrifugation at 12,000 rpm for 10 min in a microcentrifuge, and then layer 0.5 to 1.0 ml of the supernatant on pre-chilled 37-ml 5–40% sucrose gradients (prepared using Ribosome Buffer). Subject the gradients to velocity sedimentation at 32,000 rpm for 6 h (or 20,000 rpm for 14 h) and at 4° in a Beckman SW32 rotor. Fractionate the gradient while monitoring the absorbance at A_{254}, and pool all fractions containing the 80S ribosomes. Pellet the ribosomes in the 80S pool by centrifugation at 25,000 rpm for 24 h at 4° in a Type70 Ti rotor. Finally dissolve the 80S ribosome pellet in 20 mM

HEPES–KOH (pH 7.4), 50 mM potassium acetate, 2.5 mM magnesium acetate, 2 mM DTT, and 250 mM sucrose. Quantify the concentration of 80S ribosomes by measuring the A_{260} of the solution (we use the estimate that 1.0 A_{260} unit corresponds to 20 nM 80S ribosomes).

2.5.2. Purification of reassociated 80S ribosomes from yeast

For some assays, including the eIF5B ribosome-dependent GTPase assay, we have found that reassociated 80S ribosomes prepared from purified 40S and 60S subunits work better than 80S ribosomes prepared as described in the previous section. The reassociated 80S ribosomes appear to have fewer contaminants, contributing to lower background activities in the assays. Incubate the strain F353 in 6 liters of YPD medium until the $A_{600} = 1.0$ to 1.5, and then harvest and break the cells, and pellet the ribosomes through a sucrose cushion as described above for purification of crude 80S ribosomes. Dissolve the ribosomal pellet in 6 ml of Subunit Separation Buffer [50 mM HEPES–KOH (pH 7.4), 500 mM KCl, 2 mM MgCl$_2$, and 2 mM DTT] by stirring on ice for 1 h, and then clarify the solution by pelleting insoluble material using a microcentrifuge. The A_{260} of the dissolved ribosomes should be ~50 to 100. Using a 100 mM puromycin stock solution freshly prepared in DW, add puromycin to a final concentration of 1 mM and incubate the mixture on ice for 15 min, then at 37° for 10 min, and finally on ice for 10 min. Load 0.5 to 1.0 ml of the ribosome solution on 37-ml 5% to 20% sucrose gradients (prepared using Subunit Separation Buffer), and subject the gradients to centrifugation at 32,000 rpm for 6 h (or 20,000 rpm for 14 h) in a Beckman SW32 rotor. Fractionate the gradients while monitoring the absorbance at A_{260}, and separately pool the fractions containing the 40S and 60S subunits.

To reassociate the subunits, first pellet the subunits by centrifugation at 35,000 rpm for 17 h in a Beckman Type45 Ti rotor, discard the supernatant, and dissolve the ribosomal pellets in Reassociation Buffer [20 mM HEPES–KOH (pH 7.4), 20 mM magnesium acetate, 100 mM KCl, and 2 mM DTT]. Then mix the 40S and 60S subunits in a 1:1 ratio of A_{260} units (the final concentration should be 40 to 140 A_{260}) and incubate at 32° for 40 min, and then at 4° for 10 min. To separate the reassociated 80S ribosomes from the free subunits, load 0.5 to 1.0 ml of the subunit mixture on prechilled 37-ml 5% to 40% sucrose gradients (prepared using Reassociation Buffer), and subject the gradients to centrifugation at 32,000 rpm for 6 h at 4° in an SW32 rotor. Fractionate the gradients while monitoring the absorbance at A_{260}, and pool the fractions containing the 80S ribosomes. Pellet the ribosomes by centrifugation at 22,000 rpm for 24 h at 4° in a Type70 Ti rotor, discard the supernatant, and then dissolve the ribosome pellets in Reassociation Buffer and incubate at 37° for 20 min. Finally, clarify the reassociated 80S ribosome solution in a microcentrifuge, and then measure the A_{260} to quantify the concentration of 80S ribosomes.

2.5.3. Comment

1. Because the subunits are never separated during the purification steps, the crude 80S ribosomes purified by the first method are contaminated with ribosome-bound translation factors (both initiation and elongation factors), mRNAs, and tRNAs. However, the crude 80S ribosomes are useful for the eIF5B ribosome-binding assay described in the next section. In contrast, the reassociated 80S ribosomes are very clean and provide superior results in the uncoupled eIF5B ribosome-dependent GTPase assay. We also use the reassociated 80S ribosomes for hydroxyl radical mapping of the eIF5B–80S complex.

2.6. Ribosome-dependent uncoupled GTPase assay of eIF5B

The GTPase activity of eIF5B is stimulated in the presence of 80S ribosomes, but little activity is observed in the presence of either 40S or 60S subunits (Acker *et al.*, 2006; Pestova *et al.*, 2000; Shin *et al.*, 2002). In addition, kinetic analyses revealed a biphasic nature of the eIF5B GTPase activity in the context of ribosomal subunit joining. An early phase of robust GTPase activity was followed by a weaker phase that matched the rate of GTP hydrolysis observed with eIF5B and isolated 80S ribosomes (Acker *et al.*, 2006). Thus, eIF5B rapidly hydrolyzes GTP in assays coupling eIF5B GTPase activity with ribosomal subunit joining, and eIF5B hydrolyzes GTP with slower kinetics in assays uncoupled to subunit joining. Here, we described the uncoupled eIF5B ribosome-dependent GTPase assay (the protocol for the coupled assay is described by Acker *et al.*, 2006). Though unlinked to translation initiation, the uncoupled GTPase assay is valuable for examining the impact of eIF5B and ribosome mutations on eIF5B GTPase activity, and it is much simpler to perform (requiring only eIF5B and 80S ribosomes) than the coupled assay.

Reaction mixtures contain 1 μM eIF5B, 0.4 μM 80S ribosomes, and limited amounts of $[\gamma\text{-}^{33}P]$GTP (typically 50 nM) in 1× Reaction Buffer (30 mM HEPES–KOH [pH 7.4], 50 mM potassium acetate, 2.5 mM magnesium acetate, and 2 mM DTT). Prior to the assay, prepare two 6-μl solutions, the first containing 2 μM eIF5B and 0.8 μM 80S ribosomes in 1× reaction buffer and the second containing 100 nM $[\gamma\text{-}^{33}P]$GTP in 1× reaction buffer. Preincubate the two mixtures in a 30° water bath for 2 min, and then start the reaction by mixing the two solutions. Quench the reaction at various times by transferring 1-μl aliquots of the reaction mixture to tubes containing 3 μl of Stop Solution (50 mM EDTA in 90% formamide).

The hydrolysis of GTP is monitored by polyethylenimine cellulose thin-layer chromatography (PEI-TLC, Selecto Scientific, Georgia). Prerun a 10 cm × 10 cm PEI-TLC plate with DW for 10 min and then air dry.

Spot 1-μl aliquots of the quenched reactions at 1 cm from the bottom of the TLC plate. Allow the spots to dry completely, and then run the TLC for 10 min using a buffer containing 0.8 M lithium chloride and 0.8 M acetic acid. Dry the TLC plate completely, expose for phosphorimage analysis or autoradiography, and quantify the amounts of GTP and P$_i$ in each sample. To determine the rate constant for GTP hydrolysis, plot the fraction of P$_i$ released $\{[P_i]/([P_i] + [GTP])\}$ as a function of the quench time (in seconds). When calculating the fraction of P$_i$ released in the reactions, take into account the amounts of P$_i$ released in control reactions lacking ribosomes or eIF5B. We use the program KaleidaGraph (Synergy Software) to fit the results to the single exponential equation: $A[1 - \exp(-kt)]$, in which A is the amplitude and k is the rate constant. Typically in our assays, eIF5B hydrolyzes GTP in the presence of 80S ribosomes with a rate constant of $\sim$0.06 sec^{-1}.

2.6.1. Comment

1. The ribosome-dependent uncoupled GTPase activity of eIF5B is dramatically inhibited by increasing concentrations of potassium acetate or KCl. For example, the GTPase activity of WT eIF5B is decreased more than 10-fold when the concentration of KCl (or potassium acetate) is increased from 50 mM to 100 mM. At this time it is unclear whether this salt dependence reflects impacts on eIF5B binding to 80S ribosomes, eIF5B catalytic function, or the association of 40S and 60S subunits to generate 80S ribosomes. However, as 40S and 60S subunits should remain associated in 100 mM KCl, we disfavor the latter possibility.

2.7. eIF5B ribosome-binding assay

The 80S ribosome binding activity of eIF5B is regulated by the guanine nucleotide bound to the factor. In the presence of GTP, eIF5B binds to 80S ribosomes with high affinity, and the binding is decreased in the presence of GDP or no nucleotide (Pestova *et al.*, 2000; Shin *et al.*, 2002). We adapted the bacterial IF2–70S ribosome-binding assay of Moreno *et al.* (1998) to monitor eIF5B binding to 80S ribosomes pelleted through a sucrose cushion (Fig. 9.2). To conserve on reagents, and for convenience, we use a tabletop ultracentrifuge for this assay. In 50-μl reactions, mix 0.5 μM eIF5B with 0.5 μM 80S ribosomes (crude 80S) in Binding Buffer [30 mM HEPES–KOH (pH 7.4), 100 mM potassium acetate, 2.5 mM magnesium acetate, 2 mM DTT, and 2 mM guanine nucleotide], and incubate at 25° for 5 min. Aliquot 50 μl of a 10% sucrose solution (containing 2 mM GTP, GDP, GDPNP, or no nucleotide, as appropriate) to mini (1.0-ml) polycarbonate tubes (for the Beckman TLA 120.2 rotor) and then chill the tubes on ice. Load the entire (50-μl) binding reaction on top of the ice-cold sucrose

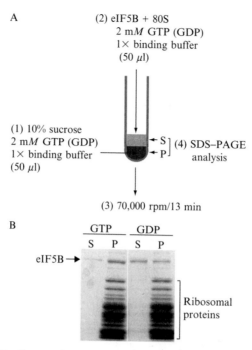

A

(2) eIF5B + 80S
2 m*M* GTP (GDP)
1× binding buffer
(50 μl)

(1) 10% sucrose
2 m*M* GTP (GDP)
1× binding buffer
(50 μl)

S ⎤ (4) SDS–PAGE
P ⎦ analysis

(3) 70,000 rpm/13 min

B

GTP   GDP

S   P   S   P

eIF5B →

Ribosomal
proteins

Figure 9.2 eIF5B ribosome binding assay. (A) Assay procedure; (B) Typical results; analysis of supernatant (S) and pellet (P) fraction by SDS–PAGE.

solution (containing guanine nucleotide). As the pellet will not be visible following centrifugation, mark the expected pellet position near the bottom of each tube prior to inserting the tubes into the rotor, and then pellet the ribosomes by centrifugation at 70,000 rpm for 13 min in a Beckman TLA 120.2 rotor.

To analyze eIF5B binding to the ribosome, sample both the supernatant and pellet (ribosome bound) fractions following centrifugation. First, collect the supernatant fraction by removing 20 μl of solution very carefully from the top of the tubes. Next, remove an additional 60 μl of solution from each tube and discard this material. Finally, suspend the ribosomal pellet in the remaining solution (~20 μl) by repeated pipetting. Analyze the supernatant and ribosome fractions by 10% sodium dodecyl sulfate polyacrylamide gel electrophoresis (SDS–PAGE), and stain the gel with Coomassie Blue. As eIF5B$^{397-1002}$ is much larger than any ribosomal proteins, the presence of eIF5B in the supernatant and pellet fractions can be easily observed on the stained gel (Fig. 9.2B). The relative amount of eIF5B in the various gel fractions can be determined by quantitative densitometry, and the fraction of total eIF5B bound to 80S ribosomes is calculated using the following equation: [eIF5B in pellet fraction]/[sum of eIF5B in both supernatant and pellet fractions]. Note that two control reactions are required for every ribosome-binding assay: first, a binding reaction without eIF5B, and second, a binding reaction without ribosomes. As the results vary between separate trials, the assay should be repeated several times (>3) to obtain reliable results.

3. FUTURE DIRECTIONS

As stated earlier, the molecular genetic techniques described here are applicable both for nonessential genes like *FUN12* and for the essential genes encoding most translation factors. Moreover, the approach to screen for intragenic suppressors will be generally applicable for proteins functioning in many cellular processes. The ability to test structural predictions in an *in vivo* model, combined with the use of unbiased genetic screens to identify important residues and domains in proteins, highlights the power of the yeast system as a tool to study the structure–function properties of proteins. With the current emphasis to acquire the 3-D structure of a large variety of proteins, we believe that the molecular genetic structure–function studies we have utilized to study eIF5B can serve as a model for the study of other yeast proteins or heterologous proteins like the mammalian kinase PKR that are functional in yeast cells (Dey *et al.*, 2005).

REFERENCES

Acker, M. G., Shin, B. S., Dever, T. E., and Lorsch, J. R. (2006). Interaction between eukaryotic initiation factors 1A and 5B is required for efficient ribosomal subunit joining. *J. Biol. Chem.* **281,** 8469–8475.

Algire, M. A., Maag, D., Savio, P., Acker, M. G., Tarun, S. Z., Jr., Sachs, A. B., Asano, K., Nielsen, K. H., Olsen, D. S., Phan, L., Hinnebusch, A. G., and Lorsch, J. R. (2002). Development and characterization of a reconstituted yeast translation initiation system. *RNA* **8,** 382–397.

Algire, M. A., Maag, D., and Lorsch, J. R. (2005). Pi release from eIF2, not GTP hydrolysis, is the step controlled by start-site selection during eukaryotic translation initiation. *Mol. Cell* **20,** 251–262.

Boeke, J. D., Trueheart, J., Natsoulis, G., and Fink, G. R. (1987). 5-Fluoroorotic acid as a selective agent in yeast molecular genetics. *Methods Enzymol.* **154,** 164–175.

Choi, S. K., Lee, J. H., Zoll, W. L., Merrick, W. C., and Dever, T. E. (1998). Promotion of met-tRNAiMet binding to ribosomes by yIF2, a bacterial IF2 homolog in yeast. *Science* **280,** 1757–1760.

Dey, M., Cao, C., Dar, A. C., Tamura, T., Ozato, K., Sicheri, F., and Dever, T. E. (2005). Mechanistic link between PKR dimerization, autophosphorylation, and eIF2alpha substrate recognition. *Cell* **122,** 901–913.

Fringer, J. M., Acker, M. G., Fekete, C. A., Lorsch, J. R., and Dever, T. E. (2007). Coupled release of eukaryotic translation initiation factors 5B and 1A from 80S ribosomes following subunit joining. *Mol. Cell Biol.* **27,** 2384–2397.

Hinnebusch, A. G. (2000). Mechanism and regulation of initiator methionyl-tRNA binding to ribosomes. In "Translational Control of Gene Expression" (N. Sonenberg, J. W. B. Hershey, and M. B. Mathews, eds.), pp. 185–243. Cold Spring Harbor Laboratory Press, Cold Spring Harbor, NY.

Iizuka, N., Najita, L., Franzusoff, A., and Sarnow, P. (1994). Cap-dependent and cap-independent translation by internal initiation of mRNAs in cell extracts prepared from *Saccharomyces cerevisiae*. *Mol. Cell Biol.* **14,** 7322–7330.

Lee, J. H., Choi, S. K., Roll-Mecak, A., Burley, S. K., and Dever, T. E. (1999). Universal conservation in translation initiation revealed by human and archaeal homologs of bacterial translation initiation factor IF2. *Proc. Natl. Acad. Sci. USA* **96,** 4342–4347.

Lee, J. H., Pestova, T. V., Shin, B. S., Cao, C., Choi, S. K., and Dever, T. E. (2002). Initiation factor eIF5B catalyzes second GTP-dependent step in eukaryotic translation initiation. *Proc. Natl. Acad. Sci. USA* **99,** 16689–16694.

Maag, D., Fekete, C. A., Gryczynski, Z., and Lorsch, J. R. (2005). A conformational change in the eukaryotic translation preinitiation complex and release of eIF1 signal recognition of the start codon. *Mol. Cell* **17,** 265–275.

Mitchell, D. A., Marshall, T. K., and Deschenes, R. J. (1993). Vectors for the inducible overexpression of glutathione S-transferase fusion proteins in yeast. *Yeast* **9,** 715–722.

Moreno, J. M., Kildsgaard, J., Siwanowicz, I., Mortensen, K. K., and Sperling-Petersen, H. U. (1998). Binding of *Escherichia coli* initiation factor IF2 to 30S ribosomal subunits: A functional role for the N-terminus of the factor. *Biochem. Biophys. Res. Commun.* **252,** 465–471.

Pestova, T. V., Lomakin, I. B., Lee, J. H., Choi, S. K., Dever, T. E., and Hellen, C. U. (2000). The joining of ribosomal subunits in eukaryotes requires eIF5B. *Nature* **403,** 332–335.

Roll-Mecak, A., Cao, C., Dever, T. E., and Burley, S. K. (2000). X-Ray structures of the universal translation initiation factor IF2/eIF5B. Conformational changes on GDP and GTP binding. *Cell* **103,** 781–792.

Sherman, F. (1991). Getting started with yeast. *Methods Enzymol.* **194,** 3–21.

Shin, B. S., Maag, D., Roll-Mecak, A., Arefin, M. S., Burley, S. K., Lorsch, J. R., and Dever, T. E. (2002). Uncoupling of initiation factor eIF5B/IF2 GTPase and translational activities by mutations that lower ribosome affinity. *Cell* **111,** 1015–1025.

Shin, B. S., Acker, M. G., Maag, D., Kim, J. R., Lorsch, J. R., and Dever, T. E. (2007). Intragenic suppressor mutations restore GTPase and translation functions of a eukaryotic initiation factor 5B switch II mutant. *Mol. Cell Biol.* **27,** 1677–1685.

Sikorski, R. S., and Boeke, J. D. (1991). *In vitro* mutagenesis and plasmid shuffling: From cloned gene to mutant yeast. *Methods Enzymol.* **194,** 302–318.

Strathern, J. N., and Higgins, D. R. (1991). Recovery of plasmids from yeast into *Escherichia coli*: Shuttle vectors. *Methods Enzymol.* **194,** 319–329.

Tarun, S. Z., Jr., and Sachs, A. B. (1995). A common function for mRNA 5′ and 3′ ends in translation initiation in yeast. *Genes Dev.* **9,** 2997–3007.

CHAPTER TEN

THE USE OF FUNGAL *IN VITRO* SYSTEMS FOR STUDYING TRANSLATIONAL REGULATION

Cheng Wu,* Nadia Amrani,[†] Allan Jacobson,[†] *and* Matthew S. Sachs*,[‡]

Contents

Abstract

The use of cell-free systems enables biochemical determination of factors and mechanisms contributing to translational processes. The preparation and use of cell-free translation systems from the fungi *Saccharomyces cerevisiae* and *Neurospora crassa* are described. Examples provided illustrate the use of these systems, in conjunction with luciferase assays, [35S]Met incorporation, and primer-extension inhibition (toeprint) analyses, to assess the translational effects of upstream open reading frames and premature termination codons.

* Department of Environmental and Biomolecular Systems, OGI School of Science and Engineering, Oregon Health and Science University, Beaverton, Oregon
[†] Department of Molecular Genetics and Microbiology, University of Massachusetts Medical School, Worcester, Massachusetts
[‡] Department of Molecular Microbiology and Immunology, School of Medicine, Oregon Health and Science University, Portland, Oregon

Methods in Enzymology, Volume 429
ISSN 0076-6879, DOI: 10.1016/S0076-6879(07)29010-X

1. INTRODUCTION

Eukaryotic cell-free *in vitro* translation systems that faithfully synthesize polypeptides when programmed with mRNA have been in wide use since the 1970s, when soluble systems derived from wheat germ (Roberts and Paterson, 1973) and rabbit reticulocytes (Pelham and Jackson, 1976) were described. Detailed protocols for the preparation of each have appeared in this series (e.g., Anderson *et al.*, 1983; Erickson and Blobel, 1983; Jackson and Hunt, 1983). Procedures adapted from those used to prepare wheat germ extracts have led to the development of cap- and poly(A)-dependent cell-free translation extracts from *Saccharomyces cerevisiae* (Iizuka and Sarnow, 1997; Iizuka *et al.*, 1994; Sachs *et al.*, 2002; Tarun and Sachs, 1995) and *Neurospora crassa* (Wang and Sachs, 1997a,b). Here we provide detailed methods for the preparation and use of extracts from these two fungi. The assays described include the measurement of luciferase enzyme activity, the mapping of ribosomes on mRNA templates by primer-extension inhibition (toeprinting), and the labeling of translation products with [^{35}S]Met.

2. METHODS AND DISCUSSION

2.1. Method set 1

Method set 1 is used in the experiments shown in Figs. 10.1 through 10.3. All the water used in reactions, and reagent solutions that do not contain Tris, are treated with diethylpyrocarbonate (DEPC) to inactivate ribonucleases. DEPC is added to a concentration of 0.1% to water or reagent solutions, and after at least 12 h has elapsed, the DEPC is inactivated by autoclaving. Unless otherwise noted, reagent solutions are stored at $-80°$.

2.1.1. Buffers

Transcription buffer (5×): 200 mM Tris–HCl, pH 7.5, 30 mM MgCl$_2$, 10 mM spermidine, 50 mM NaCl.

Buffer A: 30 mM HEPES-KOH, pH 7.6, 100 mM potassium acetate, 3 mM magnesium acetate; 2 mM dithiothreitol (DTT). This is freshly prepared before use from concentrated stock solutions that are stored at $4°$ except for DTT, which is stored at $-20°$.

0.1 M phenylmethylsulfonyl fluoride (PMSF) stock: 1 g PMSF in 57.4 ml isopropanol, stored at room temperature; it is added immediately prior to use to buffers that require it.

Protease inhibitors (20×): 500 μg/ml p-amidinophenylmethylsulfonyl fluoride, 100 μg/ml each of pepstatin A, antipain, chymostatin, and leupeptin in water.

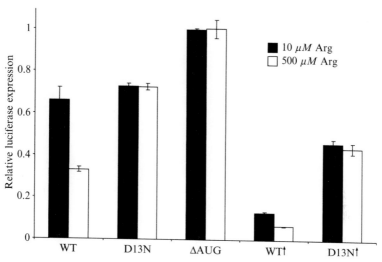

Figure 10.1 The effects of uORF regulation of translation *in vitro* can be measured by luciferase reporter activity. Micrococcal nuclease-treated *N. crassa* translation reaction mixtures (20 μl) were programmed with equal amounts (12 ng) of synthetic *S. cerevisiae* *CPA1-LUC* RNA transcripts and incubated at 25° for 20 min. The *CPA1-LUC* transcripts contained either the wild-type (WT) or D13N uORF in the wild-type initiation context or the improved initiation context (↑), or the AUG to AUU mutation that eliminates the uORF start codon (ΔAUG). All reactions contained mRNA specifying *Renilla* luciferase as an internal control. Reaction mixtures contained either 10 μM Arg or 500 μM Arg as indicated and 10 μM each of the other 19 amino acids. The results obtained from three independent translation reactions are shown. Firefly luciferase activity was normalized to *Renilla* luciferase activity and expressed as a ratio of normalized activity at a given Arg concentration relative to the normalized activity obtained from the construct lacking the *CPA1* uORF start codon (i.e., the average value of firefly enzyme activity in reaction mixtures containing 10 μM Arg is unity).

Energy mix (10×): 10 mM ATP, 2.5 mM GTP, 250 mM creatine phosphate.

Common buffer (40×): 400 mM HEPES-KOH, pH 7.6, 40 mM DTT.

Variable buffer (10×): 22.5 mM magnesium acetate, 1 M potassium acetate.

Reverse transcription buffer (5×): 250 mM Tris–HCl, pH 8.0, 375 mM KCl, 50 mM MgCl$_2$.

Cycloheximide: 10 mg/ml in water.

Edeine: Obtained from the National Cancer Institute Developmental Therapeutics Program (Compound NSC153112); we prepare a 20 mM stock from 50 mg as an oily dispersion in water.

Sodium dodecyl sulfate–polyacrylamide gel electrophoresis (SDS–PAGE) loading buffer (5×): 250 mM Tris–HCl, pH 6.8, 10% SDS, 50% glycerol, 0.5% bromophenol blue, 500 mM DTT. The DTT is added immediately prior to use from a stock solution stored at −20°; buffer lacking DTT is stored at room temperature.

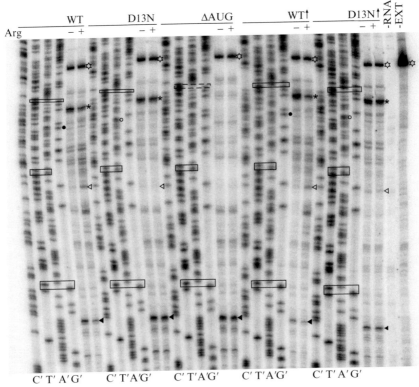

Figure 10.2 Toeprint analyses of initiation and termination of polypeptide synthesis. Micrococcal nuclease-treated *N. crassa* translation reaction mixtures (20 μl) were programmed with equal amounts (120 ng) of synthetic *S. cerevisiae CPA1-LUC* RNA transcripts and incubated at 25° for 20 min. The *CPA1-LUC* transcripts used were the same as in Fig. 10.1. Reaction mixtures contained either 10 μM Arg (–) or 2000 μM Arg (+) as indicated and 10 μM each of the other 19 amino acids. Radiolabeled primer ZW4 (19) was used for primer extension analysis and for sequencing of each *CPA1-LUC* template (sequencing was accomplished using the Thermo Sequenase Cycle Sequencing Kit [USB]). The nucleotide complementary to the dideoxynucleotide added to each sequencing reaction is indicated below the corresponding lane (C′, T′, A′, and G′) so that the sequence of the template can be directly deduced; the 5′- to-3′ sequence reads from top to bottom. Reading from top to bottom, the positions of the uORF start and stop codons, and the LUC start codon, are boxed in the nucleotide sequence (the AUU in the ΔAUG construct is boxed with broken lines), and the position of the nucleotide change at uORF codon-13 that determines whether it encodes D13 (GAC) or N13 (AAC) is indicated by solid or open circles, respectively. Primer extension products corresponding to the mRNA start 5′-end are indicated by a star, the uORF start codon by an asterisk, the duORF stop codon by an open triangle, and the LUC start codon by a solid triangle. Control samples of RNA in reaction mixtures without extract (–EXT) and reaction mixtures containing extract but not mRNA (–RNA) are indicated.

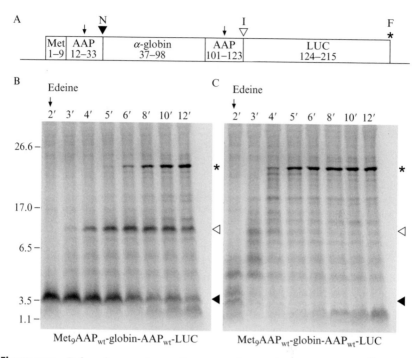

Figure 10.3 Pulse–chase analyses of nascent polypeptide synthesis using [^{35}S]Met in *N. crassa* cell-free extracts. (A) The Met$_9$-AAP-globin-AAP-LUC construct was used (Fang *et al.*, 2004). Arrows indicate where wild-type AAP Asp codons were changed to Asn (D12N mutation in the AAP). Filled arrowhead *N* indicates the C-terminus of the Met$_9$-AAP polypeptide intermediate; open arrowhead *I*, the C-terminus of the Met$_9$-AAP-globin-AAP intermediate; and asterisk *F*, the C-terminus of the completed polypeptide. Transcripts (1.2 μg) specifying (B) Met$_9$-AAP$_w$-globin-AAP$_w$-LUC or (C) Met$_9$-AAP$_m$-globin-AAP$_m$-LUC were translated in independent 200 μl *N. crassa* translation reaction mixtures supplemented with 2 mM arginine. Edeine (final concentration of 1 mM) was added at 2 min (arrow) to each and 10-μl aliquots of each reaction mixture were removed at the indicated time points for analysis by SDS–PAGE (Fang *et al.*, 2002). The arrowheads and asterisks indicate the positions where intermediate Met$_9$-AAP; *I*, the polypeptide intermediate Met$_9$-AAP-globin-AAP; and *F*, the full-length polypeptide migrate in these gels.

Toeprint gel loading buffer (2×): 0.05% bromophenol blue, 0.05% xylene cyanol FF, 20 mM ethylenediaminetetraacetic acid (EDTA), pH 8.0, 91% formamide; stored at −20°.

2.1.2. Strains used

The *N. crassa* strain is the standard laboratory wild-type designated Oak Ridge 74-OR23-IVA (Fungal Genetics Stock Center strain #2489). The *S. cerevisiae* strain is MBS [*MATa ade2-1 his3-11,15 leu2-3,112 trp1-1 ura3-1 can1-100* (rho+) L-o, M-o] (see Iizuka *et al.*, 1994).

2.1.3. *In vitro* transcription

Reactions (100 μl) to synthesize capped RNA contain 4 μg of linearized DNA template, 40 mM Tris–HCl, pH 7.5, 6 mM MgCl$_2$, 2 mM spermidine, 10 mM NaCl, 10 mM DTT, 0.5 mM each of ATP, CTP, and UTP, 2 μCi of [α-^{32}P]UTP, 0.05 mM GTP, 0.5 mM m^{7}G(5')ppp(5')G, 80 units of RNasin, and 100 units of T7, T3, or SP6 RNA polymerase as appropriate. The radiolabel is used to quantify yields of RNA. Therefore, in all experiments, a common stock of reactants without template DNA is prepared and aliquoted to separate tubes to which template DNA is then added. This ensures that RNAs prepared in parallel will be radiolabeled with [α-^{32}P]UTP to the same specific activity. Reaction mixtures are incubated for 1.5 h at 37°. The DNA template is then removed by adding 1 unit of RQ1 DNase I and incubation at 37° for an additional 15 min. Water (35 μl) and 5 M NH$_4$OAc (15 μl) are added, and each reaction mixture is extracted with 150 μl of phenol:chloroform (50:50), and then with 150 μl of chloroform. The aqueous phase is transferred to a fresh 1.6-ml Eppendorf tube and the mRNA is precipitated by adding 1 volume of isopropanol, chilling for at least 15 min at −20°, and centrifugation at 4° for 20 min in an Eppendorf 5415D minicentrifuge at 13,200 rpm. The supernatant is carefully removed and the pellet is washed once with 70% ethanol at room temperature, briefly vacuum dried, and then resuspended in 50 μl of water. mRNAs are stored in aliquots at −80°, taking into account our observation that multiple freeze–thaw cycles adversely affect the translatability of the mRNA.

The amount of RNA synthesized in each reaction is determined by measuring the incorporation of radiolabeled nucleotide. Accurate quantitation can be achieved by polyethylenimine (PEI) thin layer chromatography (Selecto PEI cellulose #11078) using 0.75 M potassium phosphate, pH 3.5 [RNA remains at the origin (Cashel *et al.*, 1969)] or TCA precipitation (Sambrook and Russell, 2001). The fraction of [α-^{32}P]UTP incorporated is used to calculate the amount of RNA synthesized. Specifically, the yields of RNA obtained in parallel reactions are determined by considering the size of each RNA, the fraction of U in each, and the fraction of radiolabeled nucleotide incorporated (the latter acts as a measure of the total amount of nucleotide incorporated). The integrity of the RNA is evaluated by ethidium bromide staining following electrophoresis of a sample (at least 180 ng of RNA) in formaldehyde agarose gels (Gaba *et al.*, 2005).

2.1.4. Preparation of *N. crassa* cell-free extracts for translation

General methods for culturing *N. crassa* can be found at the Fungal Genetics Stock Center website (www.fgsc.net/Neurospora/NeurosporaProtocol Guide.htm) (see also Davis and de Serres, 1970). Conidia are germinated in growth medium (Vogel's minimal medium + 1.5% sucrose) at a concentration of 1×10^7 conidia/ml. Specifically, a 1-liter culture in a 2-liter

Erlenmeyer flask is incubated at 32° with orbital shaking (200 rpm) until 90% of the conidia show germ tubes (typically 5 to 6 h). The germlings are harvested by vacuum filtration onto 9-cm Whatman 541 filter paper using a Büchner funnel. The filter paper with the germlings is placed in a sterile 50-ml conical screwcap tube with 40 ml of fresh growth medium; the germlings are resuspended by vigorous hand-shaking and transferred to 1 liter of fresh prewarmed growth medium in a 2-liter flask, incubated an additional 1 h at 32° with orbital shaking, and harvested again. Following harvesting, germling pads are rinsed on the filter with ice-cold buffer A, then peeled from the filter and weighed; a typical yield from the wild-type strain is 12 g. The pad is then frozen directly in liquid nitrogen. In the cold room, liquid nitrogen-frozen germlings are powdered in the presence of liquid nitrogen using a mortar (Coors #60322, 13 cm OD) and matching pestle (the mortar and pestle are precooled to −80° to help prevent breakage when liquid nitrogen is added). Ice-cold buffer A (0.5 ml/g of wet weight) is added at intervals when the liquid nitrogen vaporizes away. The mortar is refilled approximately halfway with liquid nitrogen and grinding is continued until a fine powder is obtained. The powdered mixture of cells and buffer A are carefully transferred to a 50-ml polycarbonate centrifuge tube using a wide-bore funnel, allowed to thaw on ice, and then centrifuged at 4° for 15 min at 16,000 rpm in an SS34 rotor. The supernatant is carefully collected with a sterile Pasteur pipette, avoiding both the pellet and the fatty upper layer, and placed in a fresh 15-ml conical tube that is maintained on ice. Typically, 8 ml of extract is recovered.

Small molecules are removed from the extract by centrifugation at 4° through Zeba Desalt Spin Columns (Pierce 89894) using an Eppendorf 5810R centrifuge and rotor A-4-62. Specifically, at 4°, two 10-ml columns are preequilibrated with buffer A by first centrifuging the columns at $1000 \times g$ for 2 min to remove the storage buffer; the column is then washed four times with 5 ml of buffer A, each time removing the buffer by centrifugation as above. Extract is loaded onto the columns (4 ml/column), which are then centrifuged at $1000 \times g$ for 2 min. The eluates are pooled and this material is the *N. crassa* extract used for cell-free translation experiments. Alternatively, small molecules can be removed from the extracts by chromatography through Sephadex G-25 columns (described below). In either case, protease inhibitors are added to $1\times$ final concentration to the extracts following the removal of small molecules. For storage, aliquots of extract (200 μl) are put into 1.6-ml Eppendorf tubes, frozen with liquid nitrogen, and stored at −80°.

2.1.5. Preparation of *S. cerevisiae* cell-free extracts for translation

General microbiological methods for *S. cerevisiae* cultures are as described (Burke *et al.*, 2000). A 100 ml starter culture in YPD, begun from a single colony, is grown at 30° for 24 h with orbital shaking (200 rpm). This is used to inoculate 800 ml of YPD in a 2-liter Erlenmeyer flask at an initial OD_{600} of 0.03 to 0.06. Cultures are grown at 30° and 200 rpm until they reach an $OD_{600} \approx 1.5$ (typically 8 h for a wild-type strain).

Cells are harvested by centrifugation at 4° for 5 min at 3000 rpm in a Sorvall GSA rotor using four centrifuge bottles. The supernatant is decanted and 15 ml of ice-cold buffer A/8.5% mannitol is added to each of two bottles. The pelleted cells are resuspended by shaking and each transferred to a remaining bottle containing a cell pellet, which is also resuspended. The combined cells are transferred to a preweighed 50-ml conical screwcap tube. Cells are collected by centrifugation at 4° in an Eppendorf 5810R centrifuge with rotor A-4-62 at $1470 \times g$, and washed four additional times with 10 ml buffer A/8.5% mannitol by consecutive rounds of gentle resuspension followed by centrifugation; the final centrifugation is at $2500 \times g$. The wet weight of each cell pellet is determined by weighing the tube containing the cell pellet and subtracting the predetermined weight of the tube. The cells are resuspended in 1.5 ml of buffer A/8.5% mannitol/0.5 mM PMSF per gram wet cell weight. Resuspended cells are combined with 6× the wet cell weight of cold 0.5-mm glass beads (Biospec). In the cold room, cells are lysed by manual shaking for five 1-min periods with 1-min cooling on ice between shaking periods. Shaking is performed at a rate of two cycles/sec over a 50-cm hand path. The tubes containing cell lysates and beads are centrifuged at 4° for 2 min at $650 \times g$. This supernatant is then transferred to a sterile 50-ml polycarbonate centrifuge tube using a sterile Pasteur pipette, and centrifuged at 4° for 6 min at 18,000 rpm in a Sorvall SS34 rotor. The supernatant is collected, taking care to avoid the fatty layer at the top and the cell debris at the bottom of the tube. Small molecules are removed from this clarified extract using Zeba Desalt Spin Columns (Pierce) that are preequilibrated with buffer A/0.5 mM PMSF essentially as described for the preparation of *N. crassa* extracts. Small molecules can also be removed by Sephadex G-25 chromatography (see below). Aliquots (200 μl) are pipetted into 1.6-ml Eppendorf tubes, frozen with liquid nitrogen, and stored at −80°.

2.1.6. Nuclease treatment of cell-free extracts

Immediately prior to assembly of the translation reaction mixtures, 200 μl of extract is thawed on ice, supplemented with 2 μl of 100 mM $CaCl_2$, and then treated with 10 U of micrococcal nuclease by incubation at 25° for 10 min. Then 3 μl of 170 mM ethyleneglycoltetraacetic acid (EGTA) is added to chelate the calcium and thus inhibit the activity of the Ca^{2+}-dependent nuclease, and the mixture is placed on ice. Alternatively, nuclease treatment can be accomplished prior to preparing aliquots of extracts for freezing by appropriately scaling up the nuclease treatment procedure.

2.1.7. Cell-free translation and analyses of translation products

The procedure for setting up 10 20-μl *N. crassa* or *S. cerevisiae* translation reaction mixtures for assaying luciferase activity or for toeprinting is as follows. The setup can be scaled as appropriate. All components are kept on ice until the reaction incubation is begun. A master solution, composed of 16.8 μl of energy mix, 1.2 μl of creatine phosphokinase (7.5 U/μl), 5 μl of common buffer (40×), 20 μl of variable buffer (10×), 2 μl of 1 mM 20 amino acids, and 1 μl of RNasin (Promega cat. no. 40 U/μl) is combined with DEPC-treated water to adjust the volume to 80 μl. The composition of the variable buffer used for what we define as our "standard conditions" is given; different concentrations of Mg^{2+} and K^+ can be used in the variable buffer to optimize translation of specific mRNAs (Spevak *et al.*, 2006). For measurements of luciferase activity, the internal control mRNA specifying *Renilla* (sea pansy) luciferase is added to this mixture and the amount of added water is reduced correspondingly. This mixture of reactants is aliquoted into individual tubes (8 μl/tube) and combined with 2 μl of appropriately diluted mRNA. Then 10 μl of nuclease treated cell-free extract is added, gently mixed with reactants, and reactions are incubated in a 25° water bath for 20 min and stopped by freezing in liquid N_2. Firefly luciferase activity is determined by adding 5 μl of the thawed reactions to 50 μl of the firefly luciferase assay reagent of the Dual Luciferase Reporter Assay System (Promega #E1960), and immediate measurement with a Turner TD-20e luminometer; 50 μl of *Renilla* luciferase assay reagent is added, and that enzyme's activity measured by luminometry.

For [^{35}S]Met labeling of translation products, reaction conditions are similar to those described above, except that 10 μCi of [^{35}S]methionine (MP #51001H, >1000 Ci/mmol) is used in each 20-μl reaction, and the 1 mM amino acid solution used lacks methionine. After incubation, reactions are stopped by adding SDS–PAGE loading buffer. Translation products are analyzed by electrophoresis in tricine gels containing a total concentration of 4% acrylamide in the stacking gel and 16.5% acrylamide in the separating gel (Schägger and von Jagow, 1987). Autoradiography is performed with a Phosphor Screen (Amersham).

2.1.8. Preparation of 5'-^{32}P-labeled primer for toeprinting and sequencing reactions

Oligodeoxynucleotides are labeled at their 5' termini with T4 polynucleotide kinase and [γ-^{32}P]ATP. The reaction mixture (100 μl) contains 50 pmol of oligodeoxynucleotide, 50 mM Tris–HCl, pH 8.0, 20 mM $MgCl_2$, 4 mM spermidine, 4 mM DTT, 300 μCi [γ-^{32}P]ATP (4500 Ci/ mmol; MP #38101X), and 10 units of T4 kinase. The primer is first mixed with Tris–HCl, pH 8.0, and the volume is adjusted to 44 μl; the primer is then heated at 90° for 3 min and chilled on ice. $MgCl_2$, spermidine, DTT,

$[\gamma-^{32}P]ATP$, and water (to bring the final volume to 100 μl) are added; then kinase is added and the mixture is incubated at 37° for 45 min. EDTA is added to a final concentration of 50 mM to stop the reaction. Labeled primers are purified with mini Quick Spin Oligo Columns (Roche #11814397001). An aliquot (1 μl) of purified primer is analyzed by PEI thin layer chromatography (Selecto PEI cellulose #11078) using 0.75M potassium phosphate, pH 3.5, followed by PhosphorImager analyses to confirm purification; radiolabeled oligonucleotide stays at the origin. The primer is adjusted to a predicted concentration of 0.5 μM.

2.1.9. Primer extension inhibition (toeprint) assay

The procedure for setting up 10 toeprint reaction mixtures is as follows. A 10× annealing solution contains 20 μl of 5× reverse transcription buffer, 10 μl of 0.1 M DTT, 10 μl of 2.5 mM dNTPs, 2.5 μl of RNasin (40 U/μl), and 12.5 μl of DEPC-treated water ($V_f = 55 \mu l$). Translation reactions are performed as outlined above except that instead of freezing in liquid nitrogen to stop the translation reaction, 3 μl of translation reaction is added to 5.5 μl of annealing solution on ice. The mixture is incubated at 55° for 3 min. This step is critical for the visualization of ribosomes by toeprinting. Primer (1 μl) is added and the mixture is incubated at 37° for 5 min. Then, 0.5 μl reverse transcriptase (100 U; Invitrogen Superscript III) is added and the sample is incubated at 37° for 30 min. The reactions are stopped by extraction with an equal volume of phenol:chloroform. Following extraction, the aqueous phase is combined with an equal volume of 2× Toeprint gel loading buffer. Samples are heated at 80 to 85° for 5 min and then loaded on a 6% urea–polyacrylamide gel (prerun at 75 W for 45 min) and electrophoresed at 65 W until the bromophenol blue dye runs off the gel. It may be desirable to adjust acrylamide concentrations and/or running times to optimize the resolution of products in different size ranges. Autoradiography is performed by Phosphorimaging.

2.2. Method set 2

Unless otherwise specified, protocols and the composition of the buffers are similar to those of Method set 1. Method set 2 is used in the experiments described in Figs. 10.4–10.6. The preparation of extracts described below can be done in three different stages, with the first step consisting of growing cells (overnight), collecting them the following day, and storage at −80°. In the second step, cells are lysed and can again be kept at −80° for future use. The third step involves thawing the cells and preparing the extracts. The translation and toeprinting reactions can also be done in two separate steps. Translation reactions can be aliquoted and stored at −80°, with the toeprinting reactions done on a different day using the frozen aliquots. This Method set also

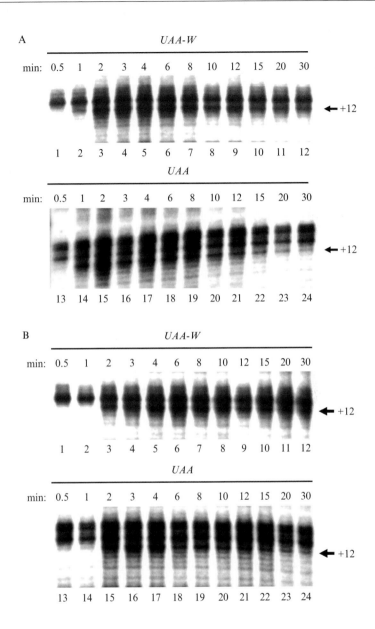

Figure 10.4 Toeprint analyses of premature termination in yeast wild-type and *upf1Δ* extracts. Premature termination codon toeprints detected during translation of *UAA-W* and *UAA* mRNAs in the absence of CHX (Amrani *et al.*, 2004). Translation reactions were incubated with the mRNA for 0.5 min to 30 min. Primer extension inhibition assays performed on aliquots of these reactions using oligoprimer #3029 (Amrani *et al.*, 2004) are shown. (A) Wild-type extracts. (B) *upf1Δ* extracts. The position of the +12 toeprint is indicated with an arrow.

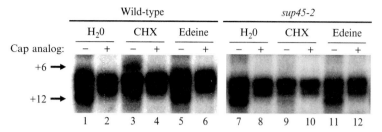

Figure 10.5 CHX addition to yeast wild-type or *sup45–2* extracts promotes dissociation of the ribosomes from the termination site. Translation reactions were incubated for 4 min, and terminated by incubation with CHX (0.6 mg/ml), edeine (0.5 μM), or water for 3 min. Positions of the toeprints are shown on the left.

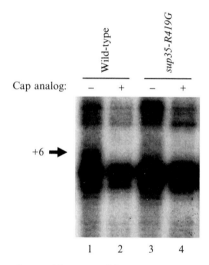

Figure 10.6 Toeprints detected in yeast wild-type extracts are absent in *sup35-R419G* extracts in the presence of CHX. The position of the +6 toeprint is indicated with an arrow.

provides a simple technique for labeling oligoprimers that will be used for both toeprinting and DNA sequencing reactions.

2.2.1. Buffers

Transcription buffer (5×): Provided by the mMessage mMachine Kit (Ambion).

Buffer C: 20 mM HEPES (pH 7.4 with KOH), 100 mM potassium acetate, 2 mM magnesium acetate, 2 mM DTT, 0.5 mM PMSF.

Buffer B: Buffer C + 20% glycerol.

Translation Buffer (6×): 32 mM HEPES (pH 7.4 with KOH), 170 mM potassium acetate, 3 mM magnesium acetate, 0.75 mM ATP, 0.1 mM GTP, 25 mM creatine phosphate, 0.04 mM complete amino-acid, 2.7 mM DTT.
Cycloheximide: 10 mg/ml in 100% ethanol, stored at −20°.

2.2.2. Strains used

The *S. cerevisiae* strains were MBS [*MATa ade2-1 his3-11,15 leu2-3,112 trp1-1 ura3-1 can1-100* (rho+) L-o, M-o], NA101 [*MATa ade2-1 his3-11,15 leu2-3,112 trp1-1 ura3-1 can1-100 upf1::HIS3* (rho+) L-o, M-o], NA207 [*MATα ade2-1 ura3-1 sup35::ADE2* (pRS316, *sup35-R419G, URA3,CEN*)], and MT552/8a (*MATα sup45-2 ade2-1 ura3-1*).

2.2.3. *In vitro* transcribed mRNAs

Synthetic, capped poly(A)-containing RNA is synthesized *in vitro* from chimeric genes cloned in a pSP65A vector that includes ∼65 dT residues for transcription of a poly(A) tail (Promega). An SP6 m*Message* m*Machine* kit (Ambion), used according to the manufacturer's protocol, generates synthetic mRNA from *Hind*III-linearized plasmids. RNA yields are quantified by spectrophotometry and their integrities are assessed, as in Method set 1, by formaldehyde agarose gel electrophoresis.

2.2.4. Preparation of extracts

2.2.4.1. *Growth of culture and cell lysis* Yeast cells are streaked onto a
YPD plate and incubated at 25° for 36 to 48 h. Starter cultures (100 ml) are grown at 25° from a single colony and 2-liter YPD cultures are grown in 6-liter Erlenmeyer flasks, starting at $OD_{600} = 0.03$ to 0.06 by inoculation from the starter culture. Growth is continued at the same temperature (110 rpm, gyratory shaker) until the cultures reach $OD_{600} = 3$ to 4.

Cultures are harvested in preweighed 0.5-liter bottles and centrifuged for 10 min at 5000 rpm in a Sorvall GS-3 centrifuge rotor. The resulting supernatants are discarded and the cell pellets are washed with 300 ml cold water and centrifuged again at 5000 rpm for 10 min. The cell pellets are resuspended and washed twice in 200 ml of freshly prepared cold buffer C. The suspensions are centrifuged for 10 min at 5000 rpm in the same rotor, the supernatants discarded, and the wet weights of the resulting cell pellets are determined. The cell pellets are resuspended in one-tenth the volume of the wet weight of the pellet using buffer C supplemented with protease inhibitor cocktail. The resulting cell suspensions are dripped into liquid N_2 to generate small frozen cell pellets in the form of tiny beads. The frozen cell pellets from each culture are transferred to two 50-ml plastic tubes for storage at −80° until the time of cell breakage.

To lyse cells, a ceramic mortar and pestle are first prechilled at −80°. A small amount of liquid N_2 is then added to the mortar. The frozen yeast pellets from one 50-ml tube are added to the liquid N_2 and the mortar is partially filled with liquid N_2. The pellets are crushed using slight pressure and a circular motion. Once most of the liquid N_2 has evaporated, the mortar is partially filled with liquid N_2 and the grinding process repeated, using much greater pressure. Liquid N_2 is added as needed and grinding is continued until a fine powder is obtained. The grinding of cells from one 50-ml tube takes about 15 min. The resulting powder is transferred to a new 50-ml tube and either stored again at $-80°$ or allowed to thaw on ice, typically for 2 to 3 h. The thawed broken cells are transferred to prechilled Nalgene 16 $\times$ 75-mm ultracentrifuge tubes and centrifuged at 18,000 rpm for 15 min in a Beckman Ti50 ultracentrifuge rotor. The supernatant is transferred to a fresh 16 $\times$ 75-mm tube and spun for an additional 15 min at 18,000 rpm. The resulting supernatant is removed with a Pasteur pipette, taking care to avoid both the lipid layer at the top and any pellet that had formed at the bottom of the centrifuge tube.

2.2.4.2. Column chromatography Sephadex G-25 superfine (Sigma, 50 ml of suspension autoclaved for 30 min, then cooled to $4°$) is poured into a 2.5 $\times$ 20-cm column and equilibrated with 50 ml buffer C in the cold room using a Rainin peristaltic pump to control the flow rate (approximately 5.5 ml/h). After equilibration, the sample (4 to 5 ml) is loaded onto the column, which is then washed with buffer B to elute the sample. Column fractions (0.5 ml) are collected in microfuge tubes. Peak fractions representing material in the void volume (which appear slightly opaque) typically elute approximately 25 min after loading the sample. The A_{260} of each fraction is determined after diluting 2 μl of sample into 998 μl of water. All fractions with a diluted A_{260} of 0.4 or higher are pooled. The protein concentration of the extract is determined by BCA assay (Pierce) using bovine serum albumin (BSA) as the standard. Aliquots (50 μl) in 0.5-ml microcentrifuge tubes are quickly frozen in liquid N_2 for subsequent storage at $-80°$. Columns can be reused after washing with buffer B.

2.2.4.3. Nuclease treatment Immediately prior to assembly of the translation reaction mixtures, 50 μl extract, 1.0 μl 50 mM $CaCl_2$, and 1 μl micrococcal nuclease (41 U/μl) are combined and incubated at $25°$ for 5 min. Then 1.0 μl of 100 mM EGTA is added to stop the nuclease reaction, and the mixture is placed on ice.

2.2.5. Analyses by translation and toeprinting
2.2.5.1. Translation reactions Translation reactions (7.5 μl) containing 2.5 μl 6$\times$ translation buffer, 1.0 μl creatine phosphokinase (4 mg/ml), 0.1 μl RNasin (40 U/μl), 1.0 μl RNA (100 ng), and 2.9 μl of water (or, in those cases where additional components are added, the amount of water required for a final volume of 7.5 μl) are assembled in 0.5-ml Eppendorf

tubes and mixed with 7.5 μl of nuclease-treated extract (different extracts are diluted to the same protein concentration, e.g., 16 mg/ml, using buffer B, after micrococcal nuclease treatment). The translation reactions are incubated at 25° for 4 min for a typical toeprinting reaction and are stopped by the addition of 1.0 μl cycloheximide for 3 min at 25°. The reactions are then kept on ice and 3-μl aliquots from each reaction are transferred into new 0.5-ml Eppendorf tubes. Routinely, samples are frozen in dry ice and kept at $-80°$ until used for toeprinting reactions. Alternatively, the toeprint procedure can follow immediately.

2.2.5.2. Preparation of ³²P-labeled primer

Labeled oligonucleotide primers are prepared in a 40 μl reaction containing 16 pmol primer, 4 μl 10× kinase buffer, 2 μl T4 polynucleotide kinase (Promega), and 8 μl ATP [γ-32P] 6000 Ci/mmol (Perkin Elmer). The reaction is incubated at 37° for 30 min and stopped by heating at 65° for 5 min. The sample is then diluted to 70 μl by the addition of water and purified with a mini Quick Spin Column (Roche) following the manufacturer's recommendations. The radiolabeled primer is then stored at $-20°$ and is suitable for use for up to 2 weeks after labeling.

2.2.5.3. Sequencing reaction protocol

This protocol is used to generate sequence ladders for toeprinting analyses and uses the DNA Sequencing Kit from USB.

1. Prepare denatured DNA in 1.6-ml Eppendorf tubes by mixing 1 μl DNA (0.5 to 1 μg), 3 μl labeled primer, 2.5 μl H$_2$O, and 1 μl dimethyl sulfoxide (DMSO). Heat the mixture at 95° for 3 min then put in dry ice.
2. Put 2.5 μl of each ddNTP per well in a microtiter plate (Nalge Nunc) and prewarm at 37° for 3 min.
3. Transfer the tube containing the denatured DNA on ice and add 2 μl of 5× Sequenase reaction buffer, 1 μl 0.1 M DTT, and 2 μl diluted Sequenase DNA polymerase (to prepare the dilution; mix 1 μl enzyme with 8 μl Sequenase dilution buffer).
4. Add 3 μl from the mixture prepared in step 3 to each well that contains the ddNTP (see step 2). Incubate for 5 min at 37°.
5. Add 4 μl of Stop Solution (from the kit) to each well.
6. Heat at 95° for 2 min, then put on ice.
7. Load 6 μl per well.

2.2.5.4. Toeprint reaction protocol

1. Prepare Annealing Solution on ice in a 0.5-ml tube [1× Annealing Solution: 1 μl radiolabeled primer, 1.25 μl water, 2 μl 5× reverse transcription

buffer, 1 μl 0.1 M DTT, 1 μl from mix of 2.5 mM each dNTP, 0.25 μl RNasin (40 U/μl)].

2. Remove the translation reaction aliquots from the freezer and thaw on ice. Add 6.5 μl annealing solution to each sample and maintain on ice.
3. Transfer the tubes to 55° for 2 min, then 37° for 5 min.
4. Add 0.5 μl (100 units) reverse transcriptase (Superscript III; Invitrogen) and incubate at 37° for 30 min.
5. Stop the reaction by adding 1 μl EDTA (5 mM) and place sample on ice.
6. Extract with an equal volume of phenol:chloroform. Vortex briefly to mix, centrifuge to separate phases, and add the extracted aqueous phase to an equal volume of toeprint gel loading buffer.
7. Heat the samples at 95° for 5 min and load on a 6% urea-polyacrylamide gel that has been prerun at 110 W for 45 min.
9. Electrophorese samples at 65 W until the bromophenol blue dye runs off the gel. If using shark's-tooth combs, load the toeprint reactions into every other lane, placing the loading buffer in the blank lanes. It may be desirable to adjust acrylamide concentrations and/or running times to optimize the resolution of products in different size ranges.

2.3. Examples of the method

2.3.1. Analysis of translational regulation by upstream open reading frames

Each of the transcripts of the homologous genes *N. crassa arg-2* and *S. cerevisiae CPA1* contains an evolutionarily conserved upstream open reading frame (uORF) encoding the arginine attenuator peptide. The nascent peptide causes ribosomes to stall in response to a high level of arginine (Hood *et al.*, 2007 and references therein). Regulation by uORFs in the *N. crassa in vitro* translation system is illustrated by analyses of synthetic reporter transcripts containing different permutations of the *S. cerevisiae CPA1* uORF (Gaba *et al.*, 2005). In these constructs, the *CPA1* mRNA 5′-leader is present upstream of firefly luciferase. Figure 10.1 shows the results of assaying firefly luciferase activity derived from translation of synthetic capped and polyadenylated mRNA. The mRNAs (1) contain the wild-type uORF, (2) contain a uORF with a missense mutation that eliminates Arg-specific regulation (D13N), (3) lack the uORF because the uORF start codon has been removed by changing the AUG to AUU (ΔAUG), (4) contain the wild-type uORF in an improved initiation context, or (5) contain the D13N uORF in an improved initiation context. Firefly luciferase activity is measured and normalized to the activity of *Renilla* luciferase that is produced from an internal control mRNA (Wang *et al.*, 1998).

The luciferase data show that the wild-type uORF, but not the D13N or ΔAUG uORF, results in arginine-specific negative regulation of luciferase synthesis. Furthermore, improving the initiation context of the uORF start codon decreases gene expression, regardless of the uORF coding sequence (Fig. 10.1), as expected, because the improved initiation context should reduce leaky scanning past the uORF start codon (Gaba *et al.*, 2005).

In the *N. crassa* cell-free system, but not in the *S. cerevisiae* cell-free system, it is possible to visualize ribosomes at start codons by toeprinting without the addition of drugs that block elongation such as cycloheximide (CHX). Toeprinting analyses of the same five *CPA1* mRNAs used for luciferase assays in *N. crassa* extracts are shown in Fig. 10.2. Controls include toeprint analyses of extract lacking mRNA (–RNA) and of mRNA incubated in the absence of extract (–EXT). For each mRNA, the oligonucleotide used for toeprinting is also used to sequence the DNA template used for mRNA transcription; this confirms the sequence of each construct and also provides an appropriate sequence ladder for high-resolution mapping of ribosomes at rate-limiting steps in translation. These data show several important aspects of *CPA1* uORF control. First, the wild-type, but not the D13N, sequence causes stalling at the uORF termination codon when high arginine is present, as evidenced by an increased toeprint signal at this site. Additional toeprint signals seen within the uORF coding region may represent additional ribosomes whose movement has been impeded by the ribosome stalled at the termination codon. Second, stalling at the termination codon in high arginine is associated with reduced loading of ribosomes at the luciferase start codon. Third, improving the context of the uORF initiation codon causes increased loading of ribosomes at that codon, relative to the wild-type context, and decreases loading of ribosomes at the downstream luciferase start codon, independent of the effects of the uORF sequence on arginine regulation; the improved uORF initiation context reduces loading for uORFs specifying wild-type or D13N peptides.

2.3.2. Use of [^{35}S]Met to examine polypeptide synthesis

The use of [^{35}S]Met to follow polypeptide synthesis in pulse–chase experiments is shown in Fig. 10.3. In this experiment, mRNAs containing the *N. crassa* AAP near the N-terminus of a polypeptide, and also internally within the same polypeptide (Fig. 10.3A), are translated in *N. crassa* extracts containing a high concentration of arginine. The results of a time-course for a construct containing two AAPs with a wild-type sequence, or a construct that contains in each of the two AAPs the Asp to Asn mutation (*N. crassa* D12N) that results in loss of arginine-specific regulation, are shown in Fig. 10.3B and C, respectively. To accomplish a pulse–chase under *in vitro* conditions, the constructs contain nine methionine residues (specified by nine consecutive AUG codons) at their N-terminus, and no internal in-frame AUG codons (these constructs also lack out-of-frame AUG codons), therefore, placing all of the [^{35}S]Met at the N-terminus. Edeine, which inhibits translation initiation but not elongation or termination, is added 2 min after the translation reaction commences, resulting in cessation of initiation and allowing the incorporation of radiolabel to be followed by taking appropriate time points for analyses by SDS–PAGE (in the experiments shown here, edeine was used at a final concentration of 1 mM). These analyses show that each wild-type AAP domain causes ribosomes to stall in high arginine, and indicate that after stalling, ribosomes resume translation. No arginine-specific stalling is observed within the polypeptide containing two D12N AAP domains.

2.3.3. Analyses of premature translation termination by toeprinting samples of yeast translation reactions

The nonsense-mediated mRNA decay (NMD) pathway eliminates mRNAs that contain premature termination codons (PTCs) and prevents suppression of genetic nonsense (Amrani *et al.*, 2006a,b; Jacobson, 1996; Pulak and Anderson, 1993). NMD substrates include transcripts that arise from genes in which a mutation or an error in transcription or processing has given rise to a premature nonsense codon, inefficiently spliced pre-mRNAs that enter the cytoplasm with their introns intact, mRNAs in which the ribosome has bypassed the initiator AUG and commenced translation further downstream, some mRNAs containing uORFs, mRNAs subject to frameshifting, bicistronic mRNAs, transcripts of pseudogenes and transposable elements, and mRNAs with abnormal extensions of their 3′-UTRs. Yeast factors that regulate NMD, identified in screens for translational suppressors or two-hybrid interactors with known factors, are principally those encoded by the *UPF1*, *NMD2(UPF2)*, and *UPF3* genes (Cui *et al.*, 1995; He and Jacobson, 1995; He *et al.*, 1997; Lee *et al.*, 1995; Leeds *et al.*, 1991, 1992). They have been characterized extensively and are conserved in all eukaryotes examined. Mutations in the yeast *UPF/NMD* genes not only lead to the stabilization of nonsense-containing mRNAs, they also promote nonsense suppression (Keeling *et al.*, 2004; Leeds *et al.*, 1992; Maderazo *et al.*, 2000; Stahl *et al.*, 2000; Weng *et al.*, 1996a,b). Mutations in the release factors, Sup35p (eRF3) and Sup45p (eRF1), also promote nonsense suppression (Keeling *et al.*, 2004; Stansfield *et al.*, 1997), and these effects are additive with those of *upf/nmd* mutations (Keeling *et al.*, 2004), indicating distinct functions in termination.

To understand the mechanism by which premature translation termination events influence mRNA decay, we have used toeprinting techniques to monitor the position of ribosomes at premature stop codons (Amrani *et al.*, 2004; Sachs *et al.*, 2002; Wang *et al.*, 1999). The yeast *can1-100* allele contains a premature UAA codon at position 47 of the *CAN1* coding region that effectively terminates translation and destabilizes the *CAN1* mRNA (Maderazo *et al.*, 2000). We constructed a gene fusion encompassing the UAA-containing segment of the *can1-100* allele and the firefly *LUC* coding region, and used this construct to generate synthetic mRNA (termed *UAA RNA*) as described previously (Amrani *et al.*, 2004). Mutagenesis techniques were used to create a *can1-100* variant with a weak terminator (CAA UAA CAA) at codon 47 (termed *UAA-W* RNA). Translation reactions in wild-type or *upf1Δ* extracts were incubated with these mRNAs for 0.5 min to 30 min without addition of CHX, and samples were taken for toeprinting analyses. As shown previously (Amrani *et al.*, 2004), toeprints corresponding to ribosomes stalled with a stop codon in their A-sites were obtained with both RNAs at the expected position, 12 to 14 nucleotides (nt) downstream of the premature nonsense codons (Fig. 10.4). However, analysis of those termination toeprints in wild-type extracts shows a lag in the release of the ribosome from a weak termination codon (CAA UAA CAA) in the *UAA-W* mRNA compared to the strong terminator in the *UAA* mRNA (AGT UAA GTC) (Fig. 10.4A, compare lanes 10, 11, and 12 with 22, 23, and 24). Likewise, toeprint reactions from *upf1Δ* extracts show a delay in the release of the ribosomes from *UAA-W* mRNA compared to *UAA* mRNA (Fig. 10.4B, compare lanes 10, 11, and 12 with 22, 23, and 24). These data show that the efficiency of translation termination indeed depends on the sequence context of the termination codon (Bonetti *et al.*, 1995). Interestingly, a comparison of the toeprint bands of wild-type extracts with those of *upf1Δ* extracts shows a delay in the appearance and disappearance of the bands in *upf1Δ* compared to the wild-type with both mRNAs (see Fig.10. 4A, lanes 1, 2 and 13, 14 and Fig. 10.4B, lanes 1, 2 and 13, 14), suggesting a weaker or slower translation reaction in the *upf1Δ* extracts. These data show that the toeprint assay can be used to monitor the kinetics of ribosome release from a termination site in both wild-type and mutant extracts.

2.3.4. Effects of cycloheximide on ribosomal toeprint at a termination site

As we showed previously (Amrani *et al.*, 2004), translation of *UAA* mRNA in wild-type extracts or *sup45-2* mutant extracts for 4 min allows detection of +12 toeprints at the premature stop codon (Fig. 10.5, lanes 1 and 7). These toeprint bands were dependent on mRNA translation, because they were sensitive to ^{7}mGpppG, a cap analog that blocks cap-dependent translation in cell-free extracts (see Fig. 10.5, lanes 2 and 8). Consistent with previous studies, we were not able to efficiently toeprint initiator AUGs on the same mRNAs in the absence of drugs that block the elongation process (Dmitriev *et al.*, 2003a,b; Kozak, 1998; Pestova and Hellen, 2003). However, in the presence of CHX, toeprints were obtained from the AUG initiator codon of the *UAA* mRNA (data not shown). These toeprints reflect 80S ribosomes, centered on AUG codons, protecting 16 to 18 nt 3' of those codons. Other CHX-dependent toeprints, sensitive to cap analog, were present in close proximity to the location of the early stop codons (see Fig. 10.5, lanes 3 and 4) and these mapped to a position 6 nt downstream of the U of the termination codon. These toeprints corresponded to ribosomes stalled at the −11 AUG after termination and retroreinitiation events (Amrani *et al.*, 2006a).

By using extracts from the *sup45-2* temperature-sensitive mutant MT552/8a, which carries an Ile-222-Ser substitution in the termination factor eRF1 (Stansfield *et al.,* 1995), we also found that the retroreinitiation band is dependent on fully functional eRF1 (see Fig. 10.5, lanes 9 and 10) (Amrani *et al.,* 2006a). Interestingly, the addition of CHX not only blocked the translocation of ribosomes from the initiator AUG (thereby allowing visualization of the toeprints), but also displaced the ribosomes from the termination site in both the wild-type and *sup45-2* extracts. The latter effect is manifested by the disappearance of the toeprints (see Fig. 10.5, lanes 3 and 9). The effect of CHX at the termination site is not solely dependent on the lack of availability of ribosomes during the 3 min of incubation with the drug since, in the presence of edeine (an inhibitor of AUG recognition that would block the formation of new 80S ribosomes during the 3 min of incubation), the +12 toeprint bands are still maintained (see Fig. 10.5, lanes 5 and 11). These data show that CHX has opposing effects on the ribosome, depending on whether the latter is positioned at an initiation site or a termination site.

2.3.5. Retroreinitiation after a premature termination requires functional Sup35p

Further elucidation of the mechanism of retroreinitiation (which led to the appearance of the +6 toeprint in wild-type extracts; Fig. 10.6, lanes 1 and 2) was obtained by analyzing toeprints in extracts made from the *sup35-R419G* mutant, which carries an Arg-419-Gly substitution in the termination factor Sup35p/eRF3 (Salas-Marco and Bedwell, 2004). Similar to *sup45-2* mutant extracts (see Fig. 10.5, lanes 9 and 10), extracts made from cells of this mutant give no toeprints comparable to those obtained in wild-type extracts (see Fig. 10.6, lanes 1 and 3). Ribosomes from *sup35-R419G* cells show no upstream +6 toeprint bands and are thus incapable of scanning after premature termination. The epistasis of the toeprints obtained in *sup45-2* and *sup35-R419G* extracts to those obtained in wild-type extracts provides additional evidence that the aberrant toeprints do not arise from leaky scanning and suggests that prior to any reinitiation event, a premature stop codon in the ribosomal A-site must be recognized by the termination machinery and trigger peptide hydrolysis in the adjoining P site (Song *et al.,* 2000; Stansfield *et al.,* 1997).

3. Summary

Cell-free translation systems enable rapid analyses of the *cis-* and *trans-* acting factors that regulate translation. The methods described herein apply this principle to unfractionated extracts from *S. cerevisiae* and *N. crassa*, two powerful genetic systems in which extracts can be prepared from wild-type or specific mutant cells and programmed with mRNAs harboring structures or mutations in need of functional characterization.

ACKNOWLEDGMENTS

Research in the authors' laboratories was supported by grants (award GM47498 to M.S.S. and award R37 GM27757 to A.J.) from the National Institutes of Health.

REFERENCES

Amrani, N., Ganesan, R., Kervestin, S., Mangus, D. A., Ghosh, S., and Jacobson, A. (2004). A faux 3′-UTR promotes aberrant termination and triggers nonsense-mediated mRNA decay. *Nature* **432,** 112–118.

Amrani, N., Dong, S., He, F., Ganesan, R., Ghosh, S., Kervestin, S., Li, C., Mangus, D. A., Spatrick, P., and Jacobson, A. (2006a). Aberrant termination triggers nonsense-mediated mRNA decay. *Biochem. Soc. Trans.* **34,** 39–42.

Amrani, N., Sachs, M. S., and Jacobson, A. (2006b). Early nonsense: mRNA decay solves a translational problem. *Nat. Rev. Mol. Cell. Biol.* **7,** 415–425.

Anderson, C. W., Straus, J. W., and Dudock, B. S. (1983). Preparation of a cell-free protein-synthesizing system from wheat germ. *Methods Enzymol.* **101,** 635–644.

Bonetti, B., Fu, L. W., Moon, J., and Bedwell, D. M. (1995). The efficiency of translation termination is determined by a synergistic interplay between upstream and downstream sequences in *Saccharomyces cerevisiae. J. Mol. Biol.* **251,** 334–345.

Burke, D., Dawson, D., and Stearns, T. (2000). "Methods in Yeast Genetics: A Cold Spring Harbor Laboratory Course Manual." Cold Spring Harbor Laboratory Press, Plainview, NY.

Cashel, M., Lazzarini, R. A., and Kalbacher, B. (1969). An improved method for thin-layer chromatography of nucleotide mixtures containing ^{32}P-labelled orthophosphate. *J. Chromatogr.* **40,** 103–109.

Cui, L., Yoon, S., Schinazi, R. F., and Sommadossi, J. P. (1995). Cellular and molecular events leading to mitochondrial toxicity of 1-(2-deoxy-2-fluoro-1-β-D-arabinofuranosyl)-5-iodouracil in human liver cells. *J. Clin. Invest.* **95,** 555–563.

Davis, R. H., and de Serres, F. J. (1970). Genetic and microbiological research techniques for *Neurospora crassa. Methods Enzymol.* **27A,** 79–143.

Dmitriev, S. E., Pisarev, A. V., Rubtsova, M. P., Dunaevsky, Y. E., and Shatsky, I. N. (2003a). Conversion of 48S translation preinitiation complexes into 80S initiation complexes as revealed by toeprinting. *FEBS Lett.* **533,** 99–104.

Dmitriev, S. E., Terenin, I. M., Dunaevsky, Y. E., Merrick, W. C., and Shatsky, I. N. (2003b). Assembly of 48S translation initiation complexes from purified components with mRNAs that have some base pairing within their 5′ untranslated regions. *Mol. Cell. Biol.* **23,** 8925–8933.

Erickson, A. H., and Blobel, G. (1983). Cell-free translation of messenger RNA in a wheat germ system. *Methods Enzymol.* **96,** 38–50.

Fang, P., Wu, C., and Sachs, M. S. (2002). *Neurospora crassa* supersuppressor mutants are amber codon-specific. *Fungal Genet. Biol.* **36,** 167–175.

Fang, P., Spevak, C. C., Wu, C., and Sachs, M. S. (2004). A nascent polypeptide domain that can regulate translation elongation. *Proc. Natl. Acad. Sci. USA* **101,** 4059–4064.

Gaba, A., Jacobson, A., and Sachs, M. S. (2005). Ribosome occupancy of the yeast *CPA1* upstream open reading frame termination codon modulates nonsense-mediated mRNA decay. *Mol. Cell* **20,** 449–460.

He, F., and Jacobson, A. (1995). Identification of a novel component of the nonsense-mediated mRNA decay pathway by use of an interacting protein screen. *Genes Dev.* **9,** 437–454.

He, F., Brown, A. H., and Jacobson, A. (1997). Upf1p, Nmd2p, and Upf3p are interacting components of the yeast nonsense-mediated mRNA decay pathway. *Mol. Cell. Biol.* **17,** 1580–1594.

Hood, H. M., Spevak, C. C., and Sachs, M. S. (2007). Evolutionary changes in the fungal carbamoyl-phosphate synthetase small subunit gene and its associated upstream open reading frame. *Fungal Genet. Biol.* **44,** 93–104.

Iizuka, N., and Sarnow, P. (1997). Translation-competent extracts from *Saccharomyces cerevisiae*: Effects of L-A RNA, $5'$ cap, and $3'$ poly(A) tail on translational efficiency of mRNAs. *Methods* **11,** 353–360.

Iizuka, N., Najita, L., Franzusoff, A., and Sarnow, P. (1994). Cap-dependent and cap-independent translation by internal initiation of mRNAs in cell extracts prepared from *Saccharomyces cerevisiae*. *Mol. Cell. Biol.* **14,** 7322–7330.

Jackson, R. J., and Hunt, T. (1983). Preparation and use of nuclease-treated rabbit reticulocyte lysates for the translation of eukaryotic messenger RNA. *Methods Enzymol.* **96,** 50–74.

Jacobson, A. (1996). Poly(A) metabolism and translation: the closed-loop model. *In* "Translational Control" (J. W. B. Hershey, M. B. Mathews, and N. Sonenberg, eds.), pp. 451–480. Cold Spring Harbor Laboratory Press, Cold Spring Harbor, NY.

Keeling, K. M., Lanier, J., Du, M., Salas-Marco, J., Gao, L., Kaenjak-Angeletti, A., and Bedwell, D. M. (2004). Leaky termination at premature stop codons antagonizes nonsense-mediated mRNA decay in *S. cerevisiae*. *RNA* **10,** 691–703.

Kozak, M. (1998). Primer extension analysis of eukaryotic ribosome-mRNA complexes. *Nucleic Acids Res.* **26,** 4853–4859.

Lee, S. I., Umen, J. G., and Varmus, H. E. (1995). A genetic screen identifies cellular factors involved in retroviral-1 frameshifting. *Proc. Natl. Acad. Sci. USA* **92,** 6587–6591.

Leeds, P., Peltz, S. W., Jacobson, A., and Culbertson, M. R. (1991). The product of the yeast *UPF1* gene is required for rapid turnover of mRNAs containing a premature translational termination codon. *Genes Dev.* **5,** 2303–2314.

Leeds, P., Wood, J. M., Lee, B. S., and Culbertson, M. R. (1992). Gene products that promote mRNA turnover in *Saccharomyces cerevisiae*. *Mol. Cell. Biol.* **12,** 2165–2177.

Maderazo, A. B., He, F., Mangus, D. A., and Jacobson, A. (2000). Upf1p control of nonsense mRNA translation is regulated by Nmd2p and Upf3p. *Mol. Cell. Biol.* **20,** 4591–4603.

Pelham, H. R., and Jackson, R. J. (1976). An efficient mRNA-dependent translation system from reticulocyte lysates. *Eur. J. Biochem.* **67,** 247–256.

Pestova, T. V., and Hellen, C. U. (2003). Translation elongation after assembly of ribosomes on the Cricket paralysis virus internal ribosomal entry site without initiation factors or initiator tRNA. *Genes Dev.* **17,** 181–186.

Pulak, R., and Anderson, P. (1993). mRNA surveillance by the *Caenorhabditis elegans smg* genes. *Genes Dev.* **7,** 1885–1897.

Roberts, B. E., and Paterson, B. M. (1973). Efficient translation of tobacco mosaic virus RNA and rabbit globin 9S RNA in a cell-free system from commercial wheat germ. *Proc. Natl. Acad. Sci. USA* **70,** 2330–2334.

Sachs, M. S., Wang, Z., Gaba, A., Fang, P., Belk, J., Ganesan, R., Amrani, N., and Jacobson, A. (2002). Toeprint analysis of the positioning of translational apparatus components at initiation and termination codons of fungal mRNAs. *Methods* **26,** 105–114.

Salas-Marco, J., and Bedwell, D. M. (2004). GTP hydrolysis by eRF3 facilitates stop codon decoding during eukaryotic translation termination. *Mol. Cell. Biol.* **24,** 7769–7778.

Sambrook, J., and Russell, D. W. (2001). "Molecular Cloning: A Laboratory Manual," 3rd ed. Cold Spring Harbor Laboratory Press, Cold Spring Harbor, NY.

Schägger, H., and von Jagow, G. (1987). Tricine-sodium dodecyl sulfate-polyacrylamide gel electrophoresis for the separation of proteins in the range from 1 to 100 kDa. *Anal. Biochem.* **166,** 368–379.

Song, H., Mugnier, P., Das, A. K., Webb, H. M., Evans, D. R., Tuite, M. F., Hemmings, B. A., and Barford, D. (2000). The crystal structure of human eukaryotic release factor eRF1—mechanism of stop codon recognition and peptidyl-tRNA hydrolysis. *Cell* **100,** 311–321.

Spevak, C. C., Park, E. H., Geballe, A. P., Pelletier, J., and Sachs, M. S. (2006). *her-2* upstream open reading frame effects on the use of downstream initiation codons. *Biochem. Biophys. Res. Commun.* **350,** 834–841.

Stahl, G., Bidou, L., Hatin, I., Namy, O., Rousset, J. P., and Farabaugh, P. (2000). The case against the involvement of the NMD proteins in programmed frameshifting. *RNA* **6,** 1687–1688.

Stansfield, I., Akhmaloka, and Tuite, M. F. (1995). A mutant allele of the *SUP45* (*SAL4*) gene of *Saccharomyces cerevisiae* shows temperature-dependent allosuppressor and omnipotent suppressor phenotypes. *Curr. Genet.* **27,** 417–426.

Stansfield, I., Kushnirov, V. V., Jones, K. M., and Tuite, M. F. (1997). A conditional-lethal translation termination defect in a *sup45* mutant of the yeast *Saccharomyces cerevisiae. Eur. J. Biochem.* **245,** 557–563.

Tarun, S. Z., Jr., and Sachs, A. B. (1995). A common function for mRNA 5′ and 3′ ends in translation initiation in yeast. *Genes Dev.* **9,** 2997–3007.

Wang, Z., and Sachs, M. S. (1997a). Arginine-specific regulation mediated by the *Neurospora crassa arg-2* upstream open reading frame in a homologous, cell-free *in vitro* translation system. *J. Biol. Chem.* **272,** 255–261.

Wang, Z., and Sachs, M. S. (1997b). Ribosome stalling is responsible for arginine-specific translational attenuation in *Neurospora crassa. Mol. Cell. Biol.* **17,** 4904–4913.

Wang, Z., Fang, P., and Sachs, M. S. (1998). The evolutionarily conserved eukaryotic arginine attenuator peptide regulates the movement of ribosomes that have translated it. *Mol. Cell. Biol.* **18,** 7528–7536.

Wang, Z., Gaba, A., and Sachs, M. S. (1999). A highly conserved mechanism of regulated ribosome stalling mediated by fungal arginine attenuator peptides that appears independent of the charging status of arginyl-tRNAs. *J. Biol. Chem.* **274,** 37565–37574.

Weng, Y., Czaplinski, K., and Peltz, S. W. (1996a). Genetic and biochemical characterization of mutations in the ATPase and helicase regions of the Upf1 protein. *Mol. Cell. Biol.* **16,** 5477–5490.

Weng, Y., Czaplinski, K., and Peltz, S. W. (1996b). Identification and characterization of mutations in the *UPF1* gene that affect nonsense suppression and the formation of the Upf protein complex but not mRNA turnover. *Mol. Cell. Biol.* **16,** 5491–5506.

CHAPTER ELEVEN

Investigating Translation Initiation Using *Drosophila* Molecular Genetics

Gritta Tettweiler *and* Paul Lasko

Contents

Abstract

Genetic tools enable insights into how translation controls development of a multicellular organism. Different genetic approaches offer the ability to manipulate the *Drosophila* genome in very precise ways, thereby allowing the investigation of how translation factors work in the context of a whole organism. We present here an overview of selected techniques used to identify genes involved in translation initiation, and quantitative methods to characterize phenotypes caused by mutations in genes encoding translation initiation or regulatory factors.

1. Introduction

The fruit fly *Drosophila melanogaster* has served as a model organism for genetic studies for over a century, and work on this organism has been instrumental in understanding various developmental and cellular processes.

Department of Biology and DBRI, McGill University, Montreal, Quebec, Canada

Methods in Enzymology, Volume 429
ISSN 0076-6879, DOI: 10.1016/S0076-6879(07)29011-1

When the genome sequence of *Drosophila melanogaster* became available, 215 transcripts were identified that encoded proteins with predicted roles in protein synthesis, and 69 were classified as translation factors (Adams *et al.*, 2000). Further analysis of the *Drosophila* genome revealed additional factors with predicted functions in translation initiation (Lasko, 2000). In the postgenomic era, subsequent studies using molecular genetic methods on individual genes uncovered in the genome sequence have provided a deeper level of information than is possible through large-scale annotation efforts.

Regulation of protein synthesis is implicated in the control of cell growth, proliferation, and differentiation. Several reports have shown that translation factors are directly involved in the regulation of normal growth and development in *Drosophila*. Galloni and Edgar (1999) identified a series of mutants with larval growth defects. One class of such mutants affected eukaryotic initiation factor 4A (eIF4A). Lachance *et al.* (2002) demonstrated that a mutation in another translation initiation factor, eIF4E, leads to larval growth arrest as well, and that phosphorylation of eIF4E is essential for normal growth. Additional studies showed that the translational inhibitor 4E-BP, and the poly(A) binding protein interactor Paip2, can have an impact on cell growth (Miron *et al.*, 2001). In addition, regulation of translation from specific maternally expressed mRNAs is essential for the establishment of an embryonic pattern (Johnstone and Lasko, 2001). Mechanisms to regulate translation of specific mRNAs in the germline are believed to target eIF4E (Nelson *et al.*, 2004; Zappavigna *et al.*, 2004), eIF5B (Johnstone and Lasko, 2004), and the $5'$ cap structure through an eIF4E-related protein, 4EHP (Cho *et al.*, 2005, 2006).

Drosophila is a valuable organism to study the biological role of translation initiation factors largely because of the broad range of genetic tools that are available. A key advantage of *Drosophila* is that the GAL4-UAS system (see below) is best elaborated in this organism. This system allows the investigator to activate or inactivate a gene in specific cells or tissues and at specific developmental times. Transgenic techniques permit the study of the effects of inducing overexpression of translation initiation factors in specific cells or tissues and/or at specific developmental stages. Such experiments enable measurement of the phenotypes produced by overexpression of a wild-type protein, or by expression of a specific mutant form of a translation factor in a wild-type or genetically mutant background. On the other hand, gene disruption, directed or random, and targeted gene silencing experiments by expressing interfering RNAs can also provide insight into the function of the gene of interest.

2. P-ELEMENTS

The most important tool for a modern *Drosophila* geneticist is the *P-element*, a naturally occurring transposon that has now been genetically engineered in various ways to conveniently tag genes or create mutations in genes. *P-element*-mediated transposition allows efficient transfer of specific DNA segments into the germline and their stable inheritance in the progeny (Rubin and Spradling, 1982). Large-scale mutagenesis projects using various *P-element* derivatives have resulted in the isolation of transposon insertions in or very near approximately two-thirds of all *Drosophila* genes (Venken and Bellen, 2005). These insertion lines are publicly available at nominal cost from various *Drosophila* stock centers (for contact information, see below). A selection of *P-element* insertion lines for translation initiation factors available from the Bloomington Stock Centre (the largest such center) is presented in Table 11.1.

The following techniques for the induced misexpression of a gene of interest are based on *P-elements*.

2.1. The Gal4-UAS system

The Gal4 upstream activating sequences (Gal4-UAS) system allows localized induced expression of a given gene under the control of a promoter of choice (Brand and Perrimon, 1993).

At the present time, two UAS constructs are most commonly used for conditional gene expression in *Drosophila*, pUAST and pUASp. By using pUAST (Brand and Perrimon, 1993), expression of a specific gene can be achieved efficiently in somatic tissue, but it is not possible to express the gene of interest in the germline during oogenesis. This problem was circumvented by creating a modified UAS vector, pUASp (Rorth, 1998). Both vectors, and other derivatives of them, are available at nominal cost from the Drosophila Genomics Resource Center.

Following cloning of the gene of interest into the appropriate UAS vector, the plasmid is injected into the pole plasm of precellular embryos (Rubin and Spradling, 1982). This procedure is carried out routinely in many *Drosophila* laboratories, but it does take some skill and significant practice to master it. Occasional producers of transgenic *Drosophila* lines without a nearby cooperating fly laboratory should consider using one of several available commercial services for this work (a list of providers is available on www.flybase.org). The plasmids contain a wild-type copy of the *white* gene and the recipient embryos are mutant for *white*. Transformant adults will therefore have red eyes (from the plasmid-borne wild-type copy of *white*) while nontransformed adults will have white eyes. Subsequently, the transgenic flies can be crossed to balancer lines to identify the chromosome on which the transgene has inserted.

To induce expression of the gene of interest, the UAS line can be crossed with flies expressing the transcriptional activator Gal4. Numerous Gal4 lines are available from the various stock centers (for contact information, see below), allowing the choice for control gene expression in time (Hsp70-Gal, for example, which allows activation of a gene at any particular

Table 11.1 Selection of transposable element insertion lines

	Gene	Transposable *P-element* insertion	Reference
Translation initiation factor			
eIF1A	CG8053	P{EP}eIF-1A^{EP935}	BDGP Project members, 2000–
eIF2 (dGcn2)	CG1609	X	
eIF2α	CG9946	X	
eIF2β	CG4153	P{EPgy2}eIF-2β^{EY08063}	Gene Disruption Project members, 2001–
eIF2γ	X	X	
eIF2Bα	CG7883	P{EPgy2}eIF2B-α^{EY03991}	Gene Disruption Project members, 2001–
eIF2Bβ	CG2677	PBac{PB}eIF2B-β^{c02002}	Gene Disruption Project members and Exelixis, 2005
eIF2Bγ	CG8190	PBac{PB}eIF2B-γ^{c01931}	Thibault *et al.*, 2004
eIF2Bδ	CG10315	P{EPgy2}eIF2B-δ^{EY03558}	Gene Disruption Project members, 2001–
eIF2Bε	CG3806	X	
eIF3-S1 (Adam)	CG12131	P{lacW}Adam^{k13906}	Spradling *et al.*, 1999
eIF3-S2 (Trip1)	CG8882	P{SUPor-P}Trip1^{KG06360}	Gene Disruption Project members, 2001–
eIF3-S3 (p40)	CG9124	P{lacW}eIF-3p40^{k09003}	Spradling *et al.*, 1999
eIF3-S4	CG10881	X	
eIF3-S5	CG8636	X	
	CG8335	P{EP}EP2003	Gene Disruption Project members, 2001–
	CG9769	P{EP}CG9769^{EP516}	BDGP Project members, 2000–
eIF3-S6	CG9677	P{PZ}Int6^{l0547}	BDGP Project members, 2000– Cunniff *et al.*, 1997
eIF3-S7 (p66)	CG10161	P{EPgy2}eIF-3p66^{EY05735}	Gene Disruption Project members, 2001–
	CG4810	X	

Gene	CG	Allele/insertion	Reference
eIF3-S8	CG4954	P{EPgy2}eIF3-S8^{EY07713}	Gene Disruption Project members, 2001–
eIF3-S9	CG4878	P{EPgy2}eIF3-S9^{EY14430}	Gene Disruption Project members, 2001–
eIF3-S10	CG9805	P{EPgy2}eIF3-S10^{EY04238}	Gene Disruption Project members, 2001–
eIF4AIII	CG7483	P{EPgy2}eIF4AIII^{EY14207}	Gene Disruption Project members, 2001–
eIF4a	CG9075	P{lacW}eIF-4a^{1013}	Gene Disruption Project members, 2001–
eIF4B	CG10837	X	Galloni and Edgar, 1999
eIF4-E-1,2	CG4035	P{PZ}eIF-4E^{07238}	Lachance *et al.*, 2002
eIF4E-3	CG8023	X	
eIF4E-4	CG10124	X	
eIF4E-5	CG8277	X	
eIF4E-6	CG1442	PBac{WH}CG1442^{f06737}	Gene Disruption Project members and Exelixis, 2005
eIF4E-7	CG32859	P{EP}EP1443	BDGP Project members, 1994–1999
eIF4E-8	CG33100	See d4EHP	
eIF4G	CG10811	PBac{RB}eIF-4G^{e02087}	Thibault *et al.*, 2004
eIF5	CG9177	X	
eIF5B (dIF2)	CG10840	P{EPgy2}eIF5B^{EY01401}	Gene Disruption Project members, 2001–
eIF6	CG17611	P{lacW}eIF6^{k13214}	Spradling *et al.*, 1999
Regulator			
d4E-HP	CG33100	P{GT1}4EHP^{BG01713}	Gene Disruption Project members, 2001–
d4E-BP	CG8846	P{lacW}Thor^{k13517}	BDGP Project members, 1994–1999
Cup	CG11181	PBac{PB}cup^{c04119}	Thibault *et al.*, 2004
PABP	CG5119	P{lacW}pAbp^{k10109}	Sigrist *et al.*, 2000
Paip2	CG12358	P{SUPor-P}KG03704	Gene Disruption Project members, 2001–

Table 11.2 Selected mutant alleles of translation initiation factors and regulators

	Allele	Viability	Method	Reference
Translation initiation factor				
eIF1A	$eIF1A^{645}$	Cell lethal	EMS screen	Collins and Cohen, 2005
eIF3-S1 (Adam)	$Adam^{14-38}$	Lethal	EMS \?, dDEB\?, X-ray \? (not clarified)	Goldstein et al., 2001
eIF3-S8	$eIF3-S8^{14F06Y-18}$	Lethal	Imprecise P{wHy} excision	Mohr and Gelbart, 2002
eIF4AIII	$eIF4AIII^{19}$	Lethal	Imprecise P{EP} excision	Palacios et al., 2004
eIF4A	$eIF-4a^{162-5}$	Larval lethal	X-ray	Dorn et al., 1993
eIF4E-1,2	$eIF4E^{67Af}$	Larval lethal	EMS screen	Leicht and Bonner, 1988
eIF5B	dIF2-null	Larval lethal	Imprecise P{lacW} excision	Carrera et al., 2000
eIF6	$eIF6^{5E24}$	Lethal	EMS screen	Sinenko et al., 2004
Regulator				
4E-HP	$d4E-HP^{CP53}$	Viable	Imprecise P{GT1} excision	Cho et al., 2005
4E-BP	$d4E-BP^{null}$; $Thor^2$	Viable	Imprecise P{lacW} excision	Bernal and Kimbrell, 2000
Cup	$cup^{\Delta212}$	Viable	Imprecise P{PZ} excision	Nakamura et al., 2004

Before the investigator starts with the crossing, which can take some months to complete, it is strongly recommended that the starting insertion line be checked by molecular means such as PCR to ensure that the insertion site has been correctly mapped. A small but significant fraction of insertion lines available from large-scale mutagenesis screens does not correspond to the annotation details in the available databases.

2.4. FLP recombinase mediated recombination

Since all of the mutant alleles of translation initiation factors described so far are lethal early in development (Table 11.2), a different approach is necessary to investigate the consequences of a mutation in the context of an intact organism. Site-specific mitotic recombination using the FLP/FRT system permits the creation of mosaic animals in which mutant clones are generated in a tissue-specific manner (Golic, 1991). After introducing FLP recombinase, a precise deletion can be obtained by the excision of two nearby transposable elements that carry FRT sites in the same orientation.

For the generation of *eIF5B*-null germline mosaic clones (Carrera *et al.*, 2000), the "FLP-DFS" (flippase-dominant female sterile) method was used (Chou and Perrimon, 1996). In this procedure, the X-linked germline-dependent dominant female sterile mutation *ovo^{D1}* acts as a selection marker for the detection of germline recombination events, because only recombinant clones escape the cell-lethal *ovo^{D1}* phenotype. The following breeding scheme applied: *y w HSFlp; FRT2A ovo^{D1}/TM3 Sb* males were mated with *eIF5B$^{\Delta 1FRT2A}$ w$^+$/TM3Ser* virgin females. The progeny larvae were heat shocked at 37° for 2 h during the second and third instar larval stages to induce mitotic recombination in proliferating germ cells. Ovaries from homozygous, non-Sb, non-Ser females were dissected and examined using Western blot analysis (Carrera *et al.*, 2000).

Mosaic clones can be generated in any tissue based on the expression pattern of the *FLP* line used. For the generation of eye-specific mosaic clones, the *eyFLP* driver is used (Newsome *et al.*, 2000). In the *eyFLP* construct, the *FLP* recombinase cDNA is under the control of four tandem copies of an eye-specific enhancer from the *eyeless* (*ey*) gene and an inducible *hsp70* heat shock promoter. Heat treatment of larvae as described above results in specific expression in eye imaginal discs. Using *white$^+$* (*w$^+$*) distal to an FRT insertion as a marker, it is possible to detect mosaicism in a *w$^-$* background.

2.5. Homologous recombination

Another method to investigate the role of a gene of interest in *Drosophila* development is the targeted knockout or replacement of a gene of interest (Rong and Golic, 2000). In this technique, a *P-element* vector containing the genomic sequence homologous to the gene of interest is constructed. The genomic sequence is mutated, and the recognition site for a restriction enzyme (*I-SceI*) that recognizes an 18-bp sequence not otherwise present in the *Drosophila* genome is inserted. The vector is introduced into the germline by *P-element*-mediated transformation. This donor construct has four key features:

1. The DNA segment is homologous to the target gene region.
2. A unique recognition site for endonuclease *I-SceI*.
3. A mini-*white* marker gene.
4. Two recognition sequences for *FLP* recombinase, FRTs.

Recombination between the FRTs results in the excision of a circular DNA that carries the marker gene and the target DNA. The circle is then cleaved by *I-SceI in vivo* producing linear molecules. The ends of these molecules are homologous to the targeted region. Homologous recombination recombines the mutated DNA from the donor plasmid with the endogenous gene to generate tandem duplication (reviewed in Adams and Sekelsky, 2002).

2.6. RNAi

Another method to characterize the function of a translation initiation factor by its loss-of-function phenotype is the targeted knockout of the gene by RNA interference (RNAi). Using transgenic techniques analogous to those described above, RNAi can be expressed in specific tissues (Lee and Carthew, 2003). In this system, a transgene is constructed containing inverted repeats of a sequence included in the target gene of interest. The RNA produced from this transgene forms a hairpin, which can lead to silencing of expression of the target gene through the RNA interference pathway. The pWIZ (for white intron zipper) vector has been the predominant one used to produce loopless hairpin RNA. To generate an RNAi-producing construct using pWIZ, the following criteria should be used to create the fragment corresponding to the *Drosophila* target gene:

1. A fragment size of 500 to 700 bp.
2. No internal restriction sites corresponding to the PCR primer sites.
3. No matching sequences to a 5' or 3' consensus splice site in either a sense or antisense direction.

After the successful creation of the pWIZ vector to knock down the target gene by RNAi, the vector is injected into recipient animals to generate stable transgenic lines. In the following step, the animals carrying the WIZ gene are crossed to flies carrying a Gal4 driver of choice. In the resulting progeny, the production of the hairpin RNA can be regulated in a temporal or spatial pattern.

A recent report described another transformation vector for RNAi experiments in *Drosophila*, pRISE (for RNAi inducing silencing effector) (Kondo *et al.*, 2006). The clear advantage with this vector is its relatively easy cloning procedure using the TOPO cloning system (Invitrogen). The resulting plasmid contains an inverted repeat of the sequence of the target gene, and a pentamer of UAS^{Gal4}. It can be transformed into flies by commonly used *P-element*-mediated transformation. When the transgenic flies are crossed to a Gal4 line, the expression of the target gene is reduced. As for pWIZ, the RNAi effect can also be controlled temporally and spatially by the choice of a Gal4 driver.

Efforts are nearly complete to establish a genome-wide library of transgenic fly lines that produce hairpin RNAs in a temporally and spatially specific pattern to enable a systematic overview of gene function in the fly (http://www.vdrc.at). There are several fly lines available in which mRNAs encoding translation initiation factors can be inducibly silenced *in vivo* using the UAS-GAL4 system. Table 11.3 provides an overview of the UAS-RNAi flies available from the National Institute of Genetics in Japan (http://www.shigen.nig.ac.jp/fly/nigfly/).

For cultured *Drosophila* cells, a library of double-stranded RNAs (dsRNAs) directed against all predicted open reading frames in the *Drosophila* genome has been created, and a high-throughput RNAi screening service is available (http://flyrnai.org; Boutros *et al.*, 2004). An excellent

Table 11.3 RNAi flies

Synonym	Gene	Phenotype
eIF2 (dGcn2)	CG1609	Viable
eIF2β	CG4153	Lethal
eIF2Bδ	CG10315	Lethal
eIF2Bγ	CG8190	Lethal
eIF3-S1	CG12131	Lethal
eIF3-S3 (p40)	CG9124	Lethal
eIF3-S4	CG10881	Lethal
eIF3-S5	CG8335	Viable
eIF3-S7 (p66)	CG10161	Lethal
eIF3-S10	CG9805	Lethal
eIF4E-3	CG8023	Viable
eIF4E-5	CG8277	Lethal
eIF4E-7	CG32859	Lethal
eIF5B (dIF2)	CG10840	Lethal
eIF6	CG17611	Lethal
Cup	CG11181	Semilethal

overview of the high-throughput RNAi screens in cultured *Drosophila* S2 cells was recently published (Armknecht *et al.*, 2005).

3. Perspectives and Conclusions

The regulation of translation initiation is a key means in controlling *Drosophila* development. To analyze the function of translation initiation factors in development, genetic approaches offer the advantage of carrying out the experiments in a whole animal. The disruption of genes encoding translation initiation factors often results in lethality of the fly, underscoring the importance of translational control for development. Since the whole genome sequence of *Drosophila* is available, the focus now shifts toward the functional analysis of the genes identified.

4. Important Sources for *Drosophila* Protocols

Ashburner, M., Golic, K. G., and Hawley, R. S. (2005). "*Drosophila*: A Laboratory Handbook." Cold Spring Harbor Laboratory Press, Cold Spring Harbor, NY.

Goldstein, L. S. B., and Fyrberg, E. A. (1994). "*Drosophila melanogaster:* Practical Uses in Cell and Molecular Biology." Academic Press, San Diego.

Greenspan, R. J. (2004). "Fly Pushing: The Theory and Practice of *Drosophila* Genetics." Cold Spring Harbor Laboratory Press, Cold Spring Harbor, NY.

Sullivan, W., Ashburner, M., and Hawley, R. S. (2000). "*Drosophila* Protocols." Cold Spring Harbor Laboratory Press, Cold Spring Harbor, NY.

http://flybase.bio.indiana.edu/ to order fly stocks, search for protocols, personal communications within the fly community, etc.

http://dgrc.cgb.indiana.edu/vectors/ *Drosophila* Genomics Resource Center.
   http://superfly.ucsd.edu/ to search for *P-element* lines.

http://portal.curagen.com/cgi-bin/interaction/flyHome.pl *Drosophila* Interaction Database [Giot, L., *et al.*, *Science* **302**(5651), 2003].

http://www.flyrnai.org/ *Drosophila* RNAi Screening Center at Harvard Medical School (Flockhart *et al.*, 2006).

5. *DROSOPHILA* STOCK CENTERS

http://flystocks.bio.indiana.edu/	Bloomington Stock Center
http://sagafly.dgrc.kit.ac.jp/en/	Kyoto Stock Center
http://kyotofly.kit.jp/cgi-bin/ehime/index.cgi	Ehime University Stock Center
http://expbio.bio.u-szeged.hu/fly/	Szeged Stock Center
http://stockcenter.arl.arizona.edu/	Tuscon Stock Center
http://drosophila.med.harvard.edu/	Harvard Stock Collection

REFERENCES

Adams, M. D., and Sekelsky, J. J. (2002). From sequence to phenotype: Reverse genetics in *Drosophila melanogaster*. *Nat. Rev. Genet.* **3**, 189–198.

Adams, M. D., Celniker, S. E., Holt, R. A., Evans, C. A., Gocayne, J. D., Amanatides, P. G., Scherer, S. E., Li, P. W., Hoskins, R. A., Galle, R. F., George, R. A., Lewis, S. E., *et al.* (2000). The genome sequence of *Drosophila melanogaster*. *Science* **287**, 2185–2195.

Armknecht, S., Boutros, M., Kiger, A., Nybakken, K., Mathey-Prevot, B., and Perrimon, N. (2005). High-throughput RNA interference screens in *Drosophila* tissue culture cells. *Methods Enzymol.* **392**, 55–73.

BDGP Project Members (1994–1999). Berkeley *Drosophila* Genome Project.

BDGP Project Members (2000–). Berkeley *Drosophila* Genome Project.

Bellen, H. J., Levis, R. W., Liao, G., He, Y., Carlson, J. W., Tsang, G., Evans-Holm, M., Hiesinger, P. R., Schulze, K. L., Rubin, G. M., Hoskins, R. A., and Spradling, A. C. (2004). The BDGP gene disruption project: Single transposon insertions associated with 40% of *Drosophila* genes. *Genetics* **167**, 761–781.

Bernal, A., and Kimbrell, D. A. (2000). *Drosophila* Thor participates in host immune defense and connects a translational regulator with innate immunity. *Proc. Natl. Acad. Sci. USA* **97**, 6019–6024.

Boutros, M., Kiger, A. A., Armknecht, S., Kerr, K., Hild, M., Koch, B., Haas, S. A., Paro, R., and Perrimon, N. (2004). Genome-wide RNAi analysis of growth and viability in *Drosophila* cells. *Science* **303**, 832–835.

Brand, A. H., and Perrimon, N. (1993). Targeted gene expression as a means of altering cell fates and generating dominant phenotypes. *Development* **118**, 401–415.

Capdevila, J., and Guerrero, I. (1994). Targetted expression of the signaling molecule decapentaplegic induces pattern duplications and growth alterations in *Drosophila* wings. *EMBO J.* **13**, 4459–4468.

Carrera, P., Johnstone, O., Nakamura, A., Casanova, J., Jackle, H., and Lasko, P. (2000). VASA mediates translation through interaction with a *Drosophila* yIF2 homolog. *Mol. Cell* **5**, 181–187.

Cho, P. F., Poulin, F., Cho-Park, Y. A., Cho-Park, I. B., Chicoine, J. D., Lasko, P., and Sonenberg, N. (2005). A new paradigm for translational control: Inhibition via 5'-3' mRNA tethering by Bicoid and the eIF4E cognate 4EHP. *Cell* **121**, 411–423.

Cho, P. F., Gamberi, C., Cho-Park, Y. A., Cho-Park, I. B., Lasko, P., and Sonenberg, N. (2006). Cap-dependent translational inhibition establishes two opposing morphogen gradients in *Drosophila* embryos. *Curr. Biol.* **16**, 2035–2041.

Chou, T. B., and Perrimon, N. (1996). The autosomal FLP-DFS technique for generating germline mosaics in *Drosophila melanogaster*. *Genetics* **144,** 1673–1679.

Collins, R. T., and Cohen, S. M. (2005). A genetic screen in *Drosophila* for identifying novel components of the hedgehog signaling pathway. *Genetics* **170,** 173–184.

Cunniff, J., Chiu, Y. H., Morris, N. R., and Warrior, R. (1997). Characterization of DnudC, the *Drosophila* homolog of an Aspergillus gene that functions in nuclear motility. *Mech. Dev.* **66,** 55–68.

Dorn, R., Morawietz, H., Reuter, G., and Saumweber, H. (1993). Identification of an essential *Drosophila* gene that is homologous to the translation initiation factor eIF-4A of yeast and mouse. *Mol. Gen. Genet.* **237,** 233–240.

Ellis, M. C., O'Neill, E. M., and Rubin, G. M. (1993). Expression of *Drosophila* glass protein and evidence for negative regulation of its activity in non-neuronal cells by another DNA-binding protein. *Development* **119,** 855–865.

Flockhart, I., Booker, M., Kiger, A., Boutros, M., Armknecht, S., Ramadan, N., Richardson, K., Xu, A., Perrimon, N., and Mathey-Prevot, B. (2006). FlyRNAi: The *Drosophila* RNAi screening center database. *Nucleic Acids Res.* **34,** D489–D494.

Galloni, M., and Edgar, B. A. (1999). Cell-autonomous and non-autonomous growth-defective mutants of *Drosophila melanogaster*. *Development* **126,** 2365–2375.

Gene Disruption Project members (2001–). BDGP Gene Disruption Project.

Gene Disruption Project members and Exelixis (2005). Genomic mapping of Exelixis insertion collection.

Goldstein, E. S., Treadway, S. L., Stephenson, A. E., Gramstad, G. D., Keilty, A., Kirsch, L., Imperial, M., Guest, S., Hudson, S. G., LaBell, A. A., O'Day, M., Duncan, C., *et al.* (2001). A genetic analysis of the cytological region 46C-F containing the *Drosophila melanogaster* homolog of the jun proto-oncogene. *Mol. Genet. Genomics* **266,** 695–700.

Golic, K. G. (1991). Site-specific recombination between homologous chromosomes in *Drosophila*. *Science* **252,** 958–961.

Greenspan, R. J. (2004). "Fly Pushing: The Theory and Practice of *Drosophila* Genetics." Cold Spring Harbor Laboratory Press, Cold Spring Harbor, NY.

Hacker, U., Nystedt, S., Barmchi, M. P., Horn, C., and Wimmer, E. A. (2003). piggyBac-based insertional mutagenesis in the presence of stably integrated P elements in *Drosophila*. *Proc. Natl. Acad. Sci. USA* **100,** 7720–7725.

Hay, B. A., Maile, R., and Rubin, G. M. (1997). P element insertion-dependent gene activation in the *Drosophila* eye. *Proc. Natl. Acad. Sci. USA* **94,** 5195–5200.

Horn, C., Offen, N., Nystedt, S., Hacker, U., and Wimmer, E. A. (2003). piggyBac-based insertional mutagenesis and enhancer detection as a tool for functional insect genomics. *Genetics* **163,** 647–661.

Johnstone, O., and Lasko, P. (2001). Translational regulation and RNA localization in *Drosophila* oocytes and embryos. *Annu. Rev. Genet.* **35,** 365–406.

Johnstone, O., and Lasko, P. (2004). Interaction with eIF5B is essential for Vasa function during development. *Development* **131,** 4167–4178.

Kondo, T., Inagaki, S., Yasuda, K., and Kageyama, Y. (2006). Rapid construction of *Drosophila* RNAi transgenes using pRISE, a P-element-mediated transformation vector exploiting an *in vitro* recombination system. *Genes Genet. Syst.* **81,** 129–134.

Lachance, P. E., Miron, M., Raught, B., Sonenberg, N., and Lasko, P. (2002). Phosphorylation of eukaryotic translation initiation factor 4E is critical for growth. *Mol. Cell. Biol.* **22,** 1656–1663.

Lasko, P. (2000). The *Drosophila* genome: Translation factors and RNA binding proteins. *J. Cell Biol.* **150,** F51–F56.

Lee, Y. S., and Carthew, R. W. (2003). Making a better RNAi vector for *Drosophila*: Use of intron spacers. *Methods* **30,** 322–329.

Leicht, B. G., and Bonner, J. J. (1988). Genetic analysis of chromosomal region 67A-D of *Drosophila melanogaster*. *Genetics* **119,** 579–593.

Miron, M., Verdu, J., Lachance, P. E., Birnbaum, M. J., Lasko, P. F., and Sonenberg, N. (2001). The translational inhibitor 4E-BP is an effector of PI(3)K/Akt signalling and cell growth in *Drosophila*. *Nat. Cell. Biol.* **3,** 596–601.

Mohr, S. E., and Gelbart, W. M. (2002). Using the P[wHy] hybrid transposable element to disrupt genes in region 54D-55B in *Drosophila melanogaster*. *Genetics* **162,** 165–176.

Nakamura, A., Sato, K., and Hanyu-Nakamura, K. (2004). *Drosophila* cup is an eIF4E binding protein that associates with Bruno and regulates oskar mRNA translation in oogenesis. *Dev. Cell* **6,** 69–78.

Nelson, M. R., Leidal, A. M., and Smibert, C. A. (2004). *Drosophila* Cup is an eIF4E-binding protein that functions in Smaug-mediated translational repression. *EMBO J.* **23,** 150–159.

Neufeld, T. P., de la Cruz, A. F., Johnston, L. A., and Edgar, B. A. (1998). Coordination of growth and cell division in the *Drosophila* wing. *Cell* **93,** 1183–1193.

Newsome, T. P., Asling, B., and Dickson, B. J. (2000). Analysis of *Drosophila* photoreceptor axon guidance in eye-specific mosaics. *Development* **127,** 851–860.

Palacios, I. M., Gatfield, D., St Johnston, D., and Izaurralde, E. (2004). An eIF4AIII-containing complex required for mRNA localization and nonsense-mediated mRNA decay. *Nature* **427,** 753–757.

Parks, A. L., Cook, K. R., Belvin, M., Dompe, N. A., Fawcett, R., Huppert, K., Tan, L. R., Winter, C. G., Bogart, K. P., Deal, J. E., Deal-Herr, M. E., Grant, D., *et al.* (2004). Systematic generation of high-resolution deletion coverage of the *Drosophila melanogaster* genome. *Nat. Genet.* **36,** 288–292.

Rong, Y. S., and Golic, K. G. (2000). Gene targeting by homologous recombination in *Drosophila*. *Science* **288,** 2013–3018.

Rorth, P. (1998). Gal4 in the *Drosophila* female germline. *Mech. Dev.* **78,** 113–118.

Roy, G., Miron, M., Khaleghpour, K., Lasko, P., and Sonenberg, N. (2004). The *Drosophila* poly(A) binding protein-interacting protein, dPaip2, is a novel effector of cell growth. *Mol. Cell. Biol.* **24,** 1143–1154.

Rubin, G. M., and Spradling, A. C. (1982). Genetic transformation of *Drosophila* with transposable element vectors. *Science* **218,** 348–353.

Sigrist, S. J., Thiel, P. R., Reiff, D. F., Lachance, P. E. D., Lasko, P., and Schuster, C. F. (2000). Postsynaptic translation affects the efficacy and morphology of neuromuscular junctions. *Nature* **405,** 1062–1065.

Sinenko, S. A., Kim, E. K., Wynn, R., Manfruelli, P., Ando, I., Wharton, K. A., Perrimon, N., and Mathey-Prevot, B. (2004). Yantar, a conserved arginine-rich protein is involved in *Drosophila* hemocyte development. *Dev. Biol.* **273,** 48–62.

Spradling, A. C., Stern, D. M., Kiss, I., Roote, J., Laverty, T., and Rubin, G. M. (1995). Gene disruptions using P transposable elements: An integral component of the *Drosophila* genome project. *Proc. Natl. Acad. Sci. USA* **92,** 10824–10830.

Spradling, A. C., Stern, D., Beaton, A., Rhem, E. J., Laverty, T., Mozden, N., Misra, S., and Rubin, G. M. (1999). The Berkeley *Drosophila* Genome Project gene disruption project: Single P-element insertions mutating 25% of vital *Drosophila* genes. *Genetics* **153,** 135–177.

Stocker, H., Radimerski, T., Schindelholz, B., Wittwer, F., Belawat, P., Daram, P., Breuer, S., Thomas, G., and Hafen, E. (2003). Rheb is an essential regulator of S6K in controlling cell growth in *Drosophila*. *Nat. Cell. Biol.* **5,** 559–565.

Struhl, G., and Basler, K. (1993). Organizing activity of wingless protein in *Drosophila*. *Cell* **72,** 527–540.

Thibault, S. T., Singer, M. A., Miyazaki, W. Y., Milash, B., Dompe, N. A., Singh, C. M., Buchholz, R., Demsky, M., Fawcett, R., Francis-Lang, H. L., Ryner, L.,

Cheung, L. M., *et al.* (2004). A complementary transposon tool kit for *Drosophila melanogaster* using P and piggyBac. *Nat. Genet.* **36,** 283–287.

Venken, K. J., and Bellen, H. J. (2005). Emerging technologies for gene manipulation in *Drosophila melanogaster*. *Nat. Rev. Genet.* **6,** 167–178.

Wolff, T. (2000). "Histological Techniques for the *Drosophila* Eye. Part II: Adult." Cold Spring Harbor Laboratory Press, Cold Spring Harbor, NY.

Zappavigna, V., Piccioni, F., Villaescusa, J. C., and Verrotti, A. C. (2004). Cup is a nucleocytoplasmic shuttling protein that interacts with the eukaryotic translation initiation factor 4E to modulate *Drosophila* ovary development. *Proc. Natl. Acad. Sci. USA* **101,** 14800–14805.

CHAPTER TWELVE

ANALYSIS OF RNA:PROTEIN INTERACTIONS IN VIVO: IDENTIFICATION OF RNA-BINDING PARTNERS OF NUCLEAR FACTOR 90

Andrew M. Parrott, Melissa R. Walsh, *and* Michael B. Mathews

Contents

Abstract

Ribonucleoprotein complexes (RNPs) perform a multitude of functions in the cell. Elucidating the composition of such complexes and unraveling their many interactions are current challenges in molecular biology. To stabilize complexes formed in cells and to preclude reassortment of their components during isolation, we employ chemical crosslinking of the RNA and protein moieties. Here we describe the identification of cellular RNAs bound to nuclear factor 90 (NF90), the founder member of a family of ubiquitous double-stranded RNA-binding proteins. Crosslinked RNA–NF90 complexes were immunoprecipitated from

Department of Biochemistry and Molecular Biology, University of Medicine and Dentistry of New Jersey, New Jersey Medical School, Newark, New Jersey

Methods in Enzymology, Volume 429
ISSN 0076-6879, DOI: 10.1016/S0076-6879(07)29012-3

stable cell lines containing epitope-tagged NF90 protein isoforms. The bound RNA was released and identified through RNase H digestion and by various gene amplification techniques. We appraise the methods used by altering crosslinking conditions, and the binding profiles of different NF90 protein isoforms in synchronized and asynchronous cells are compared. This study discovers two novel RNA species and establishes NF90 as a multiclass RNA-binding protein, capable of binding representatives of all three classes of RNA.

 1. INTRODUCTION

RNA assumes vital structural, catalytic, and regulatory roles in the cell. The recent discoveries of small interfering RNAs and microRNAs (Valencia-Sanchez *et al.*, 2006), classes of small regulatory RNA species thought to control ~30% of protein-coding genes (Lewis *et al.*, 2005), have greatly expanded our appreciation of the breadth of the biological impact of RNA. RNA molecules generally function in cells when complexed with protein as discrete ribonucleoprotein (RNP) complexes. Methods that "trap" RNA species in complexes with their physiological protein partners afford the opportunity to uncover unknown partner function and to elucidate regulatory pathways. One such method is *in vivo* crosslinking, which can stabilize weak or transient complexes and also eliminate potentially problematic partner exchange that can occur during or after cell disruption (Mili and Steitz, 2004).

Members of the nuclear factor 90 (NF90) family of double-stranded RNA-binding proteins are abundantly expressed in vertebrate tissue (Saunders *et al.*, 2001b) and participate in many aspects of RNA metabolism (reviewed by Reichman and Mathews, 2003). The two most prominent protein isoforms, NF90 and NF110, have apparent molecular masses of ~90 and ~110 kDa (Kao *et al.*, 1994; Reichman *et al.*, 2003). They are also referred to as DRBP76 or NFAR1 and ILF3, NFAR2, or TCP110, respectively (Buaas *et al.*, 1999; Patel *et al.*, 1999; Saunders *et al.*, 2001a; Xu *et al.*, 2003). Both form a heterodimeric complex with nuclear factor 45 (NF45) protein (Corthésy and Kao, 1994).

NF90 and NF110 have been shown to interact *in vitro* with several coding and noncoding RNAs of both cellular (Bose *et al.*, 2006; Larcher *et al.*, 2004; Shi *et al.*, 2005; Shim *et al.*, 2002; Tran *et al.*, 2004; Xu and Grabowski, 2005) and viral origin (Isken *et al.*, 2003; Liao *et al.*, 1998; Shin *et al.*, 2002). Although the cellular mRNAs identified from these studies are not abundant, our earlier work demonstrated that the double-stranded RNA-binding motifs (dsRBMs) of NF90 and NF110 are almost completely occupied by cellular RNA throughout the cell cycle (Parrott *et al.*, 2005). We therefore sought to identify these cellular RNA partners. To capture the cellular RNAs that are bound to NF90 and its larger splice variant NF110, we developed a strategy of *in vivo* crosslinking coupled with immunoprecipitation of epitope-tagged protein. Use of the epitope tag enhances the generality of the method and largely avoids possible complications resulting from antibody binding to critical features of the protein, such as its RNA-binding region.

2. EXPRESSION OF EPITOPE-TAGGED PROTEINS

Alternative splicing of *ILF3* transcripts generates mRNAs encoding several isoforms of NF90 and NF110, both of which may contain an NVKQ tetrapeptide insert between their two dsRBM motifs (Duchange *et al.*, 2000). The "b" isoforms contain the insert, which is absent from the "a" forms. To characterize the RNA partners of these proteins, we established stable human 293 cell lines expressing N-terminally epitope-tagged a and b isoforms of NF90 and NF110 (Fig. 12.1A). Stable cell lines were made by transfection of monolayer 293 cells with pcDNA3.1HisB plasmid (Invitrogen) constructs using Lipofectamine 2000 (Invitrogen). Single clones were selected with cloning discs (Sigma) under 500 μg/ml G418 selection pressure, then passaged several times before immunoblot screening. High levels of protein expression can lead to spurious interactions and ectopic localization within cells. Therefore, two criteria for the study of biologically meaningful interactions are that the tagged protein should not be substantially overexpressed relative to its endogenous counterpart and that it should display an identical cellular distribution. Examination of the NF90b-containing cell line showed that the tagged protein was expressed at slightly lower levels than its endogenous counterpart, and that the two proteins displayed the same subcellular distribution when analyzed by differential centrifugation (Fig. 12.1B, upper panel). Both were present in the cytoplasm, distributed evenly between the 100,000×*g* pellet and supernatant fractions, P-100 and S-100. NF45, which accompanies both NF90 and NF110, had the same cytoplasmic distribution as NF90 (see Fig. 12.1B, middle panel). The RNA content of the fractions was analyzed to validate the centrifugal fractionation: large RNA was resolved in an agarose gel and visualized by ethidium bromide staining, and 3′ end-radiolabeled small RNA was resolved in an acrylamide/urea gel. As expected, 18S and 28S rRNA were enriched in the P-100 fraction, but absent from the S-100 fraction (Fig. 12.1B, lower panel). Likewise, 5.8S rRNA was enriched in the P-100 fraction and depleted in the S-100 fraction (Fig. 12.1C, upper panel). A substantial proportion of 5S rRNA remained in the S-100 extract (see Fig. 12.1C, upper panel), presumably as RNP complexes awaiting entry to the nucleus and incorporation into nascent ribosomes (Lin *et al.*, 2001). A Western blot for S6 ribosomal protein confirmed the presence of ribosomes in the P-100 fraction (see Fig. 12.1C, lower panel). Similar results

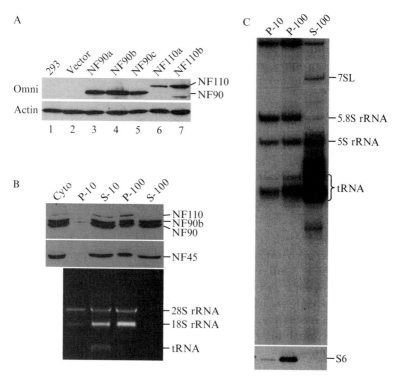

Figure 12.1 Cytoplasmic localization of ectopic NF90. (A) Immunoblot of protein extract from 293 cells (lane 1) and from stable cell lines (lanes 2 to 7) probed with anti-Omni-tag (upper panel) and anti-actin (lower panel) antibodies. (B and C) Differential centrifugation of cytoplasmic extract from 293 cells stably expressing NF90b. Extract was centrifuged at $10,000 \times g$, producing a pellet (P-10) and supernatant (S-10). S-10 was then centrifuged at $100,000 \times g$ to produce a clear pellet (P-100) and supernatant (S-100). Fractions were analyzed by Western blotting and probing with anti-DRBP76 (B, upper panel), anti-NF45 (B, middle panel), or anti-S6 antibody (C, lower panel). Extracted RNA was resolved in an ethidium bromide stained agarose gel (B, lower panel) or was 3′ end labeled and resolved in a urea/acrylamide gel (C, upper panel).

were obtained by fractionating cell lines expressing NF90a (data not shown) and by immunofluorescence of the NF90b and NF110b cell lines (Parrott *et al.*, 2005). These data indicated that the cell lines are suitable for analysis of RNAs associated with the NF90 family proteins.

3. RNP Immunoprecipitation (RIP) Assay

Next, we approached the methodology of crosslinking NF90 or NF110 to its RNA partners, and their immunoprecipitation from cell extract. Of the two established means of crosslinking, we found that formaldehyde was more effective than ultraviolet (UV) irradiation for establishing the covalent attachment of NF90 to RNA. Formaldehyde treatment also has a distinct advantage in that its covalent attachment can be reversed with heat under denaturing conditions. The overall crosslinking scheme that we employed is based on that reported by Niranjanakumari and colleagues (2002) with a significant change in the cell lysis procedure. We found that cell lysis by sonication as prescribed by these authors can be detrimental to specificity. Therefore, we employed detergent lysis as the primary means of cell lysis.

3.1. Buffers

RIPA: 50 mM Tris–HCl, pH 7.5, 1% NP-40 (v/v), 0.5% sodium deoxy-cholate (w/v), 0.05% sodium dodecyl sulfate (SDS) (w/v), 1 mM ethy-lenediaminetetraacetic acid (EDTA), 150 mM NaCl, 0.2 mM phenylmethylsulfonyl fluoride (PMSF), 0.5 mM dithiothreitol (DTT), 0.1 mg/ml yeast tRNA, and 1 U/μl RNasin (Promega).

Harsh RIPA: 50 mM Tris–Cl, pH 7.5, 1% NP-40 (v/v), 1% sodium deoxycholate (w/v) , 0.1% SDS (w/v), 1 mM EDTA, 0.4 M NaCl, 0.2 M urea, and 0.2 mM PMSF.

Crosslink reversal: 50 mM Tris–Cl, pH 7, 5 mM EDTA, 10 mM DTT, and 1% SDS (w/v).

Formamide loading buffer: 10 mM Tris, pH 8, 20 mM EDTA, pH 8, 95% formamide (v/v), 0.005% bromophenol blue, and 0.005% xylene cyanol.

RT: 20 mM Tris–HCl, pH 8.4, 50 mM KCl, 2.5 mM MgCl$_2$, 10 mM DTT, 0.4 mM dNTPs, and 1 U/μl RNasin (Promega).

Elution: 10 mM Tris–HCl, pH 8.5.

Tailing: 10 mM Tris–HCl, pH 8.4, 25 mM KCl, 1.5 mM MgCl$_2$, and 0.2 mM dCTP.

3.2. Procedure

All incubations were carried out at room temperature unless stated other-wise. Asynchronously growing 293 cells (90% confluent cells growing in four 150-mm dishes) were growth arrested with 0.1 mg/ml cycloheximide for 3 min at 37°. Cycloheximide interferes with polypeptide chain elonga-tion and effectively "freezes" the polysome profile of the cell. Cells were washed in 10 ml phosphate-buffered saline (PBS, Sigma) supplemented with 0.1 mg/ml cycloheximide and 7 mM MgCl$_2$, then incubated in 10 ml PBS with 7 mM MgCl$_2$ and 0.15 or 0.5% formaldehyde for 10 min on an orbital shaker. The crosslinking reaction was quenched with the addition of 2 M glycine (pH 7) to 0.25 M and incubating for 5 min with shaking. Cells were harvested by scraping and pelleted at $500 \times g$ for 1 min at 4°. To obtain cells synchronized at the G$_2$/M phase boundary, cells incubated at 37° were first subjected to a double-thymidine block, released for 8 h in medium supplemented with 24 μM cytidine, then incubated with 50 ng/ml nocodazole [1 mg/ml in dimethyl sulfoxide (DMSO); Sigma] for 4 h (Zieve *et al.*, 1980). G_2/M cells were manually shaken from the culture plate and then washed and crosslinked as described above.

Asynchronous and synchronous cells were lysed on ice for 5 min in 300 μl RIPA buffer and the lysate clarified at 10,000×g for 5 min at 4°. The supernatant (Lysate-S) was centrifuged again at 10,000×g for 5 min at 4°. The pellet was resuspended in 300 μl RIPA buffer and sonicated three times on ice for 5 sec, with 5-sec relaxations. The extract from the sonicated pellet (Lysate-P) was then clarified twice at 10,000×g for 5 min at 4°.

Lysate was added to 15 μl protein A-Sepharose (Amersham Biosciences) and precleared for 1 h with rocking at 4°. Protein A-Sepharose was removed and the precleared lysate was centrifuged at 10,000×g for 10 min at 4°. Anti-Omni-probe antibody (Santa Cruz Biotech.) was prepared by incubation with an equal volume of protein A-Sepharose in RIPA buffer for 2 h at 4°. The antibody–bead complex was then washed three times in 500 μl RIPA buffer. Precleared lysate was added to the antibody–bead complex and incubated with rocking for 3 h at 4°. Immunoprecipitates were washed five times in 500 μl Harsh RIPA buffer with 10 min rotations. They were resuspended in 150 μl Crosslink Reversal buffer and incubated at 70° for 45 min. Beads were agitated, pelleted at 500×g, and the RNA solution removed by aspiration. RNA was isolated by Trizol extraction according to the manufacturer's instruction (Invitrogen) and precipitated with 1 vol isopropanol in the presence of 1 M ammonium acetate and 20 μg glycogen. The RNA pellet was then washed in 75% ethanol, air-dried, and resuspended in 20 μl water.

An aliquot of isolated RNA (5.1 μl) was 3′-terminally radiolabeled by the ligation of $[5′-\alpha-^{32}P]$cytidine-3′,5′-bisphosphate (Uhlenbeck and Gumport, 1982). The reaction (7.5 μl) was catalyzed by 8 U T4 RNA ligase in the supplied buffer (New England Biolabs) supplemented with 10% DMSO (v/v) for 2 h on ice. The reaction was stopped with the addition of 1 vol Formamide Loading buffer and heating at 70° for 4 min. Radiolabeled RNA was resolved in urea/acrylamide gels.

3.3. Results: G_2/M phase cells

Preliminary cell lysis experiments showed that the nuclei of formaldehyde crosslinked cells remained largely intact even in the presence of ionic detergent. Detergent lysis might not be expected to yield a significant amount of nuclear protein to immunoprecipitate, but the majority of NF90 and NF110 are known to migrate into the cytoplasm with the onset of mitosis (Matsumoto-Taniura *et al.*, 1996). Therefore, we first coprecipitated RNA from G_2/M phase extract, when the nuclear membrane is also

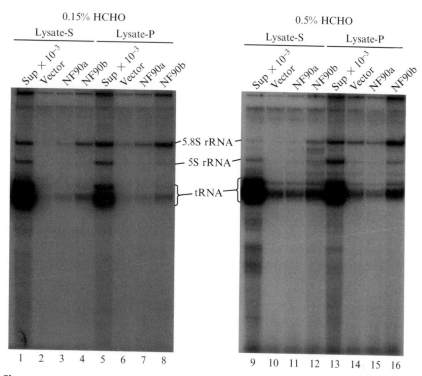

Figure 12.2 Visualization of immunoprecipitated RNA. Vector, NF90a, and NF90b stable cell lines were arrested at the G_2/M boundary, treated with 0.15 or 0.5% formaldehyde, and lysed in RIPA buffer. The lysate was centrifuged at $10,000 \times g$, producing supernatant (Lysate-S) and a pellet that was resuspended, sonicated, and centrifuged at $10,000 \times g$, and the supernatant retained (Lysate-P). RNA immunoprecipitated with Omni-probe antibody from Lysate-S and -P was 3′ end labeled and resolved in a urea/acrylamide gel. RNA extracted from the immunoprecipitation supernatant (Sup) of the vector cell line served as a marker.

relatively porous. Immunoprecipitation of NF90 from Lysate-S yielded specific small RNAs (Fig. 12.2, lanes 2 to 4 and 10 to 12). Immunoprecipitation from Lysate-P resulted in quantitative, but not qualitative, differences in the RNA band pattern when compared to the vector control (Fig. 12.2, lanes 6 to 8 and 14 to 16). The relatively high background seen with Lysate-P (compare lanes 6 and 14 with lanes 2 and 10) led us to favor Lysate-S.

We employed RNase H digestion to confirm the identity of some of the more abundant RNAs that were crosslinked and immunoprecipitated with NF90. Figure 12.3A demonstrates the RNase H digestion of 5.8S (lanes 1 and 2) and 5S rRNA (lanes 3 and 4) when annealed to their respective antisense oligonucleotides. Reverse transcription followed by polymerase chain reaction (RT-PCR) with specific primers was used to identify 7SK RNA complexed with NF90b in Lysate-P (Fig. 12.3B, upper panel).

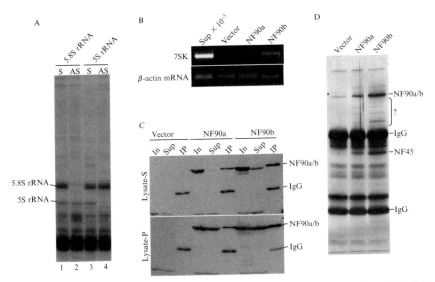

Figure 12.3 Analysis of immunoprecipitated RNA and protein. (A) 3' End-labeled RNA immunoprecipitated from NF90b Lysate-P (G$_2$/M, 0.5% HCHO) was digested with RNase H in the presence of 5.8S or 5S rRNA sense (S) or antisense (AS) oligonucleotides and resolved in a urea/acrylamide gel. (B) RT-PCR of 7SK and β-actin mRNA immunoprecipitated from Lysate-P (G$_2$/M, 0.15% HCHO). RNA extracted from the vector cell line supernatant (Sup) served as a positive control. (C) Western blot of Lysate-S and -P (G$_2$/M, 0.15% HCHO) probed with Omni-probe antibody. Input (In), supernatant (Sup), and Omni-probe antibody immunoprecipitate (IP) lanes are present and heavy IgG chain is denoted. (D) Immunoprecipitate from Lysate-S (G$_2$/M, 0.5% HCHO) was resolved in an SDS–PAGE gel and silver stained. Asterisk denotes a nonspecific protein that comigrates with tagged NF90, and IgG heavy and light chains are denoted. (See color insert.)

The high background of nonspecific binding mentioned above is evidenced by the detection of β-actin mRNA in all of the Lysate-P immunoprecipitates (Fig. 12.3B, lower panel).

The effect of varying the concentration of formaldehyde used for crosslinking was investigated. Figure 12.2 shows the different RNA profiles achieved by incubating G$_2$/M phase cells with 0.15% or 0.5% formaldehyde. It was found that 5.8S rRNA was prominent in NF90 immunoprecipitate from cells treated with a low concentration of formaldehyde (0.15%) (Fig. 12.2, lanes 4 and 8). However, the presence of 5.8S rRNA was reduced in NF90 immunoprecipitate from Lysate-S with an increase in formaldehyde concentration to 0.5% (Fig. 12.2, compare lane 4 with lane 12), suggesting that the presence of macromolecular complexes such as ribosomes in Lysate-S is reduced with increased crosslinking. We find that half of cytoplasmic NF90 fractionates in the P-100 extract (Fig. 12.1B, upper panel), and others have observed NF90 in the ribosomal salt wash fraction (Langland et al., 1999).

NF90 might then be thought to associate with ribosomes by binding directly to 5.8S rRNA. Additional bands migrating between tRNA and 5.8S rRNA also became apparent with an increase in formaldehyde concentration, possibly indicating that transient complexes are "trapped" more efficiently with increased crosslinking (Fig. 12.2, compare lane 4 with lane 12).

The efficiency of immunoprecipitation was examined by Western blotting (Fig. 12.3C). Both NF90 isoforms were effectively immunodepleted from Lysate-S supernatant (Fig. 12.3C, upper panel), but less so from Lysate-P supernatant (Fig. 12.3C, lower panel), probably because the protein is buried in macromolecular complexes in the latter. In support of this interpretation, an increased formaldehyde concentration resulted in less NF90 immunoprecipitated from Lysate-P (not shown). The presence or absence of a tetrapeptide insert in the NF90 isoform did not appear to alter its RNA-binding specificity. However, NF90a was consistently expressed at a level lower than NF90b, and this resulted in less RNA being coprecipitated (Figs. 12.2 and 12.4A). Attempts to immunoprecipitate crosslinked NF110 isoforms from Lysate-S and Lysate-P were unsuccessful (not shown), probably because this protein is involved in large nuclear complexes (Parrott *et al.*, 2005), which could mask the epitope tag or lead to insolubility after crosslinking.

Coprecipitated protein was also analyzed by silver staining (Fig. 12.3D). Both NF90a and NF90b coprecipitated NF45 in an approximately stoichiometric amount; it is therefore likely that the NF90:NF45 heterodimeric complex associates with the RNA isolated in the RIP assay, not NF90 alone. Other unidentified proteins with apparent masses of 60 to 90 kDa were enriched in the NF90 immunoprecipitates (Fig. 12.3D, bracket), and a nonspecific protein comigrated with tagged NF90 (Fig. 12.3D, asterisk). Evidently not all associated proteins were removed by harsh washing conditions, implying that these proteins, like NF45, may be crosslinked to NF90. This conclusion raises the possibility that isolated RNAs are bound indirectly to NF90, necessitating further studies to establish direct binding. We have confirmed by GST pull-down assay that the two novel RNAs identified here bind directly to the dsRBMs of NF90 (A. M. Parrott and M. B. Mathews, unpublished observations).

3.4. Results: asynchronous cells

Immunoprecipitation of NF90 from Lysate-S prepared from asynchronous cells also yielded specific RNA bands (Fig. 12.4A, lanes 5 and 6). The presence of 5S and 5.8S rRNA in Lysate-P immunoprecipitates was confirmed by RNase H digestion assay (Fig. 12.4B), as was that of 7SK from Lysate-S (not shown). The immunoprecipitation from Lysate-S of 7SL, the RNA component of the signal recognition particle, was also confirmed by RNase H digestion (not shown).

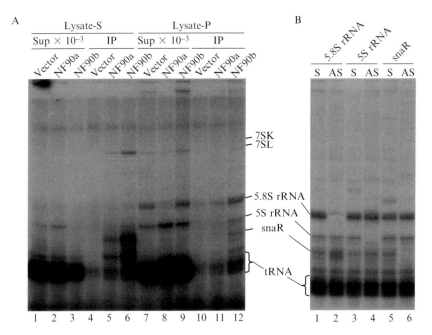

Figure 12.4 Analysis of RNA from asynchronous cells. (A) Asynchronously growing Vector, NF90a, and NF90b cell lines were treated with 0.5% formaldehyde and lysed in RIPA buffer. Lysates-S and -P were generated as before. Immunoprecipitated RNA (IP) and RNA extracted from supernatant (Sup) was 3′ end labeled and resolved in a urea/acrylamide gel. (B) 3′ End-labeled RNA immunoprecipitated from NF90b Lysate-P (asynchronous) was digested with RNase H in the presence of 5.8S, 5S rRNA, or snaR sense (S) or antisense (AS) oligonucleotides and resolved in a urea/acrylamide gel.

4. IDENTIFICATION OF UNKNOWN RNAs BY PCR AMPLIFICATION AND SEQUENCING

To identify other RNAs isolated in the RIP assay, many of which were less abundant, we employed the RT-PCR strategy outlined in Fig. 12.5A. This strategy is a modification of a 5′ rapid amplification of cDNA ends (RACE) method marketed by Invitrogen as a commercial 5′ RACE RT-PCR system (v2). Reverse transcription employs a "lock-dock" oligo(dT) primer (Borson *et al.*, 1992), which consists of an oligo(dT) primer carrying two degenerate nucleotides at its 3′ end and a specific 18-nt sequence at its 5′ end. Following RT, a tail of C residues is added. The PCR stage relies on an inosine-rich "Abridged Anchor" primer to base pair exclusively with this appended 3′-cytidine tail on the cDNA, and a Reverse primer that is identical to the 5′-terminal sequence of the Lock-dock oligo (dT) primer.

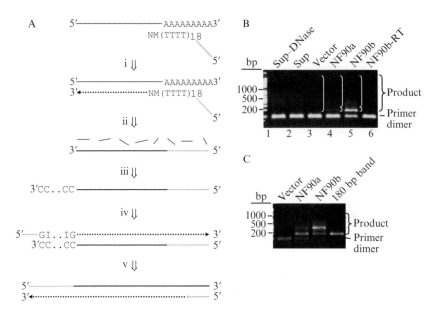

Figure 12.5 RT-PCR scheme and product. (A) Schematic of RT-PCR, indicating (i) oligo(dT) annealing to RNA and reverse transcription, (ii) RNase digestion of template RNA, (iii) dC-tailing of cDNA, and (iv) Abridged Anchor primer annealing to cDNA and positive strand transcription followed by (v) Reverse primer annealing and negative strand transcription. Primer sequences are denoted in gray. (B) RNA extracted from Lysate-S (G_2/M, 0.5% HCHO) supernatant (Sup) and immunoprecipitate was DNase I treated and then RT-PCR amplified, and the DNA was resolved in an ethidium bromide-stained agarose gel. Controls include no DNase I treatment of the supernatant (Sup-DNase; lane 2) and no reverse transcription of immunoprecipitated RNA from the NF90b cell line (NF90b-RT; lane 6). White brackets (lanes 3 to 5) denote the regions of PCR product excised from the gel. (C) PCR product from (B) was then reamplified 14-fold to yield more distinct band patterns.

4.1. Primers

Lock-dock oligo(dT): 5′-CCGCTCGAGGTATCCTAGGCATTTTTTT-TTTTTTTTTTTMN-3′ where M = A, G, or C; N = A, G, C, or T.
Reverse: 5′-CCGCTCGAGGTATCCTAGGCA-3′.
Abridged Anchor: 5′-GGCCACGCGTCGACTAGTACGGGIIGGGIIG-GGIIG-3′ where I = inosine.

4.2. Procedure

All incubations were carried out in a thermocycler. To remove residual mitochondrial DNA, RNA isolated in the RIP assay (5 μl, ~25%) was incubated with DNase I (1 U, Amplification Grade) according to the manufacturer's instructions (Invitrogen). DNase digestion was terminated with EDTA (2.3 mM final) for 10 min at 65°. Lock–dock oligo(dT) reverse transcription primer was added to the RNA solution (0.1 μM final) and annealed by heating at 70° for 10 min, then cooling. Reverse transcription (15 μl final) was performed in RT buffer at 42° for 50 min using Super-Script II Reverse Transcriptase (200 U, Invitrogen). The reaction was terminated at 70° for 15 min and RNA was digested with 0.5 μl RNase (15 U/μl RNase T1, 4 U/μl RNase H) for 30 min at 37°.

cDNA was twice purified through a QIAquick desalting column (Qia-gen), eluting in 30 μl Elution buffer. Column purification reduces (but does not eliminate) the formation of primer–dimer byproducts in the subsequent PCR step. To add the 3'-deoxycytidine tail, cDNA (10 μl) was heated at 94° for 2 min in Tailing buffer (25 μl final), then chilled, before incubating with terminal deoxynucleotidyltransferase (8 U, New England BioLabs) for 10 min at 37°. The reaction was terminated at 65° for 10 min in readiness for standard PCR. cDNA (5 μl) was amplified over 30 cycles with an annealing temperature of 63°, using Reverse primer, Abridged Anchor primer (Invitrogen), and SuperTaq polymerase (2.5 U, Ambion).

4.3. Results

RT-PCR products were resolved in agarose gels and visualized by ethidium bromide staining. RT-PCR product was generated from NF90a and NF90b immunoprecipitates, but not in the absence of reverse transcriptase or from empty vector control cells (Fig. 12.5B). Despite column purifica-tion of cDNA, primer–dimer byproducts tended to dominate subsequent cloning steps, so RT-PCR product was routinely excised from the gel, separating away the primer–dimer band. A dominant 180-bp product apparent in the NF90a and NF90b lanes was also isolated prior to cloning (Fig. 12.5B, white brackets). DNA was eluted from the excised gel and was amplified over a further 14 PCR cycles yielding a more detailed band pattern (Fig. 12.5C). PCR products were then ligated via TOPO TA Cloning technology (Invitrogen) into pCR2.1-TOPO vector (Invitrogen), and the ligation mix was used to transform competent *Escherichia coli* cells. Colonies were selected on the basis of blue–white screening and PCR inserts were sequenced by priming from the oligo(dT) end. Sequencing from the 5' end through the guanosine-rich Abridged Anchor primer often led to failed or truncated reads. The results of several cloning experiments are summarized in Table 12.1. The most abundant clones corresponded to a novel family of small RNAs represented by the 180-bp band mentioned above and termed snaR (Fig. 12.4B; A. M. Parrott and M. B. Mathews, unpublished observations), closely followed by the SINE repetitive ele-ment, dimeric *Alu*. The presence of snaR is visualized in NF90a and b immunoprecipitates from Lysate-S (Fig. 12.4A, lanes 5 and 6) and in NF90b

Table 12.1 RNA species identified

Cell cycle stage and enzymatic treatment	RNA species (number of clones)
G$_2$/M *without* DNase I	Mitochondrial DNA (2), Alu-Y (2), snaR-A, RPS19 intron containing Alu and Mir sequence
G$_2$/M	Alu-S (4), snaR-A (4), snaR-B (2), RPS9 mRNA (last exon including whole 3' UTR), mature RPSA mRNA (last two exons and part of a third), Znf131 mRNA (intron)
Asynchronous	snaR-A (13), Alu-S (9), snaR-B, Alu-Y
Asynchronous with poly(A) tailing	Pre-tRNA$_i^{met}$, hY5, U3 snRNA (180 nt fragment)
G$_2$/M and asynchronous[a]	5S rRNA, 5.8S rRNA, 7SL, 7SK

[a] RNA species were identified by RNase H digestion or gene-specific RT-PCR. All other species were identified by 5'-RACE.

immunoprecipitate from Lysate-P (Fig. 12.4A, lane 12), and was confirmed by RNase H digestion (Fig. 12.4B, lanes 5 and 6).

The RT-PCR procedure outlined above has two major limitations: first, the unknown RNA must possess a 3' oligo(A) tract to be copied into cDNA. We note that NF90 appears capable of binding RNAs such as 5.8S rRNA that do not possess a natural poly(A) tail. Second, RNA species such as *Alu* and snaR may dominate the PCR reaction and prevent the detection of rare RNAs. Hence, we modified our approach by excising gel regions devoid of *Alu* and snaR, and enzymatically appending a poly(A) tail onto the eluted RNA.

4.4. Poly(A)-tailing procedure

A region of 5% acrylamide/7 M urea gel containing 3' end-labeled RNA between the snaR and tRNA bands (Fig. 12.4A, lanes 4 to 6) was excised and the RNA eluted in 400 μl TE, pH 7.4, with 0.5% SDS (w/v) for 1.5 h at 25°. The RNA was extracted with 1 vol phenol, then twice with 1 vol chloroform:isoamyl alcohol (24:1). The RNA was precipitated with 2.5 vol ethanol in the presence of 20 μg glycogen and 1 M ammonium acetate on dry ice. The RNA pellet was dissolved in 26 μl water and treated with 1 U calf intestinal phosphatase according to the manufacturer's instructions (New England Biolabs.). The RNA was extracted and precipitated as described, then subjected to poly(A) tail addition using poly(A) polymerase (1 U) according to the manufacturer's instructions (Invitrogen; 2.5 mM

ATP final). The RNA was extracted and precipitated as described in readiness for RT-PCR.

4.5. Results

Although the poly(A) tailing procedure demands three extraction and precipitation steps, which resulted in a 60% loss in RNA quantity (as determined by scintillation counting), it led to the identification of a previously uncharacterized precursor of tRNA$_i^{met}$ and of hY5 RNA (Table 12.1).

5. SUMMARY

The methods described here employ *in vivo* chemical crosslinking coupled with immunoprecipitation to stabilize and isolate RNP complexes from cells. Several RNA species (~12) were found to associate with NF90 protein, including two previously unidentified species. RNP complexes were immunoprecipitated from two different extracts, Lysate-S and Lysate-P. RNAs purified from these complexes were radiolabeled and resolved by gel electrophoresis. The relative amount of a particular RNA bound to NF90 could then be visualized and the relative solubility of its RNP could be gauged by varying the formaldehyde concentration. RNAs that comigrated with abundant RNAs were identified by sequence-specific analysis such as RNase H digestion. Those RNAs that were too scarce to be visualized in a gel or did not comigrate with an abundant RNA were identified by 5'-RACE RT-PCR. The latter scheme was modified to identify nonpolyadenylated as well as polyadenylated RNAs.

The RNAs identified in this study (Table 12.1) are of disparate origin, between them being transcribed by all three eukaryotic RNA polymerases. Their relative abundance is also very different, ranging from mRNAs (such as that for RPS9) and previously undescribed RNAs (snaR and a precursor of tRNA$_i^{met}$) to among the most common cellular RNA species (5.8S rRNA). Care must be taken in interpreting data that include highly abundant RNAs such as 5S and 5.8S rRNA and tRNA. We observe that tRNAs are well represented in vector-alone immunoprecipitates, and conclude that these RNAs are non-specifically bound (Figs. 12.2 and 12.4A). Although 5.8S rRNA is precipitated from vector alone Lysate-P extracts, it is highly enriched in NF90 immunoprecipitations, particularly from G$_2$/M cells (Fig. 12.2, lanes 4, 8, 12, and 16), when most ribosomes are dissociated from mRNA and exist as free 40S and 60S subunits (Qin and Sarnow, 2004). The 5S rRNA is slightly enriched in NF90 immunoprecipitations, but its presence, which is considerably less than that of 5.8S rRNA, could be a consequence of NF90 binding directly to ribosomes via 5.8S rRNA. Given that a substantial proportion of cytoplasmic NF90 is found in the P-100 pellet (Fig. 12.1) and that 5.8S rRNA is a major NF90 binding partner (Fig. 12.2), it is possible that NF90 exerts some influence on translation. Interestingly, three out of four mRNAs found to crosslink with NF90 are related to the small ribosomal subunit: two encode structural components (RPS9 and RPS19) and one encodes ribosomal protein SA (RPSA), which associates with the 40S subunit (Tohgo *et al.*, 1994).

The dimeric form of *Alu* is a rare and unstable RNA polymerase III (PolIII) transcript estimated to be present in the cytosol at 100 to 1000 copies per cell (Li and Schmid, 2004; Liu *et al.*, 1994). However, dimeric *Alu* was the second most abundant clone isolated (Table 12.1). Transcription of this RNA is rapidly upregulated (within 20 min in HeLa cells) in 293 cells in response to translational inhibitors such as cycloheximide (Liu *et al.*, 1995). Therefore, it is feasible that dimeric *Alu* RNA levels are elevated in this study, but even after maximum induction (20-fold over 3 h in HeLa cells), *Alu* levels approached only 5% of the level of 7SL RNA in the cell (Liu *et al.*, 1995). There are more than one million *Alu* element copies within the human genome and the majority are inserted into untranslated regions or introns of PolII-transcribed mRNA. Strikingly, this study identified only one such *Alu* element in an intronic region of the ribosomal protein S19. Therefore, polIII-transcribed dimeric *Alu* RNA is probably a biologically meaningful partner of NF90.

A number of PolIII-transcribed RNAs found to associate with NF90 in this study are linked by secondary structure and even origin. *Alu* RNAs are considered to be derived from and share considerable sequence with 7SL RNA. These RNAs have a common secondary structure motif and both bind the cognate signal recognition particle proteins SRP9/14 (Hasler and Strub, 2006). Human Ro-associated Y5 RNA folds into a stable hairpin loop structure, reminiscent of the left arm of the *Alu* dimer (van Gelder *et al.*, 1994). Furthermore, the most abundant clone snaR-A and its close relative snaR-B probably also fold into stable hairpin loops (A. M. Parrott and M. B. Mathews, unpublished observations). This secondary structure motif could be conducive to NF90 binding and NF90 might play a role in the biology of RNA polIII transcripts and possibly in their maturation: NF90 protein specifically binds a novel precursor of $tRNA_i^{met}$, but has greatly reduced affinity for the mature form (data not shown).

The main objective of this study was to utilize *in vivo* crosslinking to find RNA partners and thereby reveal the functional roles of NF90 within the uninfected cell. A number of very different RNAs were isolated and identified, but functional interpretation remains speculative. Noting that the RNAs can be grouped into those with a shared origin (PolIII-transcribed) and those with ribosome/translation-related biology, NF90 might have at least two roles within the cell. This study places NF90 among a small group of RNA-binding proteins that recognizes different classes of cellular RNA. The La protein is a well-known member of this group. Like NF90, La is a phosphoprotein that appears to have two overlapping functions, the biogenesis of PolIII transcripts and the coordination of the production of the translational machinery (Kenan and Keene, 2004). While the NF90a and b isoforms appear to bind the same spectrum of RNA species, the present data do not altogether discount the possibility that alternative splicing and phosphorylation status may modulate partner recognition.

ACKNOWLEDGMENT

This work was supported by Grant R01 AI034552 from the National Institutes of Health.

REFERENCES

Borson, N. D., Salo, W. L., and Drewes, L. R. (1992). A lock-docking oligo(dT) primer for 5′ and 3′ RACE PCR. *PCR Methods Appl.* **2,** 144–148.

Bose, S. K., Sengupta, T. K., Bandyopadhyay, S., and Spicer, E. K. (2006). Identification of Ebp1 as a component of cytoplasmic bcl-2 mRNP (messenger ribonucleoprotein particle) complexes. *Biochem. J.* **396,** 99–107.

Buaas, F. W., Lee, K., Edelhoff, S., Disteche, C., and Braun, R. E. (1999). Cloning and characterization of the mouse interleukin enhancer binding factor 3 (Ilf3) homolog in a screen for RNA binding proteins. *Mamm. Genome* **10,** 451–456.

Corthésy, B., and Kao, P. N. (1994). Purification by DNA affinity chromatography of two polypeptides that contact the NF-AT DNA binding site in the interleukin 2 promoter. *J. Biol. Chem.* **269,** 20682–20690.

Duchange, N., Pidoux, J., Camus, E., and Sauvaget, D. (2000). Alternative splicing in the human interleukin enhancer binding factor 3 (ILF3) gene. *Gene* **261,** 345–353.

Hasler, J., and Strub, K. (2006). Alu RNP and Alu RNA regulate translation initiation *in vitro*. *Nucleic Acids Res.* **34,** 2374–2385.

Isken, O., Grassmann, C. W., Sarisky, R. T., Kann, M., Zhang, S., Grosse, F., Kao, P. N., and Behrens, S. E. (2003). Members of the NF90/NFAR protein group are involved in the life cycle of a positive-strand RNA virus. *EMBO J.* **22,** 5655–5665.

Kao, P. N., Chen, L., Brock, G., Ng, J., Kenny, J., Smith, A. J., and Corthesy, B. (1994). Cloning and expression of cyclosporin A- and FK506-sensitive nuclear factor of activated T-cells: NF45 and NF90. *J. Biol. Chem.* **269,** 20691–20699.

Kenan, D. J., and Keene, J. D. (2004). La gets its wings. *Nat. Struct. Mol. Biol.* **11,** 303–305.

Langland, J. O., Kao, P. N., and Jacobs, B. L. (1999). Nuclear factor-90 of activated T-cells: A double-stranded RNA-binding protein and substrate for the double-stranded RNA-dependent protein kinase, PKR. *Biochemistry* **38,** 6361–6368.

Larcher, J. C., Gasmi, L., Viranaicken, W., Edde, B., Bernard, R., Ginzburg, I., and Denoulet, P. (2004). Ilf3 and NF90 associate with the axonal targeting element of Tau mRNA. *FASEB J.* **18,** 1761–1763.

Lewis, B. P., Burge, C. B., and Bartel, D. P. (2005). Conserved seed pairing, often flanked by adenosines, indicates that thousands of human genes are microRNA targets. *Cell* **120,** 15–20.

Li, T. H., and Schmid, C. W. (2004). Alu's dimeric consensus sequence destabilizes its transcripts. *Gene* **324,** 191–200.

Liao, H. J., Kobayashi, R., and Mathews, M. B. (1998). Activities of adenovirus virus-associated RNAs: Purification and characterization of RNA binding proteins. *Proc. Natl. Acad. Sci. USA* **95,** 8514–8519.

Lin, E., Lin, S. W., and Lin, A. (2001). The participation of 5S rRNA in the co-translational formation of a eukaryotic 5S ribonucleoprotein complex. *Nucleic Acids Res.* **29,** 2510–2516.

Liu, W. M., Maraia, R. J., Rubin, C. M., and Schmid, C. W. (1994). Alu transcripts: Cytoplasmic localisation and regulation by DNA methylation. *Nucleic Acids Res.* **22,** 1087–1095.

Liu, W. M., Chu, W. M., Choudary, P. V., and Schmid, C. W. (1995). Cell stress and translational inhibitors transiently increase the abundance of mammalian SINE transcripts. *Nucleic Acids Res.* **23,** 1758–1765.

Matsumoto-Taniura, N., Pirollet, F., Monroe, R., Gerace, L., and Westendorf, J. M. (1996). Identification of novel M phase phosphoproteins by expression cloning. *Mol. Biol. Cell.* **7,** 1455–1469.

Mili, S., and Steitz, J. A. (2004). Evidence for reassociation of RNA-binding proteins after cell lysis: Implications for the interpretation of immunoprecipitation analyses. *Rna* **10,** 1692–1694.

Niranjanakumari, S., Lasda, E., Brazas, R., and Garcia-Blanco, M. A. (2002). Reversible cross-linking combined with immunoprecipitation to study RNA-protein interactions *in vivo. Methods* **26,** 182–190.

Parrott, A. M., Walsh, M. R., Reichman, T. W., and Mathews, M. B. (2005). RNA binding and phosphorylation determine the intracellular distribution of nuclear factors 90 and 110. *J. Mol. Biol.* **348,** 281–293.

Patel, R. C., Vestal, D. J., Xu, Z., Bandyopadhyay, S., Guo, W., Erme, S. M., Williams, B. R., and Sen, G. C. (1999). DRBP76, a double-stranded RNA-binding nuclear protein, is phosphorylated by the interferon-induced protein kinase, PKR. *J. Biol. Chem.* **274,** 20432–20437.

Qin, X., and Sarnow, P. (2004). Preferential translation of internal ribosome entry site-containing mRNAs during the mitotic cycle in mammalian cells. *J. Biol. Chem.* **279,** 13721–13728.

Reichman, T. W., and Mathews, M. B. (2003). RNA binding and intramolecular interactions modulate the regulation of gene expression by nuclear factor 110. *Rna* **9,** 543–554.

Reichman, T. W., Parrott, A. M., Fierro-Monti, I., Caron, D. J., Kao, P. N., Lee, C. G., Li, H., and Mathews, M. B. (2003). Selective regulation of gene expression by nuclear factor 110, a member of the NF90 family of double-stranded RNA-binding proteins. *J. Mol. Biol.* **332,** 85–98.

Saunders, L. R., Jurecic, V., and Barber, G. N. (2001a). The 90- and 110-kDa human NFAR proteins are translated from two differentially spliced mRNAs encoded on chromosome 19p13. *Genomics* **71,** 256–259.

Saunders, L. R., Perkins, D. J., Balachandran, S., Michaels, R., Ford, R., Mayeda, A., and Barber, G. N. (2001b). Characterization of two evolutionarily conserved, alternatively spliced nuclear phosphoproteins, NFAR-1 and -2, that function in mRNA processing and interact with the double-stranded RNA-dependent protein kinase, PKR. *J. Biol. Chem.* **276,** 32300–32312.

Shi, L., Zhao, G., Qiu, D., Godfrey, W. R., Vogel, H., Rando, T. A., Hu, H., and Kao, P. N. (2005). NF90 regulates cell cycle exit and terminal myogenic differentiation by direct binding to the 3′-untranslated region of MyoD and p21WAF1/CIP1 mRNAs. *J. Biol. Chem.* **280,** 18981–18989.

Shim, J., Lim, H., Yates, J. R., and Karin, M. (2002). Nuclear export of NF90 is required for interleukin-2 mRNA stabilization. *Mol. Cell.* **10,** 1331–1344.

Shin, H. J., Kim, S. S., Cho, Y. H., Lee, S. G., and Rho, H. M. (2002). Host cell proteins binding to the encapsidation signal epsilon in hepatitis B virus RNA. *Arch. Virol.* **147,** 471–491.

Tohgo, A., Takasawa, S., Munakata, H., Yonekura, H., Hayashi, N., and Okamoto, H. (1994). Structural determination and characterization of a 40 kDa protein isolated from rat 40 S ribosomal subunit. *FEBS Lett.* **340,** 133–138.

Tran, H., Schilling, M., Wirbelauer, C., Hess, D., and Nagamine, Y. (2004). Facilitation of mRNA deadenylation and decay by the exosome-bound, DExH protein RHAU. *Mol. Cell.* **13,** 101–111.

Uhlenbeck, O. C., and Gumport, R. I. (1982). T4 RNA ligase. *In* "The Enzymes" (Boyer, ed.), vol. 15B, pp. 31–58. Academic Press, New York.

Valencia-Sanchez, M. A., Liu, J., Hannon, G. J., and Parker, R. (2006). Control of translation and mRNA degradation by miRNAs and siRNAs. *Genes Dev.* **20,** 515–524.

van Gelder, C. W., Thijssen, J. P., Klaassen, E. C., Sturchler, C., Krol, A., van Venrooij, W. J., and Pruijn, G. J. (1994). Common structural features of the Ro RNP associated hY1 and hY5 RNAs. *Nucleic Acids Res.* **22,** 2498–2506.

Xu, Y. H., and Grabowski, G. A. (2005). Translation modulation of acid beta-glucosidase in HepG2 cells: Participation of the PKC pathway. *Mol. Genet. Metab.* **84,** 252–264.

Xu, Y. H., Leonova, T., and Grabowski, G. A. (2003). Cell cycle dependent intracellular distribution of two spliced isoforms of TCP/ILF3 proteins. *Mol. Genet. Metab.* **80,** 426–436.

Zieve, G. W., Turnbull, D., Mullins, J. M., and McIntosh, J. R. (1980). Production of large numbers of mitotic mammalian cells by use of the reversible microtubule inhibitor nocodazole. Nocodazole accumulated mitotic cells. *Exp. Cell. Res.* **126,** 397–405.

Approaches for Analyzing the Differential Activities and Functions of eIF4E Family Members

Robert E. Rhoads,* Tzvetanka D. Dinkova,[†] and Rosemary Jagus[‡]

Contents

* Department of Biochemistry and Molecular Biology, Louisiana State University Health Sciences Center, Shreveport, Louisiana
† Departamento de Bioquimica L-103, Facultad de Quimica Conjunto "E," Paseo de la Inv. Cientifica, Universidad Nacional Autonoma de Mexico, Mexico D.F.
‡ Center of Marine Biotechnology, University of Maryland Biotechnology Institute, Baltimore, Maryland

Methods in Enzymology, Volume 429
ISSN 0076-6879, DOI: 10.1016/S0076-6879(07)29013-5

Abstract

The translational initiation factor eIF4E binds to the m^7G-containing cap of mRNA and participates in recruitment of mRNA to ribosomes for protein synthesis. eIF4E also functions in nucleocytoplasmic transport of mRNA, sequestration of mRNA in a nontranslatable state, and stabilization of mRNA against decay in the cytosol. Multiple eIF4E family members have been identified in a wide range of organisms that includes plants, flies, mammals, frogs, birds, nematodes, fish, and various protists. This chapter reviews methods that have been applied to learn the biochemical properties and physiological functions that differentiate eIF4E family members within a given organism. Much has been learned to date about approaches to discover new eIF4E family members, their *in vitro* properties (cap binding, stimulation of cell-free translation systems), tissue and developmental expression patterns, protein-binding partners, and their effects on the translation or repression of specific subsets of mRNA. Despite these advances, new eIF4E family members continue to be found and new physiological roles discovered.

1. INTRODUCTION

This chapter presents investigative approaches to distinguish among family members of the mRNA cap-binding protein eIF4E. Since eIF4E acts in concert with numerous other proteins and RNAs during the initiation of protein synthesis, we begin with a brief overview of protein synthesis initiation (Kapp and Lorsch, 2004). A different class of eukaryotic initiation factors (eIF1, eIF2, etc.) catalyzes each step of initiation. A ternary complex of $eIF2 \cdot GTP \cdot Met\text{-}tRNAi$ binds to the 40S ribosomal subunit to form the 43S initiation complex. Recruitment of mRNA to the 43S initiation complex to form the 48S initiation complex requires eIF3, poly(A)-binding protein (PABP), and the eIF4 proteins. The latter consist of eIF4A, a 46-kDa RNA helicase; eIF4B, a 70-kDa RNA-binding and RNA-annealing protein; eIF4H, a 25-kDa protein that acts with eIF4B to stimulate eIF4A helicase activity; eIF4E, a 25-kDa cap-binding protein; and eIF4G, a 185-kDa protein that specifically binds to and colocalizes all of the other proteins involved in mRNA recruitment on the 40S subunit (molecular masses refer to the human proteins). In the presence of eIF1 and eIF1A, the 48S complex scans until the first AUG in good sequence context is encountered. Then the GTPase-activating protein eIF5, together with eIF5B, stimulates GTP hydrolysis by eIF2. The initiation factors are replaced by the 60S subunit, and the first peptide bond is formed.

eIF4E is probably the first of the canonical initiation factors to interact with mRNA during its recruitment to the ribosome. eIF4E has been extensively investigated in organisms that range from yeast to mammals (Dyer et al., 1998; Jankowska-Anyszka et al., 1998; Joshi et al., 2004; Rhoads et al., 1993; Robalino et al., 2004; Rodriguez et al., 1998). Sequence comparisons, coupled with deletion analyses of eIF4Es from several species, have demonstrated an evolutionarily conserved core region of 160 to 170 residues. This is represented by His-37 to His-200 of *Homo sapiens* and *Mus musculus* eIF4E-1. The consensus sequence of the core region shows amino acid residues Trp, Phe, and His in a distinctive pattern summarized as $H(x_5)W(x_2)W(x_{8-12})W(x_9)F(x_5)FW(x_{20})F(x_7)W(x_{10})W(x_{9-12})W(x_{34-35})W(x_{32-34})H$. The more variable N- and C-termini of eIF4E appear to be dispensable for translation, although they may be involved in the regulation of eIF4E activity or may affect stability of the protein.

The tertiary structures of mouse (Marcotrigiano et al., 1997), yeast (Matsuo et al., 1997), and human (Tomoo et al., 2003) eIF4Es have been solved. The specificity of eIF4E interaction with the cap results primarily from stacking of the alkylated purine base between Trp-56 and Trp-102 (amino acid positions refer to human eIF4E-1) (Marcotrigiano et al., 1997; Matsuo et al., 1997; Niedzwiecka et al., 2002). In addition, Glu-103 forms H-bonds with N1 and N2 protons of m^7G and the adjacent peptide bond. Direct interactions occur between Trp-56 and the ribose group and between Arg-157 and Lys-162 and the α- and β-phosphate oxygen atoms. A third Trp residue (Trp-166) interacts with the N^7-methyl moiety of the cap structure.

eIF4E participates in mRNA recruitment through specific and high-affinity binding to eIF4G (Keiper et al., 1999). When isolated by affinity chromatography, eIF4E from mammals and plants is found in a complex with eIF4A and eIF4G, termed eIF4F. The eIF4E–eIF4G interaction is prevented, and cap-dependent translation is inhibited, by binding of eIF4E to 4E-BPs via a canonical eIF4E-binding motif, $YXXXXL\phi$ (where ϕ is any hydrophobic amino acid). Two other conserved Trp residues in eIF4E, Trp-43 and Trp-73, are important for the interaction with eIF4G. Trp-73 is found within a phylogenetically conserved sequence $(S/T)V(e/d)(e/d)FW$. Substitution of Trp-73 of mammalian eIF4E with a nonaromatic amino acid disrupts the ability of eIF4E to interact with either eIF4G or with 4E-BPs. The function of the four other conserved Trp residues remains to be established.

In addition to translation, eIF4E functions in nucleocytoplasmic transport of mRNA, sequestration of mRNA in a nontranslatable state, and stabilization of mRNA against decay in the cytosol (Görlich and Mattaj, 1996; Richter and Sonenberg, 2005; Strudwick and Borden, 2002). eIF4E binds to the nucleocytoplasmic shuttling protein 4E-T (eIF4E-Transporter) (Dostie et al., 2000).

Both promyelocytic leukemia (PML) protein, a homeodomain protein (Cohen *et al.*, 2001), and the proline-rich homeodomain (PRH) protein (Topisirovic *et al.*, 2003) bind directly to nuclear eIF4E and inhibit its nucleo-cytoplasmic transport. Some cases of translational repression by 3′ UTR sequences involve eIF4E (Richter and Sonenberg, 2005). For instance, in meiotically arrested *Xenopus* oocytes, translation of maternal mRNAs is repressed by binding of 3′ UTR sequence elements termed CPEs by the cytoplasmic polyadenylation element binding protein (CPEB). CPEB binds Maskin, which in turn binds eIF4E, inactivating the latter for translational initiation. A signal to resume oocyte development results in phosphorylation of CPEB, release of Maskin, poly(A) lengthening, and recruitment of the maternal mRNAs to the ribosome. The cap also serves as one determinant of mRNA degradation. In the 5′ → 3′ pathway for mRNA decay, removal of the cap by the decapping enzyme Dcp1/Dcp2 exposes the transcripts to digestion by a highly processive 5′ → 3′ exonuclease, XrnI (Hsu and Stevens, 1993). Dcp1 binds to both eIF4G and PABP as free proteins as well as to the complex of eIF4E·eIF4G·PABP (Vilela *et al.*, 2000). Addition of eIF4E inhibits Dcp1/Dcp2 activity *in vitro*, but m⁷GTP restores decapping. 4E-T colocalizes with mRNA decapping factors in processing bodies (P-bodies), the sites of mRNA decay (Ferraiuolo *et al.*, 2005). Interaction of 4E-T with eIF4E represses translation, which is believed to be a prerequisite for targeting of mRNAs to P-bodies.

Intriguingly, from two to eight proteins with sequences similar to proto-typical eIF4Es have been found in plants, flies, mammals, frogs, birds, nema-todes, echinoderms, fish, and various protists (Joshi *et al.*, 2005; Morales *et al.*, 2006) (Fig. 13.1). Evolutionarily, it seems that a single early eIF4E gene underwent a series of gene duplications, generating multiple structural classes and in some cases subclasses. Today, eIF4E and its relatives comprise a family of structurally related proteins within a given organism. Sequence similarity is highest in a core region of 160 to 170 amino acid residues identified by evolutionary conservation and functional analyses (Joshi *et al.*, 2005). To distinguish prototypical eIF4E from its relatives, vertebrate eIF4E has been renamed eIF4E-1 (Keiper *et al.*, 2000) or eIF4E-1A (Robalino *et al.*, 2004). Prototypical eIF4E is considered to be eIF4E-1 of mammals, eIF4E and eIF (iso)4E of plants, and eIF4E of *Saccharomyces cerevisiae*.

With the exception of some eIF4Es from protists, all eIF4Es can be grouped into one of three classes (see Fig. 13.1). Class I members carry Trp residues equivalent to Trp-43 and Trp-56 of *H. sapiens* eIF4E and appear to be present in all eukaryotes. cDNAs encoding members of Class I can be identified in species from Viridiplantae, Metazoa, and Fungi. As judged from completed genomes, many protists also encode Class I-like family members. Evidence for gene duplication of Class I eIF4E family members can be found in certain plant species, as well as in nematodes, insects, chordates, and some fungi (Hernandez and Vazquez-Pianzola, 2005; Joshi *et al.*, 2005). Class I

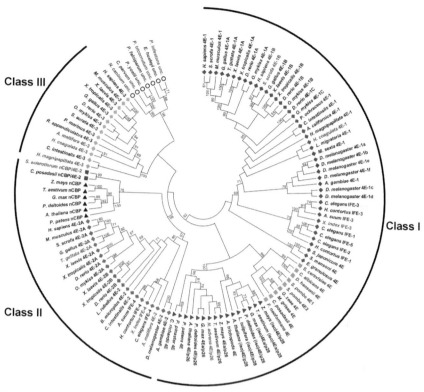

Figure 13.1 A radial cladogram describing the overall relationship of selected eIF4E family members from multiple species. The figure is taken from Joshi *et al.* (2005) and can also be viewed at http://umbicc 3–215.umbi.umd.edu/iisstart.asp. The topology of a neighbor-joining tree visualized in radial format is derived from an alignment of nucleotide sequences representing the conserved core regions of the indicated eIF4E family members. eIF4E family member names in black or red indicate whether or not the complete sequence of the conserved core region of the member could be predicted from consensus cDNA sequence data, respectively. eIF4E family member names in blue indicate that genomic sequence data were used to either verify or determine the nucleotide sequence representing the core region of the member. The shape of a "leaf" indicates the taxonomic kingdom from which the species containing the eIF4E family member derives: Metazoa (diamonds), Fungi (squares), Viridiplantae (triangles), and Protista (circles). The color of a "leaf" indicates the subgroup of the eIF4E family member: metazoan eIF4E-1 and IFE-3-like (red), fungal eIF4E-like (gold), plant eIF4E and eIF(iso) 4E-like (green), metazoan eIF4E-2-like (cyan), plant nCBP-like (blue), fungal nCBP/ eIF4E-2-like (purple), metazoan eIF4E-3-like (pink), and atypical eIF4E family members from some protists (white). eIF4E family members within structural Class I, Class II, and Class III are indicated. (See color insert.)

members include the prototypical initiation factor, but may also include eIF4Es that recognize alternative cap structures such as IFE-1, -2, and -5 of *Caenorhabditis elegans* (Jankowska-Anyszka *et al.*, 1998; Keiper *et al.*, 2000), or eIF4Es that apparently fulfill regulatory functions such as the vertebrate eIF4E-1Bs (Robalino *et al.*, 2004). Class II members possess Trp → Tyr/Phe/Leu and Trp → Tyr/Phe substitutions relative to Trp-43 and Trp-56 of *H. sapiens* eIF4E. These have been identified in Metazoa, Viridiplantae, and Fungi but are absent from the model ascomycetes *S. cerevisiae* and *Schizosaccharomyces pombe*. Class III members possess a Trp residue equivalent to Trp-43 of *H. sapiens* eIF4E but carry a Trp → Cys/Tyr substitution relative to *H. sapiens* Trp-56. They have been identified primarily in chordates with rare examples in other Coelomata and in Cnidaria (Joshi *et al.*, 2005). Many eIF4E family members from Protista, by contrast, have proven hard to characterize and can show extension or compaction relative to prototypical eIF4E family members (Joshi *et al.*, 2005).

Even more intriguing than the existence of multiple eIF4E family members within a single organism are the roles that different family members might potentially play, given that eIF4E is involved in initiation of protein synthesis, mRNA turnover, repression of mRNA translation, and export of mRNA from the nucleus. Family members exist for many if not most proteins. However, in the case of eIF4E, this carries a larger significance because eIF4E could conceivably affect the levels of every protein in the cell through one or more of these mechanisms. Physiological roles for individual eIF4E family members are now beginning to emerge, but none is understood completely. Expression of specific eIF4Es may occur at different developmental stages of an organism and affect the recruitment of a subset of mRNAs, as has been observed for *C. elegans* IFE-4 (Dinkova *et al.*, 2005; Trutschl *et al.*, 2005) and zebrafish (Robalino *et al.*, 2004). Individual eIF4Es can be expressed in a tissue- or cell-specific manner (Amiri *et al.*, 2001; Dinkova *et al.*, 2005). Pairing may occur between a specific eIF4E and subsets of mRNAs containing a common structural feature, for example, a modified cap or a sequence motif in the 3′ UTR or 5′ UTR. There can be proteins that bind differentially to eIF4E family members to affect their activity and/or mRNA selection. For instance, Class II eIF4E in *Drosophila* (eIF4E-8/4EHP) binds specifically to a region of bicoid that resembles the eIF4E–binding region within eIF4G and sequesters caudal mRNA into inactive complexes with which eIF4E-1 and ribosomes cannot interact (Cho *et al.*, 2005). Certain eIF4E family members possess only partial activities when compared to prototypical eIF4Es (Joshi *et al.*, 2004) and thus may act as inhibitors of mRNA recruitment.

Understanding the physiological roles of individual eIF4E family members will shed light on basic cellular and molecular processes, but there are also implications for human health. Since eIF4E plays a critical role in both germ cell maturation and early embryonic development (Richter and Sonenberg,

2005), a better knowledge of eIF4E family members could contribute to our understanding of normal human development and abnormalities resulting in birth defects. Also, IFE-1 is required for spermatogenesis (Amiri *et al.*, 2001), and understanding the mechanism may shed light on some infertility conditions. The involvement of eIF4E in cancer is particularly well documented. Overexpression of eIF4E-1 in cultured mammalian cells produces rapid growth in foci (De Benedetti and Rhoads, 1990), causes them to form tumors in nude mice (Lazaris-Karatzas *et al.*, 1990), and prevents apoptosis after growth factor restriction (Polunovsky *et al.*, 1996). Conversely, reduction in eIF4E-1 expression decreases protein synthesis, cell growth, and malignant transformation (De Benedetti *et al.*, 1991; Rinker-Schaeffer *et al.*, 1993), including invasiveness, metastasis, and angiogenesis in mice (Graff *et al.*, 1995; Nathan *et al.*, 1997). Naturally occurring cancers overexpress eIF4E-1 (De Benedetti and Graff, 2004). Several studies have shown that eIF4E levels provide a better indicator of cancer severity, progression, and recurrence than do histological changes (De Benedetti and Graff, 2004). The high levels of eIF4E-1 in breast cancer cell lines have been exploited to destroy experimental tumors selectively in mice as a model for gene therapy (DeFatta *et al.*, 2002a,b). Interestingly, a specific isoform of human eIF4E from Class II (*H. sapiens* 4E-2A in Fig. 13.1, also termed 4EHP) is a signature of metastasis in primary solid tumors (Ramaswamy *et al.*, 2003).

In this chapter, we outline techniques for discovering differences between eIF4E family members, beginning with the detection and analysis of eIF4E family members *in silico*. This is followed by a consideration of *in vitro* methods for assessing differential cap-binding properties of eIF4E family members. For this reason, we concentrate here on how the results of such assays can distinguish among eIF4E family members rather than the assays themselves. Next we describe how eIF4E family members can be differentiated based on their expression *in vivo* and their effect on overall translational activity. Interaction of eIF4E family members with other proteins is another feature that can be used to characterize them. Finally, we describe global microarray analysis of polysomal mRNA distribution to determine subsets of mRNAs whose levels or translational efficiency are affected by individual eIF4E family members.

2. *IN SILICO* DETECTION AND ANALYSIS OF EIF4E FAMILY MEMBERS

2.1. Sequence comparisons

The first step is to search for eIF4E family members in a species of interest. To obtain nucleotide sequences representing an accurate description of the repertoire of functional genes encoding eIF4E family members within an organism, it is most useful to begin with expressed sequence tag (EST) databases since these represent the cDNAs of expressed genes. The direct use of genomic sequences is more problematic due to the possibility of including pseudogenes and can be further complicated by deficiencies in our ability to predict intron/exon boundaries. In addition, problems can arise from errors in the assembly of genomic sequence data. However, when insufficient EST data are available to assemble and verify an eIF4E sequence, it can be helpful to use genomic sequences and whatever EST sequences are available for confirmation. The use of genomic sequences is easier for organisms whose genomes are known to lack, or contain few, introns in genes transcribed by RNA polymerase II such as some Protista and yeasts or in species for which expressed cDNAs can be identified in closely related organisms.

Initial searches for expressed nucleotide sequences encoding putative eIF4E family members can be acquired from GenBank NR, dbEST, or any other available databases by using the nucleotide and amino acid sequences encoding the conserved core regions of eIF4Es from the closest related species/order/phylum as probes for BLAST searches. Sequences encoding putative eIF4E family members can be easily identified by comparison of computed translations with the core sequences of known eIF4Es and the consensus pattern for the conserved core region, $H(x_5)W(x_2)W(x_{8-12})W(x_9)$ $F(x_5)FW(x_{20})F(x_7)W(x_{10})W(x_{9-12})W(x_{34-35})W(x_{32-34})H$ (Joshi *et al.*, 2005). The retrieved eIF4E-related sequences can be used to reprobe the databanks to identify further sequences of overlapping cDNA fragments from the same species or to obtain sequences from additional species. Overlapping sequences can be aligned and a picture of the complete core, along with amino and carboxy termini, can be assembled. Using the new eIF4E cDNA sequence, the process of iteration can be continued to obtain sequences encoding more eIF4E family members.

To facilitate searches for novel eIF4E family members, a website database has been established by Dr. Bhavesh Joshi, http://umbicc3-215.umbi.umd.edu, that contains over 400 eIF4E sequences from eukaryotes and one from a virus (Mimivirus). Access is available to all but requires registration and log-in prior to use. Nucleotide or protein sequences can be downloaded for the "core" or full sequence (where available). EST sequences corresponding to each eIF4E family member are also provided, as well as its class and subtype if apparent. The eIF4E Family Member Database allows the retrieval of sequences representing eIF4E subtypes from multiple species. The database contains EST alignments, nucleotide coding sequences, and amino acid sequences. Retrieved records can be viewed through use of the Jalview application (Clamp *et al.*, 2004). Jalview requires that the investigator's browser is capable of using and displaying JAVA applets. For MacOS users, this may be a problem if using older versions of OSX. For Windows users, it is necessary to install JAVA software, available from SunOS.

Once new eIF4E family members have been identified, they can be assessed as to class membership. This will provide an initial intimation of function. In organisms with more than three eIF4E family members, the greatest duplication has been found in Class I forms. For instance, in *C. elegans*, there are five eIF4E family members, IFE-1 to IFE-5. Four of them (IFE-1, -2, -3, and -5) carry Trp residues equivalent to Trp-43 and Trp-56 of *H. sapiens* eIF4E, which is characteristic of Class I, and one (IFE-4) has a Trp → Tyr substitution at amino acids equivalent to Trp-43 and Trp-56 of *H. sapiens* eIF4E, characteristic of Class II. In *Drosophila*, there are seven eIF4E genes encoding eight eIF4Es, seven of which belong to Class I (Hernandez *et al.*, 2005). The analysis of amino acid sequences of different eIF4E family members places plant eIF4E and eIF(iso)4E within Class I (Joshi *et al.*, 2005), although they share only 50% identity. These eIF4E family members have also shown some functional divergence in cap binding (see Section 3 below).

In contrast to Viridiplantae, Metazoa, and Fungi, some protists express atypical eIF4E family members that are difficult to place within the three defined structural classes due to extensive stretches of amino acids between units of the eIF4E core structure (Joshi *et al.*, 2005). eIF4E family members from Stramenopiles and Alveolata contain an unusual stretch of 12 to 15 amino acids between residues equivalent to Trp-73 and Trp-102 of mammalian eIF4E-1 and 4 to 9 amino acids between residues equivalent to Trp-102 and Trp-166. Extended stretches between structural units of the core, albeit in positions that differ from those found in Alveolata and Stramenopiles, are also found in the eIF4E family member termed Leish4E-1 from the trypanosome *Leishmania major* (Yoffe *et al.*, 2004). Leish4E-1 contains two areas of extended amino acid stretches between structural units of the core. Studies *in vitro* have shown that LeishIF4E-1 binds with similar affinities both m⁷GTP and an unusual cap structure found in *L. major* termed cap-4 (see Section 3). Such comparisons hint that the atypical protist eIF4Es are also likely to interact with unusual cap structures and will serve to inform future functional studies. Other differences in conserved motifs suggest alternative functions. For example, Mimivirus, the sole member of the newly proposed Mimiviridae family of large double-stranded DNA viruses, encodes its own eIF4E (Raoult *et al.*, 2004). Mimivirus eIF4E differs from other eIF4E family members in that it lacks a Trp residue equivalent to Trp-73 of mouse eIF4E-1, suggesting that the protein may not interact with eIF4G or 4E-BPs and pointing toward a role in subversion of the host translational machinery by the virus.

2.2. Relationships among sequences

Similarity searches and multiple alignments of sequences lead to the question of how they are related, which can in turn intimate similarities or differences in function. Relationships are often more apparent by tree or dendrogram analyses, using DNA or protein sequences as taxonomic characters. Looking again at the example of *C. elegans*, IFE-1, -2, -3, and -5 possess more similarity to each other in sequence than to Class I family members from other phyla of Metazoa, suggesting they arose from gene duplications of a progenitor IFE-3

(Joshi *et al.*, 2005). However, IFE-1, -2, and -5 are more like each other than IFE-3, suggestive of a divergence of function (Joshi *et al.*, 2005; Keiper *et al.*, 2000). This corresponds well with the finding that only these three subtypes bind to the unusual trimethylated caps (containing 2,2,7-trimethylguanosine 5′ triphosphate; $m_3^{2,2,7}GTP$) present on ∼70% of *C. elegans* mRNAs. Evidence in support of this hypothesis comes from recent studies of the sole eIF4E from the nematode *Ascaris suum*, an IFE-3–like protein. *A. suum* IFE-3 (also termed eIF4E-3) can bind to and stimulate the translation of mRNAs possessing mono- or trimethylated cap structures *in vitro* (Lall *et al.*, 2004). Furthermore, sequences identified from some nematodes such as the parasitic *Haemanchus contortus* suggest that they express a single form of eIF4E similar to IFE-3 and a single form related to IFE-1, -2, and -5.

2.3. Information derived from naturally occurring mutations

Clues to the function of different eIF4E family members can also be derived from the study of naturally occurring mutations in eIF4E genes. In plants, recessive mutations conferring viral resistance have been mapped to eIF4E and eIF(iso)4E genes. For example, the pepper pvr2 resistance gene against potato virus Y (PVY) corresponds to eIF4E (Ruffel *et al.*, 2002). Natural mutations in eIF4E genes from other plant species also display resistance to viral infections (Gao *et al.*, 2004; Nicaise *et al.*, 2003; Ruffel *et al.*, 2005). In all cases, mutations result in substitution of nonconserved amino acids near the cap-binding pocket or at the surface of the protein (based on tertiary structure models). Interestingly, whereas all natural mutations in plant eIF4E result in expression of a modified protein, the one reported for eIF(iso)4E results in the absence of a functional protein due to a premature stop codon after the first 29 amino acids (Ruffel *et al.*, 2006). Although the absence of an obvious mutant phenotype raises the possibility that plant eIF4E and eIF(iso)4E are functionally redundant, the changes in viral resistance suggest otherwise.

2.4. Tertiary structure modeling

The structural basis for the differences in cap-binding specificities for two *C. elegans* eIF4E family members was explored by molecular modeling of tertiary structure and amino acid (Miyoshi *et al.*, 2002). Tertiary structure models of IFE-3 and IFE-5 were created by homology modeling (Dwyer, 2001) based on the known atomic coordinates for mouse eIF4E-1. The width and depth of the cap-binding cavity were shown by molecular dynamics simulations to be larger in IFE-5 than in IFE-3. This supports a model in which IFE-3 discriminates against $m_3^{2,2,7}GTP$ by steric hindrance. Using site-directed mutagenesis, amino acid sequences of IFE-5 were substituted with homologous sequences from IFE-3. Two substitutions (N64Y/V65L) converted the cap specificity of IFE-5 to essentially that of IFE-3, as detected by affinity chromatography (see Section 3) and also narrowed the cap-binding cavity in molecular models. Computer modeling has also been used to show that the Class II *H. sapiens* 4EHP (4E-2A in Fig. 13.1) is very similar in tertiary structure to *H. sapiens* eIF4E-1 (Rom *et al.*, 1998), consistent with its ability to bind m⁷GTP. Similar computer modeling of *L. major* LeishIF4E-1 indicated that the protein lacks a region corresponding to the C-terminus of mouse eIF4E-1 and yeast eIF4E (Yoffe *et al.*, 2004).

3. Assessing Differential Cap-Binding Properties of eIF4E Family Members

The most obvious property to test when seeking differences among cap-binding proteins is the affinity for cap structures. This can be determined by a number of methods, among which fluorescence quenching and affinity chromatography are the most widely used. Since most mRNAs contain a canonical cap structure (m⁷GpppG), the place to start is with cap analogs of this type.

3.1. Fluorescence quenching

The most easily performed quantitative method for determining affinity for the cap is quenching of intrinsic Trp fluorescence in eIF4E upon binding of the cap. This biophysical measurement is carried out with a single purified protein and a single cap analog. It is therefore necessary to obtain the eIF4E family member in a pure form. Since all eIF4Es have similar molecular masses and amino acid compositions, biochemical purification and separation of family members can be difficult. Instead, expression of a recombinant protein is preferred (see Section 6.2.2). Recombinant *A. thaliana* nCBP, a Class II member (see Fig. 13.1), has been shown by fluorescence quenching to bind m⁷GTP with 5- to 20-fold higher affinity than eIF(iso)4F (Ruud *et al.*, 1998). Equilibrium binding measurements comparing recombinant *C. elegans* IFE-3, -4, and -5 revealed that IFE-5 formed specific complexes with both $m_3^{2,2,7}$G- and m⁷G-containing cap analogs, but IFE-3 and IFE-4 discriminated strongly in favor of m⁷G-containing analogs (Stachelska *et al.*, 2002) (Table 13.1).

It is also possible to measure the binding affinity of capped oligonucleotides to purified eIF4E by biophysical methods. This makes it possible to test the effect of both primary and secondary structure of the attached oligonucleotide on overall affinity to eIF4E. Capped oligonucleotides bind to mammalian eIF4E with an affinity that is up to 23-fold greater (depending on salt concentration and the nature of the oligonucleotide chain) than that of dinucleotide cap analogs, as determined by both equilibrium (Carberry *et al.*, 1992) and

Table 13.1 Binding affinities of *C. elegans* eIF4E family members to different cap analogs[a]

Cap analog	IFE-3	IFE-4	IFE-5
		$K_{AS}\,(M^{-1})^{b}$	
m^7GTP	5×10^6	4×10^6	7×10^6
m^7GpppG	3×10^6	2×10^6	3×10^6
$m_3^{2,2,7}GpppG$	$<6 \times 10^4$	$<6 \times 10^4$	2×10^6
$m_3^{2,2,7}GTP$	$<6 \times 10^4$	$<6 \times 10^4$	3×10^6

[a] The data are taken from Stachelska *et al.* (2002) with permission of the publisher.
[b] Recombinant *C. elegans* proteins were titrated with the indicated cap analogs at pH 7.2 and 20°. In each case, the equilibrium association constant, K_{AS}, was calculated from fluorescence quenching data. Estimated standard deviations are ±30%.

rapid kinetic (Slepenkov *et al.*, 2006) methods. This suggests that an interaction occurs between eIF4E and at least some of the nucleotide residues adjacent to the cap. Wheat eIF4E and eIF(iso)4E, present in their corresponding eIF4F and eIF(iso)4F complexes, were shown to have different preferences for structural features at the 5′ end of mRNA (Carberry and Goss, 1991). eIF(iso)4F prefers linear structures, whereas eIF4F can bind hairpin structures and is sensitive to their position in the mRNA sequence.

As noted in Section 2, *L. major* contains an unusual cap structure, termed cap-4, in which the first four nucleotide residues adjacent to the cap are methylated (Yoffe *et al.*, 2004). The first and fourth base moieties are methylated (6,6-dimethyadenine and 3-methyluracil) and all four ribose moieties are methylated at the 2′ O position: $m^7Gpppm_2^{6,6,2'}$ Apm$^{2'}$ Apm$^{2'}$ C$^{3,2'}$ U. The fluorescence quenching method was used to compare affinities of one *Leishmania* family member, LeishIF4E-1, and mouse eIF4E-1, both to canonical cap analogs and cap-4. Whereas LeishIF4E-1 bound m^7GTP and cap-4 with similar affinities, mouse eIF4E-1 discriminated against cap-4 by ~5-fold. Deletion of the 13 C-terminal amino acid residues from mouse eIF4E-1 diminished the difference between binding to m^7GTP and cap-4. As noted above, this region is not homologous between mouse eIF4E-1 and LeishIF4E-1, suggesting that it plays a role in cap-4 interaction and underscoring the utility of tertiary structure modeling.

3.2. Affinity chromatography

An alternative approach for characterizing the cap-binding properties of eIF4E family members is to measure retention on an affinity medium containing an immobilized cap. The major disadvantage of this method is that quantitative determination of binding affinity is not straightforward; the results tend to be an all-or-none assessment of whether the protein can bind to cap structures. The advantages, however, are numerous. It is not necessary to clone the cDNA for the protein of interest or express it in a soluble, recombinant form, which is sometimes difficult (see Section 6.2.2). It makes it possible to test binding in crude extracts in which any eIF4E-interacting proteins that modulate binding are present. The retention of all cap-binding proteins present in an extract can be observed in a single experiment, provided that there are methods for distinguishing among them, e.g., different mobilities on sodium dodecyl sulfate polyacrylamide gel electrophoresis (SDS–PAGE) or specific antibodies.

Affinity chromatography has been used to indicate cap-binding specificities of all five *C. elegans* family members (Jankowska-Anyszka *et al.*, 1998; Keiper *et al.*, 2000). m^7GTP-Sepharose can be obtained commercially (Amersham Biosciences), but $m_3^{2,2,7}GTP$-Sepharose must be synthesized (Jankowska-Anyszka *et al.*, 1998). Affinity chromatography of either crude extracts or as recombinant proteins shows that all five family members are retained on m^7GTP-Sepharose, but only IFE-1, -2, and -5 are retained on $m_3^{2,2,7}GTP$-Sepharose. The interpretation of these experiments is that *C. elegans* eIF4E family members differ in cap specificity: IFE-3 and -4 bind only m^7GTP, whereas IFE-1, -2, and -5 bind both m^7GTP and $m_3^{2,2,7}GTP$. This conclusion was confirmed by a competition assay in which either m^7GTP or $m_3^{2,2,7}GTP$ was used to compete for binding of *C. elegans* eIF4Es to the two types of affinity column. In a similar experiment, *A. suum* IFE-3, the only Class I eIF4E from this species (see Fig. 13.1), was retained on both m^7GTP-Sepharose and $m_3^{2,2,7}GTP$-Sepharose, and binding was completed by either m^7GTP or $m_3^{2,2,7}GTP$ (Lall *et al.*, 2004). By contrast, LeishIF4E-1 binds to m^7GTP-Sepharose but not $m_3^{2,2,7}GTP$-Sepharose (Yoffe *et al.*, 2004).

4. EXPRESSION OF EIF4E FAMILY MEMBERS

Expression of different eIF4E family members from a variety of organisms has been assessed throughout development and in different tissues. In most cases, eIF4E family members are differentially expressed. Expression can be measured at both the RNA and the protein levels. Below we describe the main techniques that have been used in the analysis of eIF4E expression with their corresponding advantages and disadvantages.

4.1. Determination of eIF4E mRNA levels

To measure RNA levels, Northern blotting, semiquantitative reverse transcriptase polymerase chain reaction (RT-PCR), or quantitative real-time RT-PCR can be used with either total or poly(A)-enriched RNA preparations.

When using Northern blotting, probes and hybridization conditions for each eIF4E family member should be sufficiently specific to avoid cross reaction. This method has been used for two of the plant eIF4E family members, eIF4E and eIF(iso)4E (Dinkova and Sanchez de Jimenez, 1999; Manjunath *et al.*, 1999; Rodriguez *et al.*, 1998), three from mouse, eIF4E-1, eIF4E-2, and eIF4E-3 (Joshi *et al.*, 2004), and all five IFEs from *C. elegans* (Amiri *et al.*, 2001) (T. D. Dinkova, unpublished observations). The results revealed differential expression for the mRNAs of each eIF4E family member. The members of Class III (e.g., mouse eIF4E-3) are the least abundant and may be difficult to detect by Northern blotting (Joshi *et al.*, 2004). For *C. elegans*, two of the five mRNAs, for IFE-4 and IFE-5, could not be detected by this technique. In addition, sometimes the probe against a specific eIF4E family member detects more than one mRNA species, making the results difficult to interpret (Joshi *et al.*, 2004). In general, Northern blotting is neither sufficiently sensitive nor quantitative to be very useful.

Semiquantitative RT-PCR has been used for zebrafish eIF4E-1A and eIF4E-1B (Robalino *et al.*, 2004) and for mammalian eIF4E-1, eIF4E-2, and eIF4E-3 (Joshi *et al.*, 2004). This method usually requires the use of an internal control, for example, a housekeeping gene. Sometimes it is difficult to find a good standard for all tissues or developmental stages. However, for eIF4E, the most appropriate control is probably the translational elongation factor EF1A since the aim is generally to see how the level of a particular eIF4E family member compares with those of other components of the translation machinery. The method is useful to score important differences but is not sensitive to small changes. The key is to use a low cycle number, but it is almost impossible to be sure that all samples are still in the exponential versus plateau phase of the amplification reaction. Competitive RT-PCR is an option, but it involves additional adjustments. Quantitative real-time PCR (QRT-PCR) provides the most information. It is quantitative and very sensitive, but it also requires careful primer design with appropriate programs (e.g., the Beacon Designer Tool from Bio-Rad) as well as the generation of *in vitro* transcripts for each eIF4E family member to be used for quantitation.

In small organisms for which specific tissues are difficult to isolate, or when specific cell type localization is needed, *in situ* hybridization can be used to describe eIF4E mRNA distributions. This has been done for IFE-1 in *C. elegans* (Amiri *et al.*, 2001). Also, making use of available microarray databases for a specific organism could help to identify the mRNA expression pattern of eIF4E family members. These are available for *C. elegans* (Trutschl *et al.*, 2005), mouse, and *Arabidopsis thaliana* (https://www.genevestigator.ethz.ch/). It is advisable to confirm the database expression pattern by Northern blotting, RT-PCR, or QRT-PCR.

The mRNA level of a given eIF4E family member does not necessarily correspond to its protein level and may change with changing situations, such as developmental stage (Dinkova and Sanchez de Jimenez, 1999; Duprat *et al.*, 2002). Northern blotting and RT-PCR experiments with total or poly(A)-enriched RNA do not provide information about translation of the mRNA. A better way to obtain insight about protein expression is to look at the polysomal distribution of the mRNA under investigation. Such an approach has been widely used for many organisms (Dinkova *et al.*, 2005; Kawaguchi and Bailey-Serres, 2005; Zong *et al.*, 1999). Polysomes are obtained by sucrose gradient sedimentation, with the specific gradient conditions depending on the organism and separation requirements. Polysomal fractions can be analyzed independently, pooled as light (two to five ribosomes bound per mRNA) and heavy (more than five ribosomes per mRNA) (Dinkova *et al.*, 2005), or pooled altogether (Zong *et al.*, 1999) prior to analysis for specific mRNAs. This topic is treated further in Section 7.

4.2. Determination of eIF4E protein levels

For protein levels of different eIF4E family members, Western blotting with specific antibodies can be used. Alternatively, a picture of the relative levels of expression of different eIF4E family members can be developed using reporter constructs [green fluorescent protein (GFP) fusions or other tagged versions] expressed behind the promoter for the eIF4E family member gene. Western blotting is dependent on antibody availability. Some antibodies may react with multiple eIF4E family members, in which case it may be necessary to seek a gel electrophoresis system that will separate different eIF4E family members, e.g., by vertical slap isoelectric focusing (VSIEF) or SDS–PAGE (Robalino *et al.*, 2004). In general, antibodies are most likely to work within a class and less likely to work between classes. Antibodies raised to human eIF4E-1 have been shown to interact with zebrafish eIF4E-1A and -1B, but not with human eIF4E-2 (4EHP) or zebrafish eIF4E-2 (4EHP). Similarly, antibodies raised against wheat eIF4E and eIF(iso)4E (or the corresponding 4F and iso4F complexes) work for *Arabidopsis*, maize, rice, tobacco, and other plant species, but nCBP, a Class II family member, is not recognized by these antibodies. Usually eIF4E family members have similar molecular masses, making separation difficult. However, zebrafish eIF4E-1A and -1B (24,848 and 24,742 Da, respectively) can be resolved on 17.5% high-Tris gels using extended running times (1600 V-h) (Robalino *et al.*, 2004). VSIEF also provides a reliable separation not only of different eIF4E family members but also of differently modified forms of the same eIF4E family member (such as different phosphorylation states) (Manjunath *et al.*, 1999; Robalino *et al.*, 2004).

The five IFE proteins in *C. elegans* can be distinguished with antibodies against peptides corresponding to dissimilar C-terminal sequences (Jankowska-Anyszka *et al.*, 1998; Keiper *et al.*, 2000). These antibodies work well with the recombinant proteins. However, detection of IFE-4 and IFE-5 in tissue extracts, even if enriched by m^7GTP- Sepharose chromatography, is difficult (Keiper *et al.*, 2000), making Western blotting unsuitable for developmental expression studies of these two family members. Instead, GFP fusions have been primarily used (see below).

Expression of reporter constructs encoding GFP or other tagged fusions can be especially useful when low levels of the transcript do not allow detection of the mRNA, suitable antibodies are not available, or an eIF4E family member is present at very low levels. In addition, this approach provides a powerful tool for microscopy and subcellular localization of a protein. GFP fusions have been used for all five *C. elegans* IFE proteins. In this technique, the promoter for each eIF4E family member gene along with the coding sequence is fused to the cDNA for GFP and the constructs microinjected into worms. If a promoter has not been verified experimentally, it is still possible to take a certain amount of DNA sequence upstream of the AUG (this depends on the organism; 1000 to 1500 bp works for *C. elegans*) (Dinkova *et al.*, 2005). If the GFP fusion protein is not functionally equivalent to the native protein, this could negatively influence the developmental process. Using this approach, it has been shown that IFE-1::GFP is specifically expressed in germline cells and located in P-granules (Amiri *et al.*, 2001), whereas IFE-4::GFP is expressed in muscle and neurons (Dinkova *et al.*, 2005). It is also possible to use GFP fusions to display the differential developmental expression pattern beginning from late embryos to adults. The developmental expression patterns differ for all five *C. elegans* IFEs (Amiri *et al.*, 2001; Trutschl *et al.*, 2005; B. D. Keiper and R. E. Rhoads, unpublished).

4.3. Altering expression of eIF4E family members

Clues to the function of eIF4E family members can also be derived from specific gene deletions or reductions. However, deletion of a particular gene may or may not display an obvious phenotype. Reduction of the levels of Class I members that function as initiation factors, which includes the canonical eIF4Es, is expected to have drastic consequences on global cellular translation levels and hence to display striking phenotypes. Such has been the case of *C. elegans* IFE-3 knockdown by RNA interference (RNAi) (Keiper *et al.*, 2000). However, for members of noncanonical eIF4E classes, their deletion or reduction may not correlate with an obvious phenotype.

Whenever possible, it is preferable to work with a null mutation, since technical knockouts using agents such as RNAi are not always equally efficient in all tissues or organisms (Combe *et al.*, 2005; Dinkova *et al.*, 2005). In *C. elegans*, there are differences in phenotype between the knockouts and RNAi-treated animals. The null mutation proved very useful for investigating *C. elegans* IFE-4 (Class II/4EHP). Initially IFE-4 was considered rare and of little importance, and no phenotype was apparent from RNAi knockdowns (Keiper *et al.*, 2000). However, the availability of a knockout strain for IFE-4 revealed a subtle phenotype characterized by defective egg laying and food sensing (Dinkova et al., 2005). Mutation of the orthologous gene in *Drosophila* (eIF4E-8/4EHP) also showed reduced embryo hatching and patterning defects (Cho et al., 2005).

Knockout of other genes can also change the specific expression pattern of an eIF4E family member and thereby provide suggestions of function. The *C. elegans* germline-deficient strain *glp-4* was used to show that IFE-1 is specifically expressed in the germline (Amiri et al., 2001). IFE-3 and IFE-5 were also shown by Western blotting to be enriched in germline, but not IFE-2 and IFE-4. RNAi against *ife-1* does not cause embryonic lethality (Keiper et al., 2000), but a closer look at the F1 progeny of RNAi-treated worms reveals temperature-sensitive sterility and reduced brood size in fertile animals (Amiri et al., 2001). This phenotype is the result of defective spermatogenesis in the absence of IFE-1. Therefore, the specific expression and location of the IFE-1 protein are required for proper germline development, although its mechanism of action is not known at present.

There are mutants generated by gene interruption through T-DNA insertion for each eIF4E family member in *A. thaliana* (mutant stock center at Sainsbury Laboratory, http://www.arabidopsis.org). If the insertion is at the beginning of the coding sequence, the protein is usually absent. Null mutants for plant eIF(iso)4E have no obvious phenotype, but in such organisms, the eIF4E family member is overexpressed (Duprat et al., 2002). As mentioned in Section 2, natural mutations within plant eIF4E family member genes have been identified by viral resistance. By this method, correlations between either eIF4E or eIF(iso)4E and specific viral resistance have been found (Table 13.2). A virus may require eIF4E in one plant species and eIF(iso)4E in another for infection, or it may require both family members. Such characteristics have recently found plausible explanations based on interaction between eIF4E and the VPg, a protein linked to the 5' end of the viral RNA (see Section 6). Since both eIF4E and eIF(iso)4E belong to Class I, it has been proposed that they perform redundant functions in plants. However, the fact that all plant species have both eIF4E and eIF(iso)4E, the specificity of eIF4E or eIF(iso)4E in viral resistance, and the emerging knowledge about eIF4E family members from other organisms indicate that more careful follow-up on mutant phenotypes should be performed to elucidate why plants have conserved both family members throughout evolution.

When eIF4E knockout mutants are not available, RNAi is a good option in some organisms to knock down individual family members specifically or a group of family members. The method is rapid and usually very efficient for organisms in which RNAi is systemic. However, it is necessary to take care in the design and delivery of double-stranded RNA (dsRNA). RNAi can reveal striking phenotypes, e.g., lethal versus viable, and is also useful for transient knockdown of an eIF4E family member's expression (Keiper et al., 2000). The latter feature was used to separate direct from indirect effects of an *ife-4* knockout strain (Dinkova et al., 2005).

Table 13.2 eIF4E family member mutations in different plant species that correlate with specific potyviral resistance

Virus	Host	eIF4E family member	Nature of alteration	Source
Turnip mosaic virus	*A. thaliana*	eIF(iso)4E	Knockout by T–DNA insertion	Duprat *et al.*, 2002
Tobacco etch virus	*A. thaliana*	eIF(iso)4E	Knockout by T–DNA insertion	Duprat *et al.*, 2002
	Capsicum spp.	eIF4E	Amino acids changed near the cap-binding pocket: V67; L67; D109	Ruffel *et al.*, 2002
Lettuce mosaic virus	*A. thaliana*	eIF(iso)4E	Knockout by T–DNA insertion	Duprat *et al.*, 2002
	Lactuca spp.	eIF4E	Amino acids changed near the cap-binding pocket: A70; Q108–G109–A110 replaced by H	Nicaise *et al.*, 2003
Potato virus Y	*Lycopersicon* spp.	eIF4E	Amino acids changed near the cap-binding pocket: N68; A77; M108	Ruffel *et al.*, 2005
	Capsicum spp.	eIF4E	Amino acids changed near the cap-binding pocket: V67; L67; D109	Ruffel *et al.*, 2002
Pepper veinal mottle virus	*Capsicum* spp.	eIF4E	Amino acids changed near the cap-binding pocket: V67; L67; D109	Ruffel *et al.*, 2006
		eIF(iso)4E	Natural knockout by insertion of a stop codon after aa 51	
Pea seed borne mosaic virus	*Pisum sativum*	eIF4E	Amino acids changed near the cap-binding pocket: W62; D73; D74; R107	Gao *et al.*, 2004

Possible limitations of the method are the efficiency of RNAi and dsRNA delivery. In *C. elegans*, RNAi does not work efficiently in neurons, which led to knockdown of IFE-4 by RNAi in muscle but not in neurons (Dinkova *et al.*, 2005). To circumvent this, the RNAi can be directed to neurons by the use of a neuron-specific promoter. In plants, a knockdown of tobacco eIF4E, eIF(iso)4E, or both by antisense RNA caused either no phenotype or a mild one (Combe *et al.*; 2005). However, the efficiency of knockdown was low, making it difficult to draw conclusions.

RNAi, although a commonly used technique for gene silencing, is not the approach of choice for vertebrates, since RNAi can activate the interferon pathway as well as PKR, leading to spurious phenotypes (Marques and Williams, 2005). Even short hairpin RNAs (shRNAs), which are processed to give small interfering RNAs (siRNAs) of 21 nucleotides, have been reported to trigger an interferon response. In view of this, it is not surprising that RNAi has not proven satisfactory for use in zebrafish (Nasevicius and Ekker, 2000). However, functional depletion of several genes in developing zebrafish has been successful using morpholino oligonucleotides (Corey and Abrams, 2001). Morpholinos are nonionic DNA analogues with altered backbone linkages that still bind to complementary nucleic acids by Watson–Crick base pairing. When selected to target the translational start site, they can be used to block translation of mRNA. Not all morpholinos are successful in knocking down gene expression, but the use of multiple morpholinos can make them useful tools.

5. Assessing eIF4E Family Members in Translation Systems

Examining functional activities of eIF4E family members at the level of interaction with the 5'-cap structure of mRNAs and with specific mRNA sequences is outlined in Section 3. Function can also be assessed by looking at their activities in translation, either *in vivo* in a yeast complementation system, or *in vitro* in a cell-free translation system.

5.1. Function by complementation in yeast

The most informative single assay to determine whether an eIF4E family member acts as a functional translational factor is the rescue of the lethal disruption of the sole *S. cerevisiae* eIF4E gene (*cdc33*). To accomplish this, an eIF4E family member must be able to interact with *S. cerevisiae* mRNA as well as with its eIF4G. This approach has exploited the finding that the evolutionarily distant mammalian eIF4E can rescue growth of *S. cerevisiae* lacking functional eIF4E (Altmann *et al.*, 1989). The early "eIF4E knockout-and-rescue" systems employed auxotrophic markers, such as that of the *Leu2* gene, for the eIF4E gene replacement in either diploid or haploid yeast strains that previously lacked this gene. Use of diploid strains to assess the function of an untested eIF4E is technically difficult and time consuming, requiring specialized microscopic equipment for, and expertise in, the isolation, separation, and analysis of haploid spores (tetrad analysis). However, an *S. cerevisiae* strain, JOS003, is available in which the yeast gene has been replaced with a G418-resistance cassette and that contains a vector carrying human eIF4E-1 cDNA under the control of the glucose-repressible, galactose-dependent *GAL1* promoter (Joshi *et al.*, 2002). A simple glucose-based selection is used to deplete the strain of a human eIF4E substitute and to assess the functionality of an untested eIF4E family member, provided its expression is controlled by a glucose–insensitive promoter, using media selection techniques akin to those used routinely to culture and maintain bacterial stocks. This strain has been used as a tool to assess the ability of zebrafish (Robalino *et al.*, 2004) and mammalian (Joshi *et al.*, 2004; McKendrick *et al.*, 2001) eIF4E family members to rescue growth. Complementation assays have also demonstrated that five of the *Drosophila* Class I eIF4Es function as translation factors, eIF4E-1 (a), -2 (eIF4E-1a-related), -3 (d), -4 (b), and -7 (e), consistent with their abilities to interact with eIF4G and m^7GTP-Sepharose (Hernandez *et al.*, 2005).

   While a demonstration of complementation in yeast gives a clear indication that an eIF4E family member functions as a translation factor, a negative result may arise from multiple causes. Failure to complement could signify that the eIF4E in question does not function as a translation factor. For instance, the Class II eIF4Es from mammals (eIF4E-2/4EHP) and *Drosophila melanogaster* (eIF4E-8) do not complement *S. cerevisiae* and do not function as translation factors, consistent with their inability to interact with eIF4G (Hernandez *et al.*, 2005; Joshi *et al.*, 2004). Conversely, failure to complement could indicate that the eIF4E under investigation is too evolutionarily distant to interact with *S. cerevisiae* eIF4G or can recognize only 5′-cap structures distinct from those found in yeast. For instance, two eIF4Es, termed eIF4E1 and eIF4E2, are found in the deeply rooted protist *Giardia lamblia* (Li and Wang, 2005) and fall under the classification of atypical eIF4Es (Joshi *et al.*, 2005). Of the two, eIF4E2 has been shown to be essential in *G. lamblia* and binds to m^7GTP-Sepharose, suggesting that it functions in protein synthesis. The other, eIF4E1, is not essential and binds only to $m_3^{2,2,7}$GTP-Sepharose. However, neither homologue can rescue yeast eIF4E in a complementation assay.

   Expression of an uncharacterized eIF4E family member in *S. cerevisiae* can provide additional clues. For instance, if expression reduces growth in the presence of galactose (when human eIF4E-1 is expressed), this could indicate an inhibitory function. In such a situation, an *S. cerevisiae* strain lacking an endogenous eIF4E gene can be constructed in which the human eIF4E-1 is expressed from a glucose-insensitive promoter and the putative inhibitory eIF4E family member is expressed under control of the *GAL1* promoter. Removal of glucose from medium containing both glucose and galactose would allow a determination of the effects of increasing levels of the putative inhibitory form on protein synthetic activity and growth.

5.2. Function in cell-free translation systems

Assessment of the function of eIF4E family members in cell-free translation systems has so far received little attention for a variety of reasons. Most important is the fact that eukaryotic cell-free translation systems are commercially available only from rabbit reticulocytes and wheat germ, although useful systems have been developed for other species including the nematode *A. suum* (Lall et al., 2004) and the sea urchin *Strongylocentrotus purpuratus* (Jagus et al., 1992, 1993). Another complication is that cell-free translation systems already have their own eIF4E. However, it is possible to develop eIF4E-depleted mRNA-dependent systems to demonstrate the ability of an eIF4E family member to functionally replace the endogenous eIF4E, as well as use cell-free systems to identify forms of eIF4E that inhibit or support translation of specific mRNAs.

5.2.1. Recovery of translation in eIF4E-depleted mRNA-dependent translation systems

It is possible to assess whether an uncharacterized eIF4E family member can functionally replace rabbit eIF4E-1 and act as a translation initiation factor with the ability to stimulate cap-dependent translation. eIF4E-depleted cell-free translation systems can be produced by passing the cell extract over m^7GTP-Sepharose to remove eIF4E. The complication is that eIF4E is tethered to eIF4G, eIF3, and numerous other factors (Keiper et al., 1999), and consequently many other initiation factors are also removed by such treatments. One solution to this problem has been to pass wheat germ extracts over m^7GTP-Sepharose and then add back the missing factors as recombinant proteins (Gallie and Browning, 2001). This created a system that is highly dependent on added wheat eIF4F (eIF4E in complex with eIF4G) or wheat eIF(iso)4F [eIF(iso)4E in complex with eIF(iso)4G]. The investigators found that both eIF4F and eIF(iso)4F stimulated translation of mRNAs containing little 5′ UTR secondary structure, but eIF4F supported translation of an mRNA containing 5′-proximal secondary structure substantially better than did eIF(iso)4F. This correlates with the direct binding results described in Section 3. A caveat for this study is that it was not possible to separate the effects due to eIF4E from those due to eIF4G. This is particularly relevant since eIF4A bound to eIF4G acts as the RNA helicase that is necessary to unwind 5′ UTR secondary structure, whereas a role for eIF4E per se in unwinding has not been shown. A wheat germ translation system made dependent on eIF4E and eIF4G has also been used to demonstrate that recombinant *A. thaliana* nCBP is capable of stimulating translation, but only half as well as eIF(iso)4E (Ruud *et al.*, 1998).

An eIF4E-dependent reticulocyte translation system can be prepared in which endogenous rabbit eIF4E-1 is depleted by the addition of 4E-BP1 (displacing eIF4G from the eIF4E·eIF4G complex) followed by removal of the 4E-BP1·eIF4E-1 complex by m^{7}GTP-Sepharose chromatography (McKendrick *et al.*, 2001). This technique removes eIF4E but only negligible amounts of eIF4G. Depletion of rabbit eIF4E-1 reduces translation of a capped luciferase reporter mRNA to 20% of control values, and translation can be restored by the addition of 1.4 μM recombinant eIF4E-1. Similar stimulation of translation by an uncharacterized eIF4E family member would indicate that it functions as a bona fide initiation factor. The addition of up to 100-fold more recombinant human eIF4E-1 or zebrafish eIF4E-1A than rabbit eIF4E-1 present in a reticulocyte lysate has no inhibitory effect on translation in this system, suggesting that addition of other recombinant eIF4Es should pose no problem (J. Robalino, B. Joshi, and R. Jagus, unpublished). Using appropriate luciferase reporter mRNAs, such a system could also be used to assess the ability of a particular eIF4E family member to stimulate translation of mRNAs containing different types of caps, "strong" or "weak" 5′ UTRs, poly(A) tracts, etc.

5.2.2. Function of eIF4E family members in an mRNA-dependent reticulocyte translation system *not* depleted of eIF4E

There are several ways in which an eIF4E family member could function as a competitive inhibitor of translation by mimicking only some of the activities of eIF4E-1. If titration of an eIF4E family member into an unmodified reticulocyte translation system causes an inhibition of translation, the result may indicate that the protein competes with eIF4E-1 for binding to translation components. Inhibition can also be assessed and verified in the eIF4E-depleted system described above: suboptimal amounts of recombinant human eIF4E-1 can be added to reconstitute the eIF4E-dependent translation system, and the eIF4E family members being investigated can be added to test for inhibition of translation. Recovery of inhibition by addition of more eIF4E-1 would indicate competitive inhibition, whereas failure to recover may indicate the presence of a noncompetitive inhibitor. If an eIF4E family member binds only to the mRNA cap and not to eIF4G, or only to eIF4G, it could inhibit cap-dependent translation but make eIF4G available to promote internal ribosome entry site (IRES)-driven translation. Inhibition of both cap-dependent and IRES-driven translation might indicate that the family member binds eIF4G and renders it nonfunctional. Both scenarios are testable using a bicistronic reporter mRNA with *Renilla* luciferase as the first cistron and firefly luciferase as the second cistron behind a suitable IRES. Testing to see what purified factors overcome any inhibition should provide clues to the mechanism of inhibition.

6. PROTEIN–PROTEIN INTERACTION ASSAYS AS A MEANS TO DIFFERENTIATE FUNCTIONS OF EIF4E FAMILY MEMBERS

There are many ways to look at protein–protein interactions involving eIF4E family members. Feasible experimental strategies will vary with the model organism under study. While the results of such studies will allow functional characterization of an eIF4E family member, they may not illuminate the role of an eIF4E in that organism (Joshi *et al.*, 2004; Robalino *et al.*, 2004). There are many ways to look at interactions, and it is highly advisable to confirm interactions with multiple alternative methods.

6.1. Likely candidates

Except for the known exception of *S. cerevisiae*, cap-dependent translation depends on the interaction of cap-bound eIF4E with ribosomal-subunit-associated eIF4G, making eIF4E–eIF4G interaction a hallmark of prototypical Class I eIF4Es. An increasing number of proteins, generically known as eIF4E-inhibitory proteins, modulates the eIF4G–eIF4E-1 interaction (Richter and Sonenberg, 2005). The core portion of eIF4G that interacts with eIF4E is small, probably less than 15 amino acid residues. Several other proteins contain similar peptide motifs, and it is this region that competes with eIF4G for binding to eIF4E. This canonical eIF4E-binding motif is YXXXXLϕ (where ϕ is any hydrophobic amino acid). The best characterized of these are the 4E-BPs, which compete with eIF4G for binding to Class I eIF4Es and are therefore general inhibitors of cap-dependent translation. Although found in most metazoa, the 4E-BPs are present in only a selection of protists and have not been found in plants, nematodes, or most fungi. With regard to eIF4G and 4E-BP interaction, mammalian eIF4E-2 (4EHP) and eIF4E-3 each possesses a range of partial activities. For instance, mammalian eIF4E-2 does not interact with eIF4G but does interact with 4E-BPs, whereas mammalian eIF4E-3 interacts with eIF4G but not with 4E-BPs (Joshi *et al.*, 2004).

A compilation of eIF4E-interactve proteins is given in Table 13.3. In addition to the 4E-BPs, an increasing number of eIF4E-binding proteins interacts with Class I eIF4Es only on specific mRNAs and does so through affiliations with specific mRNA-binding proteins to give translational repression (Richter and Sonenberg, 2005). These include proteins such as Maskin and Cup, which disrupt eIF4E–eIF4G interaction for mRNAs

Table 13.3 Proteins reported to interact with eIF4E

Protein	Organism	eIF4E partner	Other partners	Binding motif[a]	Function	References
4EBPs	Some protists, a few fungi, Metazoa	eIF4E-1 eIF4E-2 (4EHP)		YDRKFLM (Mm) YDRKFLL (Mm) YER-AFML (Dm)	Prevents eIF4E-1/eIF4G interaction reducing translation of all mRNAs	Gingras et al., 1999
p20	S. cerevisiae	eIF4E-1		YSMNEL	Prevents eIF4E-1/eIF4G interaction	Altmann et al., 1997
Eap1p	S. cerevisiae	eIF4E-1		YTIDEL	Prevents eIF4E-1/eIF4G interaction	Cosentino et al., 2000
Maskin	Vertebrates	eIF4E-1	CPEB	TEADFLL (Xl)	Translational repression of CPE mRNAs	Richter and Sonenberg, 2005
Cup	Drosophila	eIF4E-1	Bruno, Smaug	YTRSR LM	Translational repression of oskar and nanos mRNA	Nakamura et al., 2004
Bicoid	Drosophila	eIF4E-8	Brat	YXXXXXL	Translational repression of caudal and hunchback mRNAs	Cho et al., 2006
4E-T	Vertebrates	eIF4E-1		YXXXXLϕ	Nucleocytoplasmic shuttling protein	Dostie et al., 2000
Gemin5	H. sapiens	eIF4E-1		YEAVELL LKLPFLK	Found in P-bodies. Involvement in mRNA recruitment?	Fierro-Monti et al., 2006
PRH	Vertebrates	eIF4E-1	PML	YXXXXLϕ	Inhibits eIF4E-1 function in mRNA transport	Topisirovic et al., 2003
AtLOX2	A. thaliana			YRKEEL	Unknown	Freire et al., 2000

BTF3	*A. thaliana*	eIF4E eIF(iso)4E nCBP	Nascent polypeptide-associated complex	STLKRI	Unknown	Freire, 2005
VPg	Potyviruses, norovirus, calcivirus	eIF4E eIF(iso)4E		Cap-binding pocket	Inhibits host translation by inhibition of cap–eIF4E interaction	Goodfellow et al., 2005; Grzela et al., 2006; Miyoshi et al., 2006
PML	Vertebrates	eIF4E-1	PRH	RING domain	Decreases cap-binding activity of eIF4E-1 *in vitro*	Cohen et al., 2001

[a] In the eIF4E-binding motifs, italicized residues are those shown to be essential for binding. Mm, *Mus musculus*; Dm, *D. melanogaster*; Xl, *Xenopus laevis*.

containing certain sequence elements in the 3′ UTR. In each case, the eIF4E interaction involves the YXXXXLφ motif of the translational repressor. More than 200 homeodomain proteins from *Drosophila* and vertebrates contain at least one of these potential eIF4E-binding sites including vertebrate PRH and PML (Topisirovic *et al.*, 2003). Thus, some homeodomain proteins are potential modulators of eIF4E. Recently a Class II eIF4E family member in *Drosophila*, 4EHP (eIF4E-8), was found as part of a similar inhibitory complex for Caudal (Cad), an mRNA also involved in antero-posterior embryo patterning (Cho *et al.*, 2005). The approach was based first on phenotypic observation of 4EHP mutants. It was found that 4EHP inhibits local translation of Cad mRNA by interacting with Bicoid (a 3′ UTR-binding protein). The mechanism appears similar to the one found for eIF4E-1 and Cup. Another likely candidate for interaction with eIF4E family members is the nucleocytoplasmic transporter protein 4E-T. It is currently unknown whether 4E-T interacts with eIF4E family members other than Class I.

New eIF4E-inhibitory proteins are still being uncovered, not all of which carry the YXXXXL motif. For instance, the VPgs of some potyviruses have been shown to interact with *A. thaliana* eIF4E and eIF(iso)4E with a binding site in or near the cap-binding pocket (Khan *et al.*, 2006; Michon *et al.*, 2006; Miyoshi *et al.*, 2006). Not surprisingly, searches for naturally occurring viral resistance have identified either eIF(iso)4E or eIF4E in different plant species as the mutated genes related to viral resistance (Table 13.2). Future investigations are likely to uncover new eIF4E-inhibitory proteins, some of which will interact differentially with eIF4E family members, by methods such as those described below.

6.2. "Pull-down" methods

A demonstration of the interaction between eIF4E family members and likely candidate proteins can include copurification using appropriate affinity columns such as m^7GTP-Sepharose or $m_3^{2,2,7}$GTP-Sepharose, coimmunoprecipitation if appropriate antibodies are available, "pull-downs" using tagged recombinant protein as bait or prey, and far-Western analyses. For any of these approaches, consideration should be paid to the source of "bait" or "prey," as well as to the analytical methods used to assess the interaction.

6.2.1. Source of "bait"

Since new eIF4E family members are usually identified as the predicted translation products of cDNAs, it is most likely that some form of recombinant eIF4E will be used. Production of recombinant proteins allows the inclusion of a variety of tags that can be used to purify the protein or track its interactions with other proteins. Production of recombinant eIF4E family members is most likely to be from *Escherichia coli*, using vectors such as the pET (EMD Biosciences) or pGEX (GE Healthcare) series, since any eukaryotic expression system will contain endogenous eIF4E family members as well as interacting proteins such as eIF4G or 4E-BPs. Alternatively, recombinant eIF4Es can be generated by *in vitro* expression using translation vectors such as pCITE (EMD Biosciences), which gives the added benefit of generating radiolabeled eIF4Es if desired.

6.2.2. The vagaries of recombinant protein production

It is never possible to predict what conditions will favor the production of properly folded, active eIF4Es. Recombinant eIF4E is found primarily in inclusion body pellets, requiring either the use of the small proportion found in the bacterial supernatant or solubilization and renaturation from the pellets. A suitable solubilization/renaturation protocol is likely to require trial and error. Solubilization in 6 M guanidine–HCl and 100 mM dithiothreitol (DTT) to give a protein concentration of not more than 1 mg/ml usually works well (Joshi *et al.*, 2004). Proteins can be renatured by staged dialyses or by rapid dilution (20-fold) into buffer containing 50 mM HEPES-KOH, pH 7.2, 200 mM NaCl (Rudolph *et al.*, 1997). If a glutathione-S-transferase (GST) vector is used, renaturation can be monitored by measuring GST activity spectrophotometrically (B. Joshi, unpublished method). This is particularly useful when making a recombinant eIF4E family member of unknown function, since a test of renaturation by function cannot be easily made. The renatured fusion proteins can be concentrated and purified by glutathione-Sepharose (APBiotech) affinity chromatography and stored in liquid N_2 until use. Since a bacterially produced eIF4E family member may not be completely active, it is useful to compare its activity with the same recombinant protein produced in an *in vitro* translation system (Joshi *et al.*, 2004; Robalino *et al.*, 2004).

6.2.3. Source of "prey"

If recombinant eIF4E "bait" is used, the ideal prey would be endogenous protein from tissue or cell extracts. This may work well for analyzing the partners in a model system such as human or mouse, for which there is easy or commercial access to antibodies, but works less well in systems for which fewer resources are available, such as zebrafish, *Xenopus*, sea urchin, or protists. The easiest strategy to monitor interactions between eIF4E family members and suspected prey proteins is to cotranslate mRNAs for S-tagged eIF4E family members and a potential prey protein such as eIF4G or the 4E-BPs. This is done using translation vector constructs such as pCITE in the presence of [^{35}S]methionine in a coupled reticulocyte transcription/translation system (Joshi *et al.*, 2004; Robalino *et al.*, 2004). ^{35}S-labeled prey proteins bound to S-tagged, ^{35}S-labeled, eIF4E family members can be recovered by binding to S-protein agarose (Novagen) and visualized by

SDS–PAGE and autoradiography. However, to be confident about results using only recombinant proteins, it is advisable to seek confirmation via additional methods. For instance, zebrafish eIF4E-1A but not eIF4E-1B interacts with zebrafish 4E-BP, as assessed by pull-down of *in vitro*-synthesized, ^{35}S-labeled 4E-BP with S-tagged, *in vitro*-synthesized, ^{35}S-labeled eIF4E-1A and -1B (Robalino *et al.*, 2004). This result was confirmed by demonstrating that endogenous eIF4E-1A but not eIF4E-1B from zebrafish ovary extracts could be pulled down with His-labeled 4E-BP (Robalino *et al.*, 2004).

When developing prey constructs for use in interaction studies, it is preferable to use cDNAs encoding the potential prey (eIF4G, BPs, 4E-T, Maskin, etc.) from the same or a closely related species. This may not be important between rabbit and human, or even among vertebrates, but it could be important for more distantly related species. For instance, human eIF4G-1 interacts well with zebrafish eIF4E-1A (Robalino *et al.*, 2004). On the other hand, *Drosophila* eIF4E-5(c) can interact with *Drosophila* eIF4G but not yeast eIF4G (Hernandez *et al.*, 2005). Finally, when looking for interactions of eIF4E family members with large proteins such as eIF4G, it is technically challenging to use the full-length proteins, although expression of the eIF4E-binding domain can be useful. A 455-amino acid fragment equivalent to amino acids 159 to 614 of human eIF4G-1, which includes the eIF4E-binding domain, interacts robustly with mouse eIF4E-1 and zebrafish eIF4E-1A (Joshi *et al.*, 2004; Robalino *et al.*, 2004). Similar studies have successfully used both a short eIF4E-binding peptide and an ~100 amino acid domain, although no careful comparisons of different sized fragments have been reported.

6.2.4. Yeast two-hybrid (Y2H) analysis

If partners for eIF4E family members are identified by pull-down assays or other methods described below, they can be confirmed by Y2H analysis. For instance, the interactions of VPgs from different potyviruses with either eIF(iso)4E or eIF4E have been demonstrated by pull-down assays of recombinant tagged proteins and competition for cap binding to m^{7}GTP-Sepharose or capped mRNA analogs, with the interactions then confirmed by Y2H. Conversely, there have been potential binding partners for eIF4E family members in plants uncovered by Y2H that have been partially confirmed in other assays.

6.3. Far-Western analyses

Because far-Western analysis employs a very concentrated membrane-bound substrate, it is possible that the sensitivity of this technique allows detection of weaker interactions. This has been shown to be the case for mouse eIF4E-2 (4EHP), which shows a robust interaction with 4E-BP2 and -BP3 in far-Western assays using GST-tagged eIF4E-2 as probe, and some interaction with 4E-BP1 (Joshi *et al.*, 2004). However, the interaction is much less obvious in pull-down experiments. When identifying protein–protein interactions by the far-Western technique, it is important to always include appropriate controls to distinguish true protein–protein interaction bands from nonspecific artifactual ones. For example, experiments involving detection with recombinant GST fusion proteins should be replicated with GST alone. A bait protein with an amino acid substitution in the predicted interaction domain could also be used as a control to determine specificity as could a nonrelevant protein. Ideally, the control protein would be of similar size and charge as the protein under investigation and would not interact nonspecifically with the bait protein.

6.4. Methods to uncover new protein partners

6.4.1. Y2H

The Y2H system has been used as an approach to find eIF(iso)4E-interacting proteins in *A. thaliana* (Freire, 2005; Freire *et al.*, 2000). Interestingly, the *in vivo* partner of eIF(iso)4E, eIF(iso)4G, was not found in the screen. Y2H indicated an interaction between lipoxygenase 2 (AtLOX2; mainly a chloroplast protein) and eIF(iso)4E and an interaction between BTF3, a component of the nascent polypeptide-associated complex, and eIF(iso)4E. The Y2H findings were confirmed by coimmunoprecipitation, copurification on m^7GTP-Sepharose, and reapplication of the Y2H system. Although initially found for eIF(iso)4E, *in vitro* interactions with these proteins were demonstrated for all three plant eIF4E family members. No *in vivo* significance of such interactions is known. In principle, the interaction of prey proteins with eIF4E could be mediated by yeast eIF4G, but this is unlikely for two reasons. First, the eIF4E under investigation would need to have high affinity for yeast eIF4G, which does not appear to be the case for human eIF4E (Joshi *et al.*, 2002). Second, the interaction between the bait and prey proteins needs to be close enough to bring together the DNA-binding and the activation domains; if the interaction were through yeast eIF4G, it would increase this distance.

6.4.2. Mass spectrometry

Proteins from cells or tissue extracts obtained by pull-downs using eIF4E family members as bait proteins can be identified by liquid chromatography and both MALDI and electrospray tandem mass spectrometric methods. Such analyses can only be usefully done for organisms for which a protein database has been established. An example of a novel eIF4E-1-interacting protein found in this way is Gemin5 (Fierro-Monti *et al.*, 2006). Gemin5 is present in human cell lines and binds human eIF4E-1 in a GST pull-down assay.

The protein contains the YXXXXLϕ characteristic of eIF4E-interacting proteins and colocalizes to P-bodies.

6.4.3. Clues from interactome databases

The availability of genome-scale sets of cloned open reading frames has facilitated systematic efforts at creating proteome-scale data sets of protein–protein interactions. These are represented as complex networks, or "interactome" maps. Currently, two experimental methodologies are used for generating genome-scale protein interaction maps: high-throughput yeast two-hybrid analysis (HT-Y2H) and analysis of protein complexes by affinity purification and mass spectrometry (AP-MS). Y2H, being a binary assay, captures direct protein–protein interactions, whereas AP-MS identifies components of stable complexes. Although far from complete, such maps are a useful resource to predict the function(s) of thousands of genes. These large-scale systematic surveys of protein–protein interactions are available for an increasing number of species, giving databases that can be mined for interacting partners of eIF4E family members. At present, comprehensive databases are available only for a handful of model organisms: *S. cerevisiae* [http://yeast-complexes.embl.de; http://tap.med.utoronto.edu (Legrain and Selig, 2000); *C. elegans* (http://vidal.dfci.harvard.edu/interactomedb/ i-View/interactomeCurrent.pl [Li *et al.*, 2004]); and *D. melanogaster* (http:// gifts.univ-mrs.fr/FlyNets/FlyNets; http://gifts.univ-mrs.fr/GIFTS_home_ page.html [Mohr *et al.*, 1998]). They are currently being developed for *H. sapiens* (http://www.himap.org:80/main/index.jsp) and *A. thaliana* (http://www.assocriomics.org/). So far, the worm and fly interactome maps each contains more than 5000 high-quality putative interactions, derived primarily from HT-Y2H screens. Deductions on the dynamic nature of interactions can be obtained when HT-Y2H and AP-MS are combined or when interaction data are supplemented with expression profiling data and phenotypic analyses (Ge *et al.*, 2003).

7. Global Microarray Studies of Polysomal mRNA Distribution

When no obvious phenotype is observable in a mutant strain for an eIF4E family member, and when there are no clues from specific protein interactions, global expression analysis of mutants becomes very useful. Microarrays are available for completely or partially sequenced organisms as well as cell lines and tissues. Such an approach is feasible for several organisms in which eIF4E family members have been found. Since eIF4E is a translation factor, the approach would initially consist of a search for translationally affected genes when a specific eIF4E family member is knocked out.

Such an analysis was undertaken in *C. elegans* to study the function of IFE-4, a Class II member (Dinkova *et al.*, 2005). Translationally active mRNAs were purified from "heavy" and "light" polysomal fractions, that is, having fast and slow sedimentation rates on a sucrose gradient, respectively. To assess the steady-state level of each mRNA, total RNA was also prepared from the same samples. Hybridization of each mRNA pool to Affymetrix microarrays containing the whole genome of *C. elegans* ($\sim$19,000 genes) revealed that a small subset of transcripts changed in polysomal distribution in the absence of IFE-4 (microarray results were compared between the Δ*ife-4* and wild-type strains). Some of the affected mRNAs correlated with specific phenotypic traits of the Δ*ife-4* worms.

In addition to the polysomal distribution changes, the microarray analysis showed that there were important changes in the steady-state levels of some mRNAs. The mechanisms underlying such regulation by IFE-4 in *C. elegans* remain unknown. As noted above, cap recognition by eIF4E plays a role not only in translation initiation but also in mRNA transport from the nucleus and mRNA degradation. Therefore, microarrays can be an important tool to find targets of eIF4E function at several levels of mRNA metabolism. However, conclusions from microarray data on gene expression should be further confirmed by other techniques such as Northern blotting, QRT-PCR, Western blotting, etc.

Microarray experiments yield a huge amount of information that is difficult to analyze with standard biochemical tools. In the past 5 years, a wealth of information has become available for transcriptomes of sequenced organisms such as *S. cerevisiae*, *C. elegans*, *D. melanogaster*, *M. musculus*, and *A. thaliana*. As described in Section 4 , such information can be used to find the mRNA expression pattern of different eIF4E family members within the life cycle or individual tissues in the organism by simply typing the gene ID entry of the gene of choice (e.g., https://www.genevestigator.ethz.ch/). In addition, microarray data obtained for mutants of any eIF4E family member could be analyzed using bioinformatics tools applied to available databases on expression profiles, polysomal profiles, or UTR sequences (Kawaguchi and Bailey-Serres, 2005; Trutschl *et al.*, 2005).

We combined information from public databases on expression profiles for *C. elegans* (http://cmgm.stanford.edu/~kimlab/dev/) with the results from our Δ*ife-4* mutant polysomal microarray experiments (http://genome.cs.lsus.edu/mRNA/PG2005/) to search for a relationship between developmental profile of expression and sensitivity of translation to IFE-4 (Trutschl *et al.*, 2005). We developed a method that utilized two algorithms for clustering datasets according to the expression profile during development and polysomal distribution in the Δ*ife-4* mutant. The outputs are linked using a two-dimensional color scale for visualization of any correlations. The result of this analysis indicated the mRNAs affected in polysome distribution by the loss of IFE-4 display a specific developmental expression pattern, similar to that observed for the IFE-4::GFP fusion protein. However, not all the mRNAs with such an expression pattern are affected by the absence of IFE-4. Therefore, additional characteristics of the mRNAs, besides being expressed with the same developmental pattern as IFE-4, are needed to make them dependent on this eIF4E family member for translation. The characteristics are yet to be discovered.

The basis for mRNA specificity by individual eIF4E family members is only beginning to be understood. As mentioned in Sections 3.1 and 6.1, eIF4E family members can have preferences for specific mRNA sequences in the 5' UTR or can bind proteins that recognize specific sequences in the 3' UTR. Therefore, it may be worthwhile to undertake bioinformatics approaches such as the one described above to explore the relationships between sequences in the UTRs and dependence of an mRNA's translation on a specific eIF4E family member. A limitation is the paucity of reliable UTR sequences for genes that are present in a given microarray. However, if relevant databases are available for the organism of choice, UTR sequences could be extracted with bioinformatics tools. For example, the affinity purification of capped mRNAs from *A. thaliana* has allowed the capture of 5' UTR sequences for over 14,000 full-length cDNAs (Kawaguchi and Bailey-Serres, 2005). These and other collections of high-quality cDNA sequence data are publicly available and provide a valuable resource for bioinformatics analysis of features in the 5' UTR, coding sequence, and 3' UTR that are relevant for translational regulation .

ACKNOWLEDGMENTS

The authors gratefully acknowledge the assistance of Dr. Bhavesh Joshi for help with Fig. 13.1. This work was supported by Grant MCB-0134013 from the National Science Foundation (to R. J.) and 2 R01-GM020818 from the National Institute of General Medical Sciences (to R. E. R.).

REFERENCES

Altmann, M., Müller, P. P., Pelletier, J., Sonenberg, N., and Trachsel, H. (1989). A mammalian translation initiation factor can substitute for its yeast homologue *in vivo*. *J. Biol. Chem.* **264,** 12145–12147.

Altmann, M., Schmitz, N., Berset, C., and Trachsel, H. (1997). A novel inhibitor of cap-dependent translation initiation in yeast: P20 competes with eIF4G for binding to eIF4E. *EMBO J.* **16,** 1114–1121.

Amiri, A., Keiper, B. D., Kawasaki, I., Fan, Y., Kohara, Y., Rhoads, R. E., and Strome, S. (2001). An isoform of eIF4E is a component of germ granules and is required for spermatogenesis in *C. elegans. Development* **128,** 3899–3912.

Carberry, S. E., and Goss, D. J. (1991). Wheat germ initiation factors 4F and (iso)4F interact differently with oligoribonucleotide analogues of rabbit a-globin mRNA. *Biochemistry* **30,** 4542–4545.

Carberry, S. E., Friedland, D. E., Rhoads, R. E., and Goss, D. J. (1992). Binding of protein synthesis initiation factor 4E to oligoribonucleotides: Effects of cap accessibility and secondary structure. *Biochemistry* **31,** 1427–1432.

Cho, P., Poulin, F., Cho-Park, Y., Cho-Park, I., Chicoine, J., Lasko, P., and Sonenberg, N. (2005). A new paradigm for translational control: Inhibition via 5'-3' mRNA tethering by Bicoid and the eIF4E cognate 4EHP. *Cell* **121,** 411–423.

Cho, P., Gamberi, C., Cho-Park, Y., Cho-Park, I., Lasko, P., and Sonenberg, N. (2006). Cap-dependent translational inhibition establishes two opposing morphogen gradients in *Drosophila* embryos. *Curr. Biol.* **16,** 2035–2041.

Clamp, M., Cuff, J., Searle, S. M., and Barton, G. J. (2004). The Jalview Java alignment editor. *Bioinformatics* **20,** 426–427.

Cohen, N., Sharma, M., Kentsis, A., Perez, J. M., Strudwick, S., and Borden, K. L. (2001). PML RING suppresses oncogenic transformation by reducing the affinity of eIF4E for mRNA. *EMBO J.* **20,** 4547–4559.

Combe, J., Petracek, M., van Eldik, G., Meulewaeter, F., and Twell, D. (2005). Translation initiation factors eIF4E and eIFiso4E are required for polysome formation and regulate plant growth in tobacco. *Plant Mol. Biol.* **57,** 749–760.

Corey, D., and Abrams, J. (2001). Morpholino antisense oligonucleotides: Tools for investigating vertebrate development. *Genome Biol.* **2,** reviews1015.1–1015.3.

Cosentino, G. P., Schmelzle, T., Haghighat, A., Helliwell, S. B., Hall, M. N., and Sonenberg, N. (2000). Eap1p, a Novel eukaryotic translation initiation factor 4E-associated protein in *Saccharomyces cerevisiae*. *Mol. Cell. Biol.* **20,** 4604–4613.

De Benedetti, A., and Graff, J. R. (2004). eIF-4E expression and its role in malignancies and metastases. *Oncogene* **23,** 3189–3199.

De Benedetti, A., and Rhoads, R. E. (1990). Overexpression of eukaryotic protein synthesis initiation factor 4E in HeLa cells results in aberrant growth and morphology. *Proc. Natl. Acad. Sci. USA* **87,** 8212–8216.

De Benedetti, A., Joshi-Barve, S., Rinker-Schaeffer, C., and Rhoads, R. E. (1991). Expression of antisense RNA against initiation factor eIF-4E mRNA in HeLa cells results in lengthened cell division times, diminished translation rates, and reduced levels of both eIF-4E and the p220 component of eIF-4F. *Mol. Cell. Biol.* **11,** 5435–5445.

DeFatta, R. J., Chervenak, R. P., and De Benedetti, A. (2002a). A cancer gene therapy approach through translational control of a suicide gene. *Cancer Gene Ther.* **9,** 505–512.

DeFatta, R. J., Li, Y., and De Benedetti, A. (2002b). Selective killing of cancer cells based on translational control of a suicide gene. *Cancer Gene Ther.* **9,** 573–578.

Dinkova, T. D., and Sanchez de Jimenez, E. (1999). Differential expression and regulation of translation initiation factors -4E and -iso4E during maize germination. *Physiol. Plant.* **107,** 419–425.

Dinkova, T. D., Keiper, B. D., Korneeva, N. L., Aamodt, E. J., and Rhoads, R. E. (2005). Translation of a small subset of *Caenorhabditis elegans* mRNAs is dependent on a specific eIF4E isoform. *Mol. Cell. Biol.* **25,** 100–113.

Dostie, J., Ferraiuolo, M., Pause, A., Adam, S. A., and Sonenberg, N. (2000). A novel shuttling protein, 4E-T, mediates the nuclear import of the mRNA 5' cap-binding protein, eIF4E. *EMBO J.* **19,** 3142–3156.

Duprat, A., Caranta, C., Revers, F., Menand, B., Browning, K., and Robaglia, C. (2002). The *Arabidopsis* eukaryotic initiation factor (iso)4E is dispensable for plant growth but required for susceptibility to potyviruses. *Plant J.* **32,** 927–934.

Dwyer, D. S. (2001). Model of the 3-D structure of the GLUT3 glucose transporter and molecular dynamics simulation of glucose transport. *Proteins* **42,** 531–541.

Dyer, J. R., Pepio, A. M., Yanow, S. K., and Sossin, W. S. (1998). Phosphorylation of eIF4E at a conserved serine in *Aplysia. J. Biol. Chem.* **273,** 29469–29474.

Ferraiuolo, M. A., Basak, S., Dostie, J., Murray, E. L., Schoenberg, D. R., and Sonenberg, N. (2005). A role for the eIF4E-binding protein 4E-T in P-body formation and mRNA decay. *J. Cell Biol.* **170,** 913–924.

Fierro-Monti, I., Mohammed, S., Matthiesen, R., Santoro, R., Burns, J., Williams, D., Proud, C., Kassem, M., Jensen, O., and Roepstorff, P. (2006). Quantitative proteomics identifies Gemin5, a scaffolding protein involved in ribonucleoprotein assembly, as a novel partner for eukaryotic initiation factor 4E. *J. Proteome Res.* **5,** 1367–1378.

Freire, M. (2005). Translation initiation factor (iso) 4E interacts with BTF3, the beta subunit of the nascent polypeptide-associated complex. *Gene* **345,** 271–277.

Freire, M., Tourneur, C., Granier, F., Camonis, J., El Amrani, A., Browning, K., and Robaglia, C. (2000). Plant lipoxygenase 2 is a translation initiation factor-4E-binding protein. *Plant Mol. Biol.* **44,** 129–140.

Gallie, D. R., and Browning, K. S. (2001). eIF4G functionally differs from eIFiso4G in promoting internal initiation, cap-independent translation, and translation of structured mRNAs. *J. Biol. Chem.* **276,** 36951–36960.

Gao, Z., Johansen, E., Eyers, S., Thomas, C., Noel Ellis, T., and Maule, A. (2004). The potyvirus recessive resistance gene, sbm1, identifies a novel role for translation initiation factor eIF4E in cell-to-cell trafficking. *Plant J.* **40,** 376–385.

Ge, H., Walhout, A., and Vidal, M. (2003). Integrating 'omic' information: A bridge between genomics and systems biology. *Trends Genet.* **19,** 551–560.

Gingras, A.-C., Raught, B., and Sonenberg, N. (1999). eIF4 initiation factors: Effectors of mRNA recruitment to ribosomes and regulators of translation. *Annu. Rev. Biochem.* **68,** 913–963.

Goodfellow, I., Chaudhry, Y., Gioldasi, I., Gerondopoulos, A., Natoni, A., Labrie, L., Laliberte, J., and Roberts, L. (2005). Calicivirus translation initiation requires an interaction between VPg and eIF 4 E. *EMBO Rep.* **6,** 968–972.

Görlich, D., and Mattaj, I. W. (1996). Nucleocytoplasmic transport. *Science* **271,** 1513–1518.

Graff, J. R., Boghaert, E. R., De Benedetti, A., Chan, S. K., and Zimmer, S. G. (1995). Reduction of the levels of initiation factor 4E mediates the malignant phenotype of *ras*-transformed rat fibroblasts. *Int. J. Cancer* **60,** 255–263.

Grzela, R., Strokovska, L., Andrieu, J., Dublet, B., Zagorski, W., and Chroboczek, J. (2006). Potyvirus terminal protein VPg, effector of host eukaryotic initiation factor eIF4E. *Biochimie* **88,** 887–896.

Hernandez, G., and Vazquez-Pianzola, P. (2005). Functional diversity of the eukaryotic translation initiation factors belonging to eIF4 families. *Mech. Dev.* **122,** 865–876.

Hernandez, G., Altmann, M., Sierra, J. M., Urlaub, H., del Corral, R. D., Schwartz, P., and Rivera-Pomar, R. (2005). Functional analysis of seven genes encoding eight translation initiation factor 4E (eIF4E) isoforms in *Drosophila. Mech. Dev.* **122,** 529–543.

Hsu, C., and Stevens, A. (1993). Yeast cells lacking 5′ → 3′ exoribonuclease 1 contain mRNA species that are poly(A) deficient and partially lack the 5′ cap structure. *Mol. Cell. Biol.* **13,** 4826–4835.

Jagus, R., Huang, W.-I., Hansen, L. J., and Wilson, M. A. (1992). Changes in rates of protein synthesis and eukaryotic initiation factor-4 inhibitory activity in cell-free translation systems of sea urchin eggs and early cleavage stage embryos. *J. Biol. Chem.* **267,** 15530–15536.

Jagus, R., Huang, W.-I., Hiremath, L. S., Stern, B. D., and Rhoads, R. E. (1993). Mechanism of action of developmentally regulated sea urchin inhibitor of eIF-4. *Dev. Genet.* **14,** 412–423.

Jankowska-Anyszka, M., Lamphear, B. J., Aamodt, E. J., Harrington, T., Darzynkiewicz, E., Stolarski, R., and Rhoads, R. E. (1998). Multiple isoforms of eukaryotic protein synthesis initiation factor 4E in *C. elegans* can distinguish between mono- and trimethylated mRNA cap structures. *J. Biol. Chem.* **273,** 10538–10542.

Joshi, B., Robalino, J., Schott, E. J., and Jagus, R. (2002). Yeast "knockout-and-rescue" system for identification of eIF4E-family members possessing eIF4E-activity. *Biotechniques* **33,** 392–398.

Joshi, B., Cameron, A., and Jagus, R. (2004). Characterization of mammalian eIF4E-family members. *Eur. J. Biochem.* **271,** 2189–2203.

Joshi, B., Lee, K., Maeder, D., and Jagus, R. (2005). Phylogenetic analysis of eIF4E-family members. *BMC Evol. Biol.* **5,** 48.

Kapp, L. D., and Lorsch, J. R. (2004). The molecular mechanics of eukaryotic translation. *Ann. Rev. Biochem.* **73,** 657–704.

Kawaguchi, R., and Bailey-Serres, J. (2005). mRNA sequence features that contribute to translational regulation in *Arabidopsis. Nucl. Acids Res.* **33,** 955–965.

Keiper, B. D., Gan, W., and Rhoads, R. E. (1999). Molecules in focus: Protein synthesis initiation factor 4G. *Int. J. Biochem. Cell Biol.* **31,** 37–41.

Keiper, B. D., Lamphear, B. J., Deshpande, A. M., Jankowska-Anyszka, M., Aamodt, E. J., Blumenthal, T., and Rhoads, R. E. (2000). Functional characterization of five eIF4E isoforms in *Caenorhabditis elegans. J. Biol. Chem.* **275,** 10590–10596.

Khan, M. A., Miyoshi, H., Ray, S., Natsuaki, T., Suehiro, N., and Goss, D. J. (2006). Interaction of genome-linked protein (VPg) of turnip mosaic virus with wheat germ translation initiation factors eIFiso4E and eIFiso4F. *J. Biol. Chem.* **281,** 28002–28010.

Lall, S., Friedman, C. C., Jankowska-Anyszka, M., Stepinski, J., Darzynkiewicz, E., and Davis, R. E. (2004). Contribution of *trans*-splicing, 5′-leader length, cap-poly(A) synergism, and initiation factors to nematode translation in an *Ascaris suum* embryo cell-free system. *J. Biol. Chem.* **279,** 45573–45585.

Lazaris-Karatzas, A., Montine, K. S., and Sonenberg, N. (1990). Malignant transformation by a eukaryotic initiation factor subunit that binds to mRNA 5′ cap. *Nature* **345,** 544–547.

Legrain, P., and Selig, L. (2000). Genome-wide protein interaction maps using two-hybrid systems. *FEBS Lett.* **480,** 32–36.

Li, L., and Wang, C. C. (2005). Identification in the ancient protist *Giardia lamblia* of two eukaryotic translation initiation factor 4E homologues with distinctive functions. *Eukaryot. Cell* **4,** 948–959.

Li, S., Armstrong, C. M., Bertin, N., Ge, H., Milstein, S., Boxem, M., Vidalain, P.-O., Han, J.-D. J., Chesneau, A., Hao, T., Goldberg, D. S., Li, N., *et al.* (2004). A map of the interactome network of the metazoan *C. elegans. Science* **303,** 540–543.

Manjunath, S., Williams, A., and Bailey-Serres, J. (1999). Oxygen deprivation stimulates Ca2+-mediated phosphorylation of mRNA cap-binding protein eIF4E in maize roots. *Plant J.* **19,** 21–30.

Marcotrigiano, J., Gingras, A.-C., Sonenberg, N., and Burley, S. K. (1997). Cocrystal structure of the messenger RNA 5′ cap-binding protein (eIF4E) bound to 7-methyl-GDP. *Cell* **89,** 951–961.

Marques, J., and Williams, B. (2005). Activation of the mammalian immune system by siRNAs. *Nat. Biotechnol.* **23,** 1399–1405.

Matsuo, H., Li, H., McGuire, A. M., Fletcher, C. M., Gingras, A.-C., Sonenberg, N., and Wagner, G. (1997). Structure of translation factor eIF4E bound to m^{7}GDP and interaction with 4E-binding protein. *Nature Struct. Biol.* **4,** 717–724.

McKendrick, L., Morley, S. J., Pain, V. M., Jagus, R., and Joshi, B. (2001). Phosphorylation of eukaryotic initiation factor 4E (eIF4E) at Ser209 is not required for protein synthesis *in vitro* and *in vivo. Eur. J. Biochem.* **268,** 5375–5385.

Michon, T., Estevez, Y., Walter, J., German-Retana, S., and Le Gall, O. (2006). The potyviral virus genome-linked protein VPg forms a ternary complex with the eukaryotic initiation factors eIF4E and eIF4G and reduces eIF4E affinity for a mRNA cap analogue. *FEBS J.* **273,** 1312–1322.

Miyoshi, H., Dwyer, D. S., Keiper, B. D., Jankowska-Anyszka, M., Darzynkiewicz, E., and
Rhoads, R. E. (2002). Discrimination between mono- and trimethylated cap structures
by two isoforms of *Caenorhabditis elegans* eIF4E. *EMBO J.* **21,** 1–11.

Miyoshi, H., Suehiro, N., Tomoo, K., Muto, S., Takahashi, T., Tsukamoto, T., Ohmori, T.,
and Natsuaki, T. (2006). Binding analyses for the interaction between plant virus
genome-linked protein (VPg) and plant translational initiation factors. *Biochimie* **88,**
329–340.

Mohr, E., Horn, F., Janody, F., Sanchez, C., Pillet, V., Bellon, B., Roder, L., and Jacq, B.
(1998). FlyNets and GIF-DB, two internet databases for molecular interactions in
Drosophila melanogaster. Nucleic Acids Res. **26,** 89–93.

Morales, J., Mulner-Lorillon, O., Cosson, B., Morin, E., Belle, R., Bradham, C.,
Beane, W., and Cormier, P. (2006). Translational control genes in the sea urchin
genome. *Dev. Biol.* **300,** 293–307.

Nakamura, A., Sato, K., and Hanyu-Nakamura, K. (2004). *Drosophila* cup is an eIF4E
binding protein that associates with Bruno and regulates oskar mRNA translation in
oogenesis. *Dev. Cell* **6,** 69–78.

Nasevicius, A., and Ekker, S. (2000). Effective targeted gene 'knockdown' in zebrafish. *Nat.
Genet.* **26,** 216–220.

Nathan, C., Carter, P., Liu, L., Li, B., Abreo, F., Tudor, A., Zimmer, S., and De
Benedetti, A. (1997). Elevated expression of eIF4E and FGF-2 isoforms during vascular-
ization of breast carcinomas. *Oncogene* **15,** 1087–1094.

Nicaise, V., German-Retana, S., Sanjuan, R., Dubrana, M.-P., Mazier, M.,
Maisonneuve, B., Candresse, T., Caranta, C., and LeGall, O. (2003). The eukaryotic
translation initiation factor 4E controls lettuce susceptibility to the potyvirus lettuce
mosaic virus. *Plant Physiol.* **132,** 1272–1282.

Niedzwiecka, A., Marcotrigiano, J., Stepinski, J., Jankowska-Anyszka, M., Wyslouch-
Cieszynska, A., Dadlez, M., Gingras, A.-C., Mak, P., Darzynkiewicz, E.,
Sonenberg, N., Burley, S. K., and Stolarski, R. (2002). Biophysical studies of eIF4E
cap-binding protein: Recognition of mRNA 5' cap structure and synthetic fragments of
eIF4G and 4E-BP1 proteins. *J. Mol. Biol.* **319,** 615–635.

Polunovsky, V. A., Rosenwald, I. B., Tan, A. T., White, J., Chiang, L., Sonenberg, N., and
Bitterman, P. B. (1996). Translational control of programmed cell death: Eukaryotic
translation initiation factor 4E blocks apoptosis in growth-factor-restricted fibroblasts
with physiologically expressed or deregulated Myc. *Mol. Cell. Biol.* **16,** 6573–6581.

Ramaswamy, S., Ross, K., Lander, E., and Golub, T. (2003). A molecular signature of
metastasis in primary solid tumors. *Nat. Genet.* **33,** 49–54.

Raoult, D., Audic, S., Robert, C., Abergel, C., Renesto, P., Ogata, H., La Scola, B.,
Suzan, M., and Claverie, J.-M. (2004). The 1.2-megabase genome sequence of mimivirus. *Science* **306,** 1344–1350.

Rhoads, R. E., Joshi-Barve, S., and Rinker-Schaeffer, C. (1993). Mechanism of action and
regulation of protein synthesis initiation factor 4E: Effects on mRNA discrimination,
cellular growth rate, and oncogenesis. *Prog. Nucl. Acid Res. Mol. Biol.* **46,** 183–219.

Richter, J., and Sonenberg, N. (2005). Regulation of cap-dependent translation by eIF4E
inhibitory proteins. *Nature* **433,** 477–480.

Rinker-Schaeffer, C. W., Graff, J. R., De Benedetti, A., Zimmer, S. G., and Rhoads, R. E.
(1993). Decreasing the level of translation initiation factor 4E with antisense RNA causes
reversal of *ras*-mediated transformation and tumorigenesis of cloned rat embryo fibroblasts. *Int. J. Cancer* **55,** 841–847.

Robalino, J., Joshi, B., Fahrenkrug, S. C., and Jagus, R. (2004). Two zebrafish eIF4E family
members are differentially expressed and functionally divergent. *J. Biol. Chem.* **279,**
10532–10541.

Rodriguez, C., Freire, M., Camilleri, C., and Robaglia, C. (1998). The *Arabidopsis thaliana* cDNAs coding for eIF4E and eIF(iso)4E are not functionally equivalent for yeast complementation and are differentially expressed during plant development. *Plant J.* **13,** 465–473.

Rom, E., Kim, H. C., Gingras, A.-C., Marcotrigiano, J., Favre, D., Olsen, H., Burley, S. K., and Sonenberg, N. (1998). Cloning and characterization of 4EHP, a novel mammalian eIF4E-related cap-binding protein. *J. Biol. Chem.* **273,** 13104–13109.

Rudolph, R., Bohm, G., Lilie, H., and Jaenicke, R. (1997). Folding proteins. *In* "Protein Function: A Practical Approach" (T. E. Creighton, ed.), pp. 57–100. Oxford University Press, Oxford.

Ruffel, S., Dussault, M., Palloix, A., Moury, B., Bendahmane, A., Robaglia, C., and Caranta, C. (2002). A natural recessive resistance gene against potato virus Y in pepper corresponds to the eukaryotic initiation factor 4E (eIF4E). *Plant J.* **32,** 1067–1075.

Ruffel, S., Gallois, J., Lesage, M., and Caranta, C. (2005). The recessive potyvirus resistance gene pot-1 is the tomato orthologue of the pepper pvr2-eIF4E gene. *Mol. Genet. Genomics* **274,** 346–353.

Ruffel, S., Gallois, J.-L., Moury, B., Robaglia, C., Palloix, A., and Caranta, C. (2006). Simultaneous mutations in translation initiation factors eIF4E and eIF(iso)4E are required to prevent pepper veinal mottle virus infection of pepper. *J. Gen. Virol.* **87,** 2089–2098.

Ruud, K. A., Kuhlow, C., Goss, D. J., and Browning, K. S. (1998). Identification and characterization of a novel cap-binding protein from *Arabidopsis thaliana*. *J. Biol. Chem.* **273,** 10325–10330.

Slepenkov, S. V., Darzynkiewicz, E., and Rhoads, R. E. (2006). Stopped-flow kinetic analysis of eIF4E and phosphorylated eIF4E binding to cap analogs and capped oligoribonucleotides: Evidence for a one-step binding mechanism. *J. Biol. Chem.* **281,** 14927–14938.

Stachelska, A., Wieczorek, Z., Ruszczynska, K., Stolarski, R., Pietrzak, M., Lamphear, B. J., Rhoads, R. E., Darzynkiewicz, E., and Jankowska-Anyszka, M. (2002). Interaction of three *Caenorhabditis elegans* isoforms of translation initiation factor eIF4E with mono- and trimethylated mRNA 5′ cap analogues. *Acta Biochim. Polon.* **49,** 671–682.

Strudwick, S., and Borden, K. L. (2002). The emerging roles of translation factor eIF4E in the nucleus. *Differentiation* **70,** 10–22.

Tomoo, K., Shen, X., Okabe, K., Nozoe, Y., Fukuhara, S., Morino, S., Sasaki, M., Taniguchi, T., Miyagawa, H., Kitamura, K., Miura, K., and Ishida, T. (2003). Structural feature of human factor 4E, studied by X-ray crystal analysis and molecular dynamics simulations. *J. Mol. Biol.* **328,** 365–383.

Topisirovic, I., Culjkovic, B., Cohen, N., Perez, J. M., Skrabanek, L., and Borden, K. L. B. (2003). The proline-rich homeodomain protein, PRH, is a tissue-specific inhibitor of eIF4E-dependent cyclin D1 mRNA transport and growth. *EMBO J.* **22,** 689–703.

Trutschl, M., Dinkova, T. D., and Rhoads, R. E. (2005). Application of machine learning and visualization of heterogeneous datasets to uncover relationships between translation and developmental stage expression of *C. elegans* mRNAs. *Physiol. Genom.* **21,** 264–273.

Vilela, C., Velasco, C., Ptushkina, M., and McCarthy, J. E. G. (2000). The eukaryotic mRNA decapping protein Dcp1 interacts physically and functionally with the eIF4F translation initiation complex. *EMBO J.* **19,** 4372–4382.

Yoffe, Y., Zuberek, J., Lewdorowicz, M., Zeira, Z., Keaser, C., Orr-Dahan, I., Jankowska-Anyszka, M., Stepinski, J., Darzynkiewicz, E., and Shapira, M. (2004). Cap-binding activity of an eIF4E homolog from *Leishmania*. *RNA* **10,** 1764–1775.

Zong, Q., Schummer, M., Hood, L., and Morris, D. R. (1999). Messenger RNA translation state: The second dimension of high-throughput expression screening. *Proc. Natl. Acad. Sci. USA* **96,** 10632–10636.

TETHERED FUNCTION ASSAYS: AN ADAPTABLE APPROACH TO STUDY RNA REGULATORY PROTEINS

Jeff Coller* *and* Marv Wickens[†]

Contents

* Center for RNA Molecular Biology, Case Western Reserve University, Cleveland, Ohio
[†] Department of Biochemistry, University of Wisconsin, Madison, Wisconsin

Methods in Enzymology, Volume 429
ISSN 0076-6879, DOI: 10.1016/S0076-6879(07)29014-7

Abstract

Proteins and protein complexes that regulate mRNA metabolism must possess two activities. They bind the mRNA, and then elicit some function, that is, regulate mRNA splicing, transport, localization, translation, or stability. These two activities can often reside in different proteins in a complex, or in different regions of a single polypeptide. Much can be learned about the function of the protein or complex once it is stripped of the constraints imposed by RNA binding. With this in mind, we developed a "tethered function" assay, in which the mRNA regulatory protein is brought to the 3′ UTR of an mRNA reporter through a heterologous RNA–protein interaction. In this manner, the functional activity of the protein can be studied independent of its intrinsic ability to recognize and bind to RNA. This simple assay has proven useful in dissecting numerous proteins involved in posttranscriptional regulation. We discuss the basic assay, consider technical issues, and present case studies that exemplify the strengths and limitations of the approach.

1. INTRODUCTION AND RATIONALE

In studying proteins that regulate mRNA metabolism, it often is useful to experimentally separate function from mRNA binding. In many instances, the natural mRNA target for a given protein is unknown; any assay of function must therefore be performed independent of the natural RNA–protein interaction. In addition, because posttranscriptional regulatory steps often are coupled, genetic analysis of functions *in vivo* can be complicated by indirect effects. Lastly, mutations in many critical RNA-binding proteins have pleiotropic effects on the cell and make it impossible to deduce which functions are direct. To circumvent these problems, we have developed a useful technique that allows the function of a protein to be analyzed, unconstrained by that protein's natural ability to interact with its mRNA target. We commonly refer to the technique as a "tethered function assay." The approach is adaptable and overcomes multiple complications in the study of mRNA-binding proteins.

In tethered function assays, the polypeptide of interest is tethered to a reporter mRNA through a heterologous RNA–protein interaction

(Fig. 14.1). Usually, the tethering site lies in the 3′ untranslated region (UTR) of the mRNA; this region is relatively unconstrained evolutionarily, and the natural site of action of many mRNA regulators. Tethered function assays have been used to show the role of proteins in control of mRNA transport, translation, localization, and stability (Coller and Wickens, 2002). Different reporters need to be used to assay each of these processes.

The tethered function assay takes advantage of the observation that many nucleic acid-binding proteins are modular. For example, many DNA transcription factors are bipartite, with separate DNA-binding and transcriptional activation domains (Hope and Struhl, 1986; Keegan *et al.*, 1986). Often the activities of these two domains are autonomous and separable; in other instances, they reside in distinct members of a multipolypeptide complex. RNA-binding proteins display similar modularity. The rationale of the tethered function approach is to examine solely the "functional" activity of an RNA-binding protein tethered artificially to an mRNA, circumventing the constraints imposed by natural RNA binding.

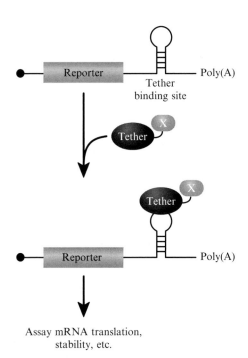

Figure 14.1 Tethered function assays using the 3′ UTR. A protein (X) is brought to a reporter mRNA through an artificial RNA–protein interaction (tether). In this example, the tethered binding site has been shown in the 3′ UTR of the reporter, but other locations have been used. The function of the tethered protein in any aspect of the mRNA's metabolism or function can then be assayed by conventional methodology.

In some cases, RNA binding and function may not be readily separable. For example, in nucleases and helicases, the nucleic acid-binding site is also the active site of the protein. Moreover, the interaction of a protein with its natural RNA-binding site can regulate the protein's activity; in these instances, it may be impossible to assay the function of the tethered protein in the absence of its cognate site.

2. THE BASIC DESIGN OF THE TETHERED FUNCTION ASSAY

The design of the tethered function assay is relatively straightforward. To determine the effects of a protein X on mRNA metabolism, a chimeric protein is expressed *in vivo* in which protein X is continuous with a tethering polypeptide (see Fig. 14.1). The tethering protein is an RNA-binding protein that recognizes an RNA tag sequence with high specificity and affinity. The effect of the fusion protein on mRNA metabolism is determined by coexpressing the chimera with an mRNA reporter (such as lacZ or luciferase) into which a tag RNA sequence has been embedded. The fusion protein's effects on mRNA metabolism are assayed by conventional means [i.e., Western blot, Northern blot, reverse transcriptase polymerase chain reaction (RT-PCR), etc.]. While the assay is relatively straightforward, several issues discussed in the following sections should be considered at the outset in designing a tethering experiment.

The assay, though powerful, is artificial. Only positive results are meaningful: lack of effects cannot be interpreted. Some RNA-binding proteins may require other proteins or their cognate RNA-binding sites to function, or be inactive as chimeras, or require appropriate positioning on the mRNA.

2.1. Position of the tethering site

A first consideration when designing a tethered function assay is the position in the mRNA of the tag sequence (i.e., the tethering site). While different laboratories have used tethered function assays and placed tag sequences within all regions of the mRNA, the most useful and common site is the $3'$ UTR (Coller and Wickens, 2002). The tethering of proteins to the $3'$ UTR has particular biological and experimental advantages. Importantly, many sites that regulate diverse steps in an mRNA's life, including its transport, cytoplasmic localization, stability, and translational activity, often reside in the $3'$ UTR. Thus, tethering to that region places regulators where they might well function. In addition, it is known that the exact location of several $3'$ UTR regulators is not critical for their function, implying that precise spatial positioning is not critical. Lastly, the 3′ UTR has fewer constraints than either the 5′ UTR (which can affect translational initiation frequency) or the open reading frame. The intercistronic region of bicistronic mRNAs also is relatively unconstrained and has been used for tethered function experiments using the same rationale (De Gregorio *et al.*, 1999, 2001; Furuyama and Bruzik, 2002; Shen and Green, 2006; Spellman *et al.*, 2005; Wang *et al.*, 2006).

3. THE TETHER

In choosing which protein to use as the tether, it is necessary to consider affinity and specificity for the RNA tag, subcellular localization, and impact of the tether on the activity of the test protein. The most common tether is the bacteriophage MS2 coat protein (Beach *et al.*, 1999; Bertrand *et al.*, 1998; Coller *et al.*, 1998; Collier *et al.*, 2005; Dickson *et al.*, 2001; Dugre-Brisson *et al.*, 2005; Gray *et al.*, 2000; Kim *et al.*, 2005; Long *et al.*, 2000; Lykke-Andersen *et al.*, 2000, 2001; Minshall and Standart, 2004; Minshall *et al.*, 2001; Ruiz-Echevarria and Peltz, 2000). However, the iron response element binding protein (IRP), a derivative of bacteriophage λ N-protein (De Gregorio *et al.*, 1999, 2001), and the spliceosomal U1A protein have been used successfully (Brodsky and Silver 2000; Finoux and Seraphin, 2006). In the following sections we will discuss each of these specific tethers and their merits and drawbacks.

3.1. The MS2 bacteriophage coat protein as a tether

The MS2 coat protein has been a popular choice for several reasons. First, this protein is relatively small (14 kDa), thus minimizing potential disruptions to the test protein. Second, the biochemistry of the MS2 coat's binding to its target sequence has been well established. Specifically, the MS2 coat is known to bind with high specificity and selectivity to a 21-nucleotide RNA stem–loop ($K_d = 1$ nM; Carey and Uhlenbeck, 1983). In addition, mutations in the binding site are available that increase or decrease affinity. In particular, the substitution of a single U within the stem–loop to a C increases affinity 50-fold over wild type (Lowary and Uhlenbeck, 1987). Moreover, use of MS2 allows a high dosage of tethered proteins to be present on the mRNA: the MS2 coat interacts with its target sequence as a dimer; thus for every stem–loop present in the mRNA reporter, two tethered proteins are present. Lastly, MS2 binds cooperatively to two stem–loops, further increasing the occupancy of sites (Witherell *et al.*, 1990). In some applications, the more protein that is bound, the better; each of these factors contribute to a strong signal in the functional assay.

On the other hand, the MS2 coat protein is not the simplest option when it is necessary to carefully control the number of tethered protein molecules bound. Since the MS2 coat protein binds as a dimer to a single site, and interacts with adjacent sites cooperatively, a large (and not trivial to determine) number of protein molecules may be bound to the targeted mRNA.

3.2. N-peptide as a tether

The bacteriophage λ N protein is often used in the tethered function assay (Baron-Benhamou et al., 2004). N-protein regulates bacterial transcriptional antitermination by binding to a 19-nucleotide RNA hairpin within early phage operons called boxB (Scharpf et al., 2000). Importantly, the N-peptide/boxB interaction occurs with high affinity ($K_d = 1.3$ nM). The particular advantage of the N-peptide in tethering assays is the result of its extremely small size; only 22 amino acids are required for the high affinity interaction with boxB RNA. Because of this, many laboratories have opted to use the N-peptide rather than MS2 coat protein, reasoning that it minimizes potential interference with the fusion protein's function (Baron-Benhamou et al., 2004). Another desirable feature of N-peptide is that unlike the MS2 coat, the protein binds 1:1 to its RNA target.

3.3. U1A protein and IRP as tethers

Both the U1A protein and IRP have been used successfully as tethers (De Gregorio et al., 1999; Finoux and Seraphin, 2006). U1A is a U1 small nuclear ribonucleoprotein (snRNP)-specific protein that binds with high specificity and affinity to a 30-nt RNA hairpin ($K_d = 5$ nM; van Gelder et al., 1993). IRP also binds to a 30-nt RNA hairpin that normally resides within the UTRs of target mRNAs with high affinities ($K_d = 90$ pM; Barton et al., 1990). Like N-peptide, the concentration of both U1A and IRP on the reporter mRNA is theoretically 1:1 (protein:RNA tag). Unlike N-peptide, however, both of these proteins are relatively large: 38 kDa for U1A and 97 kDa for IRP. As a result, they have not commonly been used in tethered function assays.

In general, the MS2 coat provides the highest concentration of tethered proteins to be bound to the reporter per binding site. This may allow phenotypes to be observed without greatly increasing the overall length of the mRNA reporter, an undesired situation in some applications. N-peptide, on the other hand, allows the delivery of a single tethered protein per binding site. The cost of this control of stoichiometry can be a need to introduce multiple tandem binding sites (more than four) in order to observe a robust phenotype (see below); the trade-off is an increase in reporter length. Nonetheless, the relative merits of MS2 coat protein, N-peptide, U1A, or IRP are situation specific. All have been successfully used to measure effects on mRNA translation, turnover, and transport. Direct comparisons between different tethers have not been made.

3.4. N-terminal or C-terminal fusions

The relative positions of the tethering protein and the protein of interest can be important. For example, in our own experience, tethering the MS2 coat protein to the N-terminus of the poly(A)-binding protein (PAB) resulted in much more activity than if the tether was located at the C-terminus (data not shown). This will have to be determined on a case-by-case basis; both orientations should be tested.

3.5. Trans-effects

A third important issue to consider is that the fusion protein may have *trans-acting* effects. Often, the tethered function assay is performed in a wild-type background with the endogenous copy of the test protein present. The presence of the tethering moiety may create a dominant negative allele that blocks the function of the normal protein *in vivo*, seriously complicating analysis. As a result, controls to ensure that any observed effects occur only in *cis* with respect to the mRNA reporter are important (see below).

4. THE REPORTER MRNA

The tethered function assay can be adapted to measure the effect of a tethered protein on many steps in mRNA metabolism and function. The adaptability comes mainly from the choice of reporter mRNA and the final assay performed. We will discuss only some of the reporters and assays that have been put into practice.

The choice of reporter mRNA obviously is dictated by the effect to be assayed. For example, translational activity can be measured in yeast using the LacZ, HIS3, and *CUP1* mRNAs, while in metazoans, luciferase, CAT, and epitope tags are most common (De Gregorio *et al.*, 1999, 2001; Gray *et al.*, 2000; Pillai *et al.*, 2004). In determining the effects of a tethered protein on mRNA stability, *MFA2, PGK1*, and *YAP1* have been used as reporter mRNAs in yeast, and β-globin and luciferase have been used in mammalian systems (Amrani *et al.*, 2006; Chou *et al.*, 2006; Coller *et al.*, 1998; Finoux and Seraphin, 2006; Kim *et al.*, 2005; Lykke-Andersen *et al.*, 2001, 2001; Ruiz-Echevarria and Peltz, 2000).

The intrinsic behavior of the reporter mRNA is an important consideration. To determine whether a tethered protein stabilizes an mRNA, the mRNA must be unstable in the absence of the protein; conversely, to determine whether a tethered protein destabilizes the mRNA, the

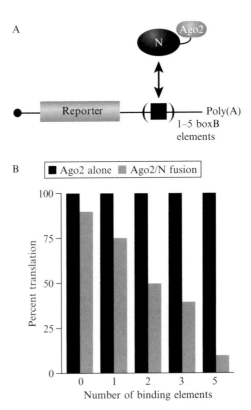

Figure 14.2 The number of tethered binding sites can influence phenotypic read-out. (A) Shown is the effect of increasing the number of tethered binding sites on translational repression mediated by tethered Ago2 (Pillai *et al.* 2004). Specifically, 0, 1, 2, 3, or 5 boxB elements were introduced into the 3′ UTR of a reporter gene expressing *Renilla* luciferase (RL). (B) The reporters were transfected into HeLa cells expressing either Ago2 (black bars) or an N-peptide Ago2 fusion (gray bars) and translation measured by enzymatic assay. As shown, increasing the number of tethered binding sites dramatically influences the repression observed.

mRNA reporter must be stable without the protein. The same reasoning applies to effects on other aspects of mRNA metabolism such as translation and subcellular localization.

4.1. The number and location of tethered binding sites

The number and location of tether binding sites are important variables. First, it should be decided where the tethered sites should be positioned, i.e., the 5′ UTR, 3′ UTR, or coding region. This depends on the suspected role of the protein in mRNA metabolism. For example, a protein thought to regulate polyadenylation might logically be placed in the 3′ UTR. It is important that the placement of the tethered binding sites not interfere on its own with the mRNA. For example, in testing the role of tethered PAB on mRNA stability, sites were placed in a region of the *MFA2 3'* UTR that was known not to affect the mRNA's half-life (Coller *et al.*, 1998; Muhlrad and Parker, 1992). Placement elsewhere would have dramatically altered the normal turnover rate of this message. It is helpful, therefore, to select as a reporter an mRNA whose *cis*-acting sequences are well characterized. Obviously these issues make it important that the behavior of the reporter mRNA with and without tethering sites be compared in the absence of the chimeric protein (see below, and Fig. 14.2).

A second issue in designing a reporter concerns the number of tethered binding sites. In many cases using the MS2 bacteriophage coat as the tether, two stem–loops have been sufficient to observe an effect (Coller *et al.*, 1998; Gray *et al.*, 2000; Minshall *et al.*, 2001; Ruiz-Echevarria and Peltz, 2000). However, many more sites have been used, ranging from 6 to 24 (Bertrand *et al.*, 1998; Fusco *et al.*, 2003; Lykke-Andersen *et al.*, 2000, 2001; Pillai *et al.*, 2004). The effect of the number of binding sites has been evaluated systematically in two reports (Lykke-Andersen *et al.*, 2000; Pillai *et al.*, 2004). Increasing the number of binding sites can increase the signal and enhance the assay's sensitivity. In Fig. 14.2, the extent of translational repression achieved by a tethered protein is proportional to the number of binding sites (Pillai *et al.*, 2004).

5. A Priori Considerations About the Logic of the Assay

5.1. Multiprotein complexes

mRNA regulatory events often occur through multiprotein complexes formed via protein–protein and protein–RNA interactions. In such cases, RNA binding may occur via one critical protein, which tethers the activity of another protein to the mRNA. Thus, the "active" protein may not directly contact the RNA. One strength of the tethered approach is its ability to assay the "activity" independent of RNA binding.

5.2. The role of RNA binding in function

The interaction between RNA and protein in some cases is essential for activity. RNA–protein interactions can change the conformation of the RNA, the protein, or both; not surprisingly, some complexes are biologically active, while the free RNAs or proteins are not (Williamson, 2000). Certain RNA ligands likely can influence activation or repression activity, much as in DNA-induced allosteric effects on transcription factors (Lefstin and Yamamoto, 1998; Scully *et al.*, 2000). In addition, the context of the natural binding site may be important for the protein's activity because essential factors are bound in the neighborhood.

These considerations have two implications. First, negative results in a tethered function assay are meaningless, even if the RNA and protein do interact on the reporter. Second, the outcome seen—for example, mRNA stabilization by a particular tethered protein—may differ when the protein is associated with its natural RNA-binding site. The same issues apply to DNA-binding transcription factor complexes, which have been powerfully dissected via comparable tethering approaches.

5.3. Analyzing function without knowing the target

In many cases, putative RNA-binding proteins have been identified, but their respective RNA target is unknown. One asset of the tethering approach is that a protein's activity can be determined without knowing the natural RNA target.

5.4. Analyzing the function of essential genes

In some cases, the RNA-binding protein under investigation is essential for cell viability; as a result, traditional genetic techniques are complicated by pleiotropic effects. The tethered function assay allows the function of the protein to be examined on just one mRNA species in an otherwise wild-type cell.

6. Important Controls

Several controls are critical in tethered function assays, and should always be performed (Fig. 14.3). It is necessary to ensure that (1) the tethered binding site does not affect the mRNA on its own, (2) the tethering protein alone (e.g., MS2 coat protein) does not have an impact, and (3) any observed effects should occur only in *cis* (that is, when the protein is bound to the mRNA). To control for possible *trans*-acting effects, the chimeric protein should be expressed alongside a reporter that lacks binding sites. This set of controls can ensure that an observed effect is specific to the protein of interest, and occurs only when it is associated with the mRNA in *cis* (see Fig. 14.3).

This concludes the general discussion of the design of a basic tethered function assay. In the following section we discuss a few specific examples with the aforementioned general principles considered. These case studies are not meant to be comprehensive of the literature but rather provide a sample of the uses of the tethered function assay to address certain biological issues. An overview is provided in Table 14.1.

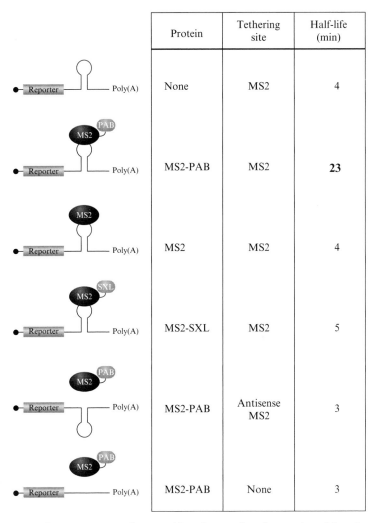

Protein	Tethering site	Half-life (min)
None	MS2	4
MS2-PAB	MS2	**23**
MS2	MS2	4
MS2-SXL	MS2	5
MS2-PAB	Antisense MS2	3
MS2-PAB	None	3

Figure 14.3   Important controls to consider when performing a tethered function assay. Shown is a representation of experiments we performed to demonstrate the effects of PAB on mRNA stability (Coller *et al.*, 1998). First, the effect of the tether was evaluated by determining half-lives of the reporter in cells expressing just the MS2 coat protein alone or MS2 fused to Sxl-lethal, a distinct RNA-binding protein of similar size to PAB (MS2-SXL). Second, we determined that the observed increase in mRNA stability was a consequence of tethering PAB in *cis*, by measuring reporter half-life when the mRNA cannot bind MS2-PAB; either the tethering sites were not present or the sites were in the antisense orientation. This latter experiment also controlled for the contribution of the tethering sites to the stability of the reporter. From these controls it was possible to conclude that the observed reporter stabilization was specific to PAB and occurred only when it was tethered.

7. EXAMPLES OF THE TETHERED FUNCTION ASSAY IN THE LITERATURE

7.1. Analyzing essential genes

Tethered function assays allow the presence of essential RNA-binding proteins to be modulated on a target mRNA without affecting cell viability. For example, in *Saccharomyces cerevisiae,* PAB is an essential gene involved in many different aspects of mRNA metabolism. Studies of *PAB1* function using conditional alleles or genetic suppressors have shown that this protein is required for efficient mRNA translation, coupled deadenylation and decay, and polyadenylation. Detailed analysis of these functions *in vivo* is complicated by the breadth of PAB's roles and the fact that it is essential. Tethered function assays were used to circumvent these pleiotropic effects. Using this approach, PAB was shown to stabilize an mRNA to which it was tethered (Coller *et al.*, 1998). The activities of mutant forms of PAB (as tethered proteins) have been determined, and the active regions identified, even though yeast carrying the equivalent mutants would not be viable (Coller *et al.*, 1998; Gray *et al.*, 2000).

Tethered function assays have also facilitated analysis of essential translation initiation factors. For example, eukaryotic initiation factor (eIF)4G, a critical member of the cap-binding complex, is thought to recruit the 40S ribosome to the mRNA by simultaneously binding both cap-binding factors (eIF4E) and a 40S ribosome-associated complex (eIF3). A wealth of biochemical data has illuminated the contribution of eIF4G to translation *in vitro*. De Gregorio *et al.* (1999) used a tethered function approach to reveal mechanisms of eIF4G action *in vivo*. They first determined that eIF4G tethered to the intergenic region of a bicistronic reporter mRNA was sufficient to drive mRNA translation independent of the cap. This enabled identification of a conserved core domain of eIF4G that is required for translational stimulation (De Gregorio *et al.*, 1999). Similar studies with translational initiation factor eIF4E demonstrated that it stimulates translation independent of its ability to bind the cap (De Gregorio *et al.*, 2001). This latter study pioneered the use of N-peptide as a tethering device (Baron-Benhamou *et al.*, 2004).

7.2. Separation of multiple functions that reside within the same protein

Many posttranscriptional events are coupled. For example, splicing and 3′ polyadenylation influence one another and these events influence transport, degradation, and translation of the mRNA. In several cases, proteins involved in an upstream event can also have a dramatic role in a downstream event. This complicates the use of conventional mutational analysis in pinpointing the protein's direct effects. In such cases, tethered function assays can help determine which of many affected steps are due directly to the activity of the protein.

In one example of this approach, SR proteins were shown to directly affect both splicing and translation (Sanford *et al.*, 2004). SR proteins are a large family of nuclear phosphoproteins required for constitutive and alternative splicing. A subset of SR proteins is known to shuttle between the nucleus and cytoplasm, suggesting that these proteins play important cytoplasmic roles in mRNA metabolism. Since many alterations in SR proteins *in vivo* impact splicing, it was difficult to determine whether any observed effects on translation were a direct effect of the SR defect or an indirect consequence of the splicing defect. To overcome this limitation, Sanford *et al.* (2004) used a tethered function assay in which they injected reporter mRNA bearing the MS2-RNA binding element with an MS2-SF2/ASF (an SR protein) protein fusion into the cytoplasm of *Xenopus* oocytes. The data demonstrated that tethered SF2/ASF stimulated translation by approximately 6-fold over the appropriate controls. This was also shown to be a general property of SF2/ASF by demonstrating that similar phenotypes were observed in HeLa cell-free translation extracts.

These findings resulted in the conclusion that SR proteins can promote mRNA translation after they are deposited on the mRNA via splicing. From the standpoint of this review, the important point is that the tethered function assay allowed the elucidation of a role for SR proteins in mRNA translation by removing the complication of the upstream event, i.e., splicing.

7.3. Dissecting complexes

Tethered function assays can be particularly useful when genetics is complex or unsuited to the problem. Many regulatory events are controlled by multiprotein complexes. Discrete components of the complex provide RNA binding and recognition, which in turn recruit the functional activity to the site of regulation.

7.3.1. Protein complexes: NMD

Analysis of non-sense-mediated decay (NMD) is exemplary. Mammalian mRNAs are targeted for rapid turnover when they contain a stop codon that is greater than 50 nucleotides upstream of the last exon–exon boundary, a process termed NMD. A group of proteins binds to the exon–exon (E/E) junction of mammalian mRNA subject to NMD (Le Hir *et al.*, 2000a,b; Singh and Lykke-Andersen, 2003). Although this complex is primarily found on NMD substrates, it was unclear if their presence was a cause or effect of the transcript being targeted for NMD. Lykke-Andersen *et al.* (2001)

used a tethered function approach to test whether the placement of any of these proteins on a normal mRNA would elicit an NMD response. While the E/E complex consists of at least five proteins, only tethered RNP S1 elicited NMD. In this case, the tethered function approach revealed a role of a specific protein in eliciting the function of a multiprotein complex (E/E complex), and showed it was a cause, rather than an effect, of the NMD process.

7.3.2. RNA–protein complexes: miRNAs

The tethered function assay has helped identify key components in the RNA protein complex associated with miRNA-mediated gene silencing. Ten years ago, a small, noncoding RNA of approximately 21 nucleotides, lin-4, was shown to bind the 3′ UTR of lin-14 mRNA in the nematode *Caenorhabditis elegans*, and to silence its translation (Pasquinelli *et al.*, 2005). Since that initial discovery, miRNAs have emerged as ubiquitous regulators of mRNA translation and stability.

Numerous factors are required for miRNA maturation and for the assembly of the miRNA into a ribonucleoprotein (RNP) complex that represses translation of the target mRNA. The RNA interference silencing complex (RISC) has been shown to be necessary for cessation of mRNA translation by an miRNA (Filipowicz, 2005; Sontheimer, 2005). Tethered function assays made it possible to dissect the repression function of RISC from the miRNA: specific components of RISC, namely Ago1–2, are sufficient to translationally repress reporter mRNAs to which they are artificially bound (Behm-Ansmant *et al.*, 2006; Pillai *et al.*, 2004; Rehwinkel *et al.*, 2005).

7.4. Mutagenesis of tethered proteins can also be useful in identifying unique gain-of-function alleles

Because the effects of a tethered protein are examined on a single reporter mRNA, the effects of many manipulations of the protein sequence can be examined readily and conclusively. This can reveal novel molecular properties in the protein.

This general approach has been applied to the Dhh1p/RCK1/p54 family of RNA helicases (Minshall and Standart, 2004; Minshall *et al.*, 2001). The *Xenopus* homolog, Xp54, is sufficient to repress the translation of an mRNA to whose 3′ UTR it is tethered. Interestingly, mutants within the putative DEAD box motif of this protein transform this helicase from a translational repressor into a translational stimulator. These results may indicate that Xp54 may serve two roles in mRNA metabolism that are dependent on modulation of its conformation or helicase activity.

7.5. Tethering of proteins to different areas of the reporter can have different effects

It should be noted that the tethered function assay measures the effect of an mRNP complex in its nonnative context and thus may induce emergent properties of the protein. Moreover, the protein of interest may have distinct functions when positioned differently on the mRNA reporter. Indeed, it has been documented that similar proteins when tethered to different areas of an mRNA can have distinct outcomes.

For example, the conserved mRNA-binding protein Staufen is important during early embryonic development in *Drosophila* and has been identified as an important regulator of mammalian mRNA processes. Tethering of mammalian Staufen to the 5′ UTR of reporter mRNAs stimulates translation without impacting mRNA stability in HEK293T cells and rabbit reticulocyte lysates (Dugre-Brisson *et al.*, 2005). Interestingly, tethering mammalian Staufen to the 3′ UTR in HeLa cells does not stimulate translation, but instead destabilizes the mRNA (Kim *et al.*, 2005). These two reports are from distinct cells types, and so require further analysis. However, it may be that Staufen possesses different activities, dependent on its location in the mRNA. This property would echo that of IRP; bound to the 5′ UTR of ferritin mRNA, it inhibits translation; bound to the 3′ UTR of transferrin mRNA, it inhibits mRNA decay (Hentze *et al.*, 2004). It may turn out to be important to compare the effects of proteins tethered to different locales to reveal region-specific differences.

7.6. Identifying mRNA localization functions and visualizing tagged mRNAs *in vivo*

Proteins that cause an mRNA to move to a particular location within a cell can be assayed using the tethered function approach. For example, yeast She2p and She3p are present in a complex on the *ASH1* 3′ UTR. Tethering either She2p or She3p to the 3′ UTR of a reporter gene was sufficient to stimulate that mRNA's localization to the bud tip (Long *et al.*, 2000). These findings directly demonstrate a localization function, and should enable its genetic dissection away from formation of the complex or binding to RNA.

Several adaptations of the tethered function assay have been developed to tag an mRNA for further analysis, rather than study a particular protein's effects. Although these are not strictly tethered function assays (as the protein is merely a tag), we mention them here because they are so closely related technically. They now are widely used, and have been reviewed in their own right (Beach *et al.*, 1999; Singer *et al.*, 2005); we discuss only a single, early pioneering example.

Bertrand *et al.* (1998) used the tethered function approach to facilitate the study of *ASH1* mRNA localization in living yeast cells. *ASH1* mRNA is distributed into daughter cells during budding, regulating asymmetric switching of yeast mating type. To determine how various mutants affect *ASH1* mRNA localization, MS2 sites were inserted into the 3′ UTR of a LacZ reporter containing the *ASH1* 3′ UTR. The localization of this RNA was then monitored in living cells by tethering an MS2/green fluorescent protein (GFP) fusion to the MS2 sites (Fig. 14.4). Tethered GFP allows for simple detection of the RNA and provides a unique perspective of *ASH1* mRNA localization in real time (Bertrand *et al.*, 1998). This assay has also been successfully used to identify the factors involved in the process. For example, certain mutants (*she2* and *she3*) perturb localization monitored by tethered GFP (Bertrand *et al.*, 1998).

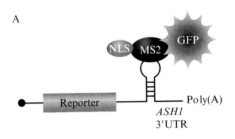

Figure 14.4   mRNA localization and tethered assays. (A) Tethered GFP can be used to monitor mRNA localization in living cells: GFP is tethered to the 3′ UTR or elsewhere in the mRNA, as a means of "tagging" the mRNA. Localization of the GFP fluorescence, and hence the mRNA, can then be monitored by microscopy. (B) Often the MS2–GFP fusion is tagged with a nuclear localization signal (NLS) as a means to reduce cytoplasmic noise. In this example, Bertrand *et al.* (1998) monitored the localization of the *ASH1* mRNA in yeast to the bud tip. Importantly, this ASH1 mRNP particle was observed only when the tethering sites were present in the reporter, and GFP was fused to the MS2 coat.

7.7. Tethered function can be used to detect both stimulatory and inhibitory events

As mentioned, the tethered function assay is highly adaptable. Tethered function assays have been used to monitor stimulatory and inhibitory effects of mRNA metabolism factors. For instance, in *Xenopus* it was demonstrated that tethered DAZL stimulates translation (Collier *et al.*, 2005), while using the same reporters others have shown that tethered Xp54 inhibits mRNA translation in *Xenopus* (Minshall and Standart, 2004; Minshall *et al.*, 2001). Similar results have been seen for assaying effects on mRNA stability. Certain classes of AU-rich binding proteins will stabilize mRNA when tethered, while others destabilize the mRNA (Barreau *et al.*, 2006; Chou *et al.*, 2006). Thus, tethered function assays provide flexibility in allowing a range of phenotypes to be observed.

7.8. Analyzing mRNA modifying enzymes

Tethered function assays have been used to identify enzymes involved in mRNA processing. Sequences near the $3'$ end of an mRNA recruit a complex of proteins that promotes $3'$ end cleavage and polyadenylation. By tethering the relevant poly(A) polymerase directly to the $3'$ end of the reporter, that enzyme was shown to be sufficient for the elongation of poly (A) tails in oocytes and to stimulate translation as a result (Dickson *et al.*, 2001). Sites for interaction with other components of the complex are dispensable (Dickson *et al.*, 2001). The same general approach has been used to identify other divergent poly(A) adding enzymes, termed the GLD-2 family, from *C. elegans,* flies, frogs, mice, and humans (Kwak *et al.*, 2004; J. E. Kwak *et al.*, unpublished observations; Wang *et al.*, 2002).

A strength of the tethered approach is that many candidate open reading frames (ORFs) can be tested rapidly. A limitation is that false negatives arise. For example, two *Saccharomyces cerevisiae* proteins, Trf4p and Trf5p, that are known to be poly(A) polymerases, differ dramatically as tethered proteins. Trf5p is active, and Trf4p is not (J. E. Kwak *et al.*, unpublished observations). This may reflect a difference in their substrate specificity, requirements for RNA or protein partners, or be an artifactual consequence of an inactive conformation in one chimeric protein.

Tethering assays can reveal unanticipated biochemical activities. In the same group of tethering experiments that identified the GLD-2 family, certain relatives of these PAPs turn out not to add poly(A) at all, but to add poly(U) instead (J. E. Kwak *et al.*, unpublished observations). Investigations into the biological role of these newly discovered poly(U) polymerases are currently underway. The key point here is that tethered assays enabled facile biochemical identification of the RNA modifications they catalyze.

8. PROSPECTS

Tethered function assays provide a simple means to address the role of specific RNA-binding proteins on mRNA metabolism and function. Their use is certainly not limited to the few examples mentioned here and in Table 14.1. The tethered function approach provides a unique platform for the study of suspect regulators of mRNA metabolism that have unknown target specificity and/or functional activity. Of particular interest are simple phenotypic screens that allow the rapid identification of tethered proteins on the metabolism of a given reporter.

As the genome sequences of more species become available, methods to analyze function beyond familial sequence resemblance are needed. Tethered function assays may provide a rapid screen to sort proteins into functional families.

ACKNOWLEDGMENTS

We thank many individuals who have contributed their thoughts and ideas to this review, most notably, Drs. Jens Lykke-Anderson, Scott Ballantyne, Kristian Baker, Kris Dickson, Niki Gray, Stan Fields, Mattias Hentze, Allan Jacobson, Roy Parker, Stu Peltz, Daniel Seay, Rob Singer, Nancy Standart, and Joan Steiz. We also thank Drs. Wenqian Hu and Thomas J. Sweet for critical reading of the manuscript. Work in the Wickens laboratory is supported by grants from the National Institutes of Health (NIH). Dr. Coller is supported by a grant from the American Cancer Society and the NIH.

REFERENCES

Amrani, N., Dong, S., He, F., Ganesan, R., Ghosh, S., Kervestin, S., Li, C., Mangus, D. A., Spatrick, P., and Jacobson, A. (2006). Aberrant termination triggers nonsense-mediated mRNA decay. *Biochem. Soc. Trans.* **34,** 39–42.

Baron-Benhamou, J., Gehring, N. H., Kulozik, A. E., and Hentze, M. W. (2004). Using the lambdaN peptide to tether proteins to RNAs. *Methods Mol. Biol.* **257,** 135–154.

Barreau, C., Watrin, T., Beverley Osborne, H., and Paillard, L. (2006). Protein expression is increased by a class III AU-rich element and tethered CUG-BP1. *Biochem. Biophys. Res. Commun.* **347,** 723–730.

Barton, H. A., Eisenstein, R. S., Bomford, A., and Munro, H. N. (1990). Determinants of the interaction between the iron-responsive element-binding protein and its binding site in rat L-ferritin mRNA. *J. Biol. Chem.* **265,** 7000–7008.

Beach, D. L., Salmon, E. D., and Bloom, K. (1999). Localization and anchoring of mRNA in budding yeast. *Curr. Biol.* **9,** 569–578.

Behm-Ansmant, I., Rehwinkel, J., Doerks, T., Stark, A., Bork, P., and Izaurralde, E. (2006). mRNA degradation by miRNAs and GW182 requires both CCR4:NOT deadenylase and DCP1:DCP2 decapping complexes. *Genes Dev.* **20,** 1885–1898.

Bertrand, E., Chartrand, P., Schaefer, M., Shenoy, S. M., Singer, R. H., and Long, R. M. (1998). Localization of ASH1 mRNA particles in living yeast. *Mol. Cell* **2,** 437–445.

Brodsky, A. S., and Silver, P. A. (2000). Pre-mRNA processing factors are required for nuclear export. *RNA* **6,** 1737–1749.

Carey, J., and Uhlenbeck, O. C. (1983). Kinetic and thermodynamic characterization of the R17 coat protein-ribonucleic acid interaction. *Biochemistry* **22,** 2610–2615.

Chou, C. F., Mulky, A., Maitra, S., Lin, W. J., Gherzi, R., Kappes, J., and Chen, C. Y. (2006). Tethering KSRP, a decay-promoting AU-rich element-binding protein, to mRNAs elicits mRNA decay. *Mol. Cell. Biol.* **26,** 3695–3706.

Coller, J., and Wickens, M. (2002). Tethered function assays using 3' untranslated regions. *Methods* **26,** 142–150.

Coller, J. M., Gray, N. K., and Wickens, M. P. (1998). mRNA stabilization by poly(A) binding protein is independent of poly(A) and requires translation. *Genes Dev.* **12,** 3226–3235.

Collier, B., Gorgoni, B., Loveridge, C., Cooke, H. J., and Gray, N. K. (2005). The DAZL family proteins are PABP-binding proteins that regulate translation in germ cells. *EMBO J.* **24,** 2656–2666.

De Gregorio, E., Preiss, T., and Hentze, M. W. (1999). Translation driven by an eIF4G core domain *in vivo*. *EMBO J.* **18,** 4865–4874.

De Gregorio, E., Baron, J., Preiss, T., and Hentze, M. W. (2001). Tethered-function analysis reveals that eIF4E can recruit ribosomes independent of its binding to the cap structure. *RNA* **7,** 106–113.

Dickson, K. S., Thompson, S. R., Gray, N. K., and Wickens, M. (2001). Poly(A) polymerase and the regulation of cytoplasmic polyadenylation. *J. Biol. Chem.* **276,** 41810–41816.

Dugre-Brisson, S., Elvira, G., Boulay, K., Chatel-Chaix, L., Mouland, A. J., and DesGroseillers, L. (2005). Interaction of Staufen1 with the 5' end of mRNA facilitates translation of these RNAs. *Nucleic Acids Res.* **33,** 4797–4812.

Filipowicz, W. (2005). RNAi: The nuts and bolts of the RISC machine. *Cell* **122,** 17–20.

Finoux, A. L., and Seraphin, B. (2006). *In vivo* targeting of the yeast Pop2 deadenylase subunit to reporter transcripts induces their rapid degradation and generates new decay intermediates. *J. Biol. Chem.* **281,** 25940–25947.

Furuyama, S., and Bruzik, J. P. (2002). Multiple roles for SR proteins *in trans* splicing. *Mol. Cell. Biol.* **22,** 5337–5346.

Fusco, D., Accornero, N., Lavoie, B., Shenoy, S. M., Blanchard, J. M., Singer, R. H., and Bertrand, E. (2003). Single mRNA molecules demonstrate probabilistic movement in living mammalian cells. *Curr. Biol.* **13,** 161–167.

Gray, N. K., Coller, J. M., Dickson, K. S., and Wickens, M. (2000). Multiple portions of poly(A)-binding protein stimulate translation *in vivo*. *EMBO J.* **19,** 4723–4733.

Hentze, M. W., Muckenthaler, M. U., and Andrews, N. C. (2004). Balancing acts: Molecular control of mammalian iron metabolism. *Cell* **117,** 285–297.

Hope, I. A., and Struhl, K. (1986). Functional dissection of a eukaryotic transcriptional activator protein, GCN4 of yeast. *Cell* **46,** 885–894.

Keegan, L., Gill, G., and Ptashne, M. (1986). Separation of DNA binding from the transcription-activating function of a eukaryotic regulatory protein. *Science* **231,** 699–704.

Kim, Y. K., Furic, L., Desgroseillers, L., and Maquat, L. E. (2005). Mammalian Staufen1 recruits Upf1 to specific mRNA 3'UTRs so as to elicit mRNA decay. *Cell* **120,** 195–208.

Kwak, J. E., Wang, L., Ballantyne, S., Kimble, J., and Wickens, M. (2004). Mammalian GLD-2 homologs are poly(A) polymerases. *Proc. Natl. Acad. Sci. USA* **101,** 4407–4412.

Lefstin, J. A., and Yamamoto, K. R. (1998). Allosteric effects of DNA on transcriptional regulators. *Nature* **392,** 885–888.

Le Hir, H., Izaurralde, E., Maquat, L. E., and Moore, M. J. (2000a). The spliceosome deposits multiple proteins 20–24 nucleotides upstream of mRNA exon-exon junctions. *EMBO J.* **19,** 6860–6869.

Le Hir, H., Moore, M. J., and Maquat, L. E. (2000b). Pre-mRNA splicing alters mRNP composition: Evidence for stable association of proteins at exon-exon junctions. *Genes Dev.* **14,** 1098–1108.

Long, R. M., Gu, W., Lorimer, E., Singer, R. H., and Chartrand, P. (2000). She2p is a novel RNA-binding protein that recruits the Myo4p-She3p complex to ASH1 mRNA. *EMBO J.* **19,** 6592–6601.

Lowary, P. T., and Uhlenbeck, O. C. (1987). An RNA mutation that increases the affinity of an RNA-protein interaction. *Nucleic Acids Res.* **15,** 10483–10493.

Lykke-Andersen, J., Shu, M. D., and Steitz, J. A. (2000). Human Upf proteins target an mRNA for nonsense-mediated decay when bound downstream of a termination codon. *Cell* **103,** 1121–1131.

Lykke-Andersen, J., Shu, M. D., and Steitz, J. A. (2001). Communication of the position of exon-exon junctions to the mRNA surveillance machinery by the protein RNPS1. *Science* **293,** 1836–1839.

Minshall, N., and Standart, N. (2004). The active form of Xp54 RNA helicase in translational repression is an RNA-mediated oligomer. *Nucleic Acids Res.* **32,** 1325–1334.

Minshall, N., Thom, G., and Standart, N. (2001). A conserved role of a DEAD box helicase in mRNA masking. *RNA* **7,** 1728–1742.

Muhlrad, D., and Parker, R. (1992). Mutations affecting stability and deadenylation of the yeast MFA2 transcript. *Genes Dev.* **6,** 2100–2111.

Pasquinelli, A. E., Hunter, S., and Bracht, J. (2005). MicroRNAs: A developing story. *Curr. Opin. Genet. Dev.* **15,** 200–205.

Pillai, R. S., Artus, C. G., and Filipowicz, W. (2004). Tethering of human Ago proteins to mRNA mimics the miRNA-mediated repression of protein synthesis. *RNA* **10,** 1518–1525.

Rehwinkel, J., Behm-Ansmant, I., Gatfield, D., and Izaurralde, E. (2005). A crucial role for GW182 and the DCP1:DCP2 decapping complex in miRNA-mediated gene silencing. *RNA* **11,** 1640–1647.

Ruiz-Echevarria, M. J., and Peltz, S. W. (2000). The RNA binding protein Pub1 modulates the stability of transcripts containing upstream open reading frames. *Cell* **101,** 741–751.

Sanford, J. R., Gray, N. K., Beckmann, K., and Caceres, J. F. (2004). A novel role for shuttling SR proteins in mRNA translation. *Genes Dev.* **18,** 755–768.

Scharpf, M., Sticht, H., Schweimer, K., Boehm, M., Hoffmann, S., and Rosch, P. (2000). Antitermination in bacteriophage lambda. The structure of the N36 peptide-boxB RNA complex. *Eur. J. Biochem.* **267,** 2397–2408.

Scully, K. M., Jacobson, E. M., Jepsen, K., Lunyak, V., Viadiu, H., Carriere, C., Rose, D. W., Hooshmand, F., Aggarwal, A. K., and Rosenfeld, M. G. (2000). Allosteric effects of Pit-1 DNA sites on long-term repression in cell type specification. *Science* **290,** 1127–1131.

Shen, H., and Green, M. R. (2006). RS domains contact splicing signals and promote splicing by a common mechanism in yeast through humans. *Genes Dev.* **20,** 1755–1765.

Singer, R. H., Lawrence, D. S., Ovryn, B., and Condeelis, J. (2005). Imaging of gene expression in living cells and tissues. *J. Biomed. Opt.* **10,** 051406.

Singh, G., and Lykke-Andersen, J. (2003). New insights into the formation of active nonsense-mediated decay complexes. *Trends Biochem. Sci.* **28,** 464–466.

Sontheimer, E. J. (2005). Assembly and function of RNA silencing complexes. *Nat. Rev. Mol. Cell. Biol.* **6,** 127–138.

Spellman, R., Rideau, A., Matlin, A., Gooding, C., Robinson, F., McGlincy, N., Grellscheid, S. N., Southby, J., Wollerton, M., and Smith, C. W. (2005). Regulation of alternative splicing by PTB and associated factors. *Biochem. Soc. Trans.* **33,** 457–460.

van Gelder, C. W., Gunderson, S. I., Jansen, E. J., Boelens, W. C., Polycarpou-Schwarz, M., Mattaj, I. W., and van Venrooij, W. J. (1993). A complex secondary structure in U1A pre-mRNA that binds two molecules of U1A protein is required for regulation of polyadenylation. *EMBO J.* **12,** 5191–5200.

Wang, L., Eckmann, C. R., Kadyk, L.C, Wickens, M., and Kimble, J. (2002). A regulatory cytoplasmic poly(A) polymerase in *Caenorhabditis elegans. Nature* **419,** 312–316.

Wang, Z., Xiao, X., Van Nostrand, E., and Burge, C. B. (2006). General and specific functions of exonic splicing silencers in splicing control. *Mol. Cell* **23,** 61–70.

Williamson, J. R. (2000). Induced fit in RNA-protein recognition. *Nat. Struct. Biol.* **7,** 834–837.

Witherell, G. W., Wu, H. N., and Uhlenbeck, O. C. (1990). Cooperative binding of R17 coat protein to RNA. *Biochemistry* **29,** 11051–11057.

CHAPTER FIFTEEN

Analysis of Ribosomal Shunting During Translation Initiation in Eukaryotic mRNAs

Vincent P. Mauro,* Stephen A. Chappell,* *and* John Dresios*,†

Contents

Abstract

In eukaryotes, translation initiation involves recruitment of ribosomal subunits at either the $5'$ m7G cap structure or at an internal ribosome entry site (IRES). For most mRNAs, the initiation codon is located some distance downstream,

* Department of Neurobiology, The Scripps Research Institute, and The Skaggs Institute for Chemical Biology, La Jolla, California
† Science Applications International Corporation, San Diego, California

Methods in Enzymology, Volume 429
ISSN 0076-6879, DOI: 10.1016/S0076-6879(07)29015-9

necessitating ribosomal movement to this site. Although the mechanistic details of this movement remain to be fully resolved, it appears to be nonlinear for some mRNAs (i.e., ribosomal subunits appear to bypass [shunt] segments of the 5′ leader as they move to the initiation codon). This chapter describes various experimental approaches to assess ribosomal shunting and to identify mRNA elements (shunt sites) that facilitate shunting. In addition, we provide an overview of approaches that can be used to investigate the mechanism used by individual shunt sites, along with a detailed protocol for investigating putative base pairing interactions between shunt sites and 18S rRNA.

1. Introduction

For eukaryotic mRNAs, ribosomal recruitment occurs some distance upstream of the initiation codon, necessitating ribosomal movement for initiation to occur. In theory, this movement may occur by linear scanning of the intervening nucleotides (Fig. 15.1A), scanning some of the intervening nucleotides while shunting others (Fig. 15.1B), not scanning at all, but moving between so-called "shunt sites" (Fig. 15.1C), or completely bypassing the intervening nucleotides by shunting directly to the initiation codon (Fig. 15.1D).

Ribosomal shunting may explain why the translation of some mRNAs proceeds efficiently even though these mRNAs contain sequence or structural elements in their 5′ leaders that should either divert or block scanning ribosomes. These "obstacles" include upstream AUG codons and stable RNA conformations, respectively. To assess shunting in such candidate mRNAs, it is necessary to identify the site or sites at which ribosomal subunits are recruited, determining which mRNA segments are shunted, defining shunt sites if they occur, and elucidating the underlying shunting mechanism.

2. Defining the Site or Sites of Ribosomal Recruitment

The first step in assessing ribosomal shunting involves determining the site of ribosomal recruitment, which is most frequently thought to occur at either the m7G cap structure or at an internal ribosome entry site (IRES; reviewed in Hellen and Sarnow, 2001; Vagner *et al.*, 2001). The former is a modified nucleotide, which is present at the 5′ ends of all mRNAs that are transcribed by RNA polymerase II. The cap structure facilitates translation initiation by interacting with a specific set of initiation factors that links it to

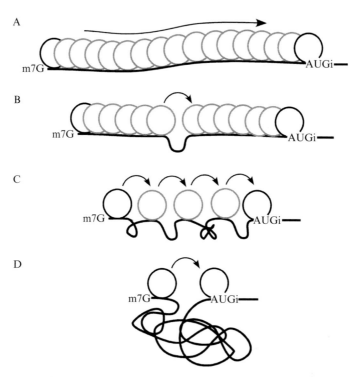

Figure 15.1 Schematic representation of possible mechanisms of ribosomal movement during translation initiation. In this figure, ribosomal recruitment is shown occurring at the m7G cap structure; however, it could also occur at an IRES contained within the mRNA. (A) Linear scanning of the 5′ leader by 40S ribosomal subunits recruited at the cap structure. Ribosomal subunits are indicated as black open circles at the site of ribosomal recruitment and at the initiation codon (AUGi); they are indicated as gray open circles where movement is indicated. The arrow indicates the 5′ to 3′ direction of movement; the black line represents the mRNA. (B) Ribosomal shunting that involves scanning of segments of the 5′ leader and bypassing of other segments. The curved arrow indicates ribosomal subunits bypassing a segment of the 5′ leader. This movement may be facilitated by shunt donor and acceptor sites located at the 5′ and 3′ ends of the shunted sequence, respectively. (C) Ribosomal shunting that does not involve scanning, but that involves ribosomal subunits moving between shunt sites. (D) Ribosomal shunting directly from the ribosomal recruitment site to the initiation codon.

the 40S ribosomal subunit. In contrast, IRESs appear to be contained only within a subset of mRNAs. IRESs have generally been found within the 5′ leader regions of mRNAs, and comprise a heterogeneous group of sequence elements that mediates translation initiation. For individual IRESs, these mechanisms differ in their requirements for initiation factors or other factors.

2.1. Assessment of cap-dependent translation using hairpin structures

To assess whether the cap structure is the site of ribosomal recruitment, it is necessary to block this site and look at the effects on translation. One way to accomplish this is to introduce a stable stem–loop (hairpin) structure near the 5' end of a reporter mRNA containing the 5' leader sequence. The mechanism by which such structures block cap-dependent translation may involve masking of the cap structure; however, it appears more likely that the structures prevent recruitment of 40S ribosomal subunits by interfering with the formation of the eukaryotic initiation factor (eIF)4F complex or with the eIF4F–eIF3 association (discussed in Kapp and Lorsch, 2004).

In our studies, we have blocked cap-dependent translation using a 60-nt palindromic sequence that is predicted to form a hairpin structure with a calculated stability of approximately −55 kcal/mol (Stoneley *et al.*, 1998). In various studies, the presence of this hairpin structure at the 5' end of a reporter mRNA inhibited cap-dependent translation by approximately 75% to 85% in transiently transfected mammalian cells, when compared to an mRNA lacking the hairpin structure (Chappell *et al.*, 2000, 2001; Stoneley *et al.*, 1998). This same hairpin structure inhibited cap-dependent translation by approximately 100% in yeast (Zhou *et al.*, 2001).

There are several ways to generate a construct encoding an mRNA with a hairpin structure. Such constructs can be generated by insertion of a hairpin structure at the 5' end of the 5' leader, either in the context of a natural or synthetic mRNA. If there is a naturally occurring restriction site at or near the 5' end of the mRNA, this site can be used to insert a hairpin structure. If such a site is not present, it can be introduced by site-directed mutagenesis, either at the transcription start site or immediately downstream of it.

Although a DNA fragment encoding a hairpin structure can be generated by polymerase chain reaction (PCR) amplification of an oligonucleotide template, we have found that it is difficult to amplify oligonucleotides containing such palindromic sequences. An easier approach involves cloning two fragments into the DNA construct, each corresponding to one-half of the hairpin structure. For this cloning, each fragment can be cloned independently to facilitate its sequencing, and then combined into the same construct. An advantage of this approach is that it introduces a restriction site into the loop of the hairpin structure; the presence of this restriction site can be used to rapidly identify plasmids containing the hairpin structure. The thermodynamic stability of a hairpin structure can be predicted using the MFold algorithm (Zuker *et al.*, 1999).

A consideration with the use of hairpin structures is the possibility of their cleavage. In our earlier studies, we observed that a hairpin structure contained within a recombinant mRNA was cleaved when this mRNA was expressed in transfected cells (Chappell *et al.*, 2006a). This observation suggests that any observed translation that relies on sequences located upstream of the hairpin structure must be due to ribosome movement across noncovalently linked RNAs. It is therefore important to perform appropriate RNA analyses to determine whether the presence of the hairpin structure affects mRNA integrity or levels. A protocol describing how to assess the integrity of a hairpin structure is included below in the section on using RNA hairpin structures as barriers to translation to assess ribosomal shunting.

A caveat associated with the use of a hairpin structure to assess cap-dependent translation is that such a structural element may inhibit translation in ways other than impeding cap-dependent ribosomal recruitment. For example, a hairpin structure may inhibit translation by affecting the conformation and activity of an IRES element that may be present downstream of the hairpin structure. Alternatively, it may affect the accessibility of the initiation codon (Chappell *et al.*, 2006b). It is therefore advisable to further assess cap-dependent translation using additional approaches (e.g., by blocking the activities of individual initiation factors involved in cap-dependent translation initiation). This can be accomplished in various ways (e.g., by expressing a hypophosphorylated 4E-BP in cells to sequester the cap-binding protein eIF4E) (Gingras *et al.*, 1999; Pinkstaff *et al.*, 2001), or by expressing the picornavirus 2A protease to cleave initiation factor eIF4G and preventing it from simultaneously interacting with both the mRNA and with the ribosomal subunit (via eIF3). While the latter approach has been used successfully to block the translation of globin mRNA when both the mRNA and enzyme were injected into *Xenopus* oocytes (Keiper and Rhoads, 1997), it should be kept in mind that cap-dependent translation may not be completely inhibited. For example, the extent to which translation is blocked will depend on various factors, including the levels of these inhibitory proteins (see Pinkstaff *et al.*, 2001). To obtain an indication of the extent to which cap-dependent translation is blocked, a cap-dependent mRNA (e.g., a recombinant mRNA containing the β-globin 5' leader) should be included in these experiments.

2.1.1. Procedure: transient transfection of mammalian cells with *Photinus* luciferase reporter constructs

The evaluation of cap-dependent translation described here, as well as other studies described in this chapter, uses the pGL3c plasmid (Promega), which contains *Photinus* luciferase as a reporter gene, and assays its levels in transiently transfected mammalian cells. However, the same strategies apply when performing such studies using other reporter cistrons or authentic coding sequences.

The experimental details of transient transfection, as typically performed in our laboratory in mouse neuroblastoma N2a cells, are detailed below. Note that we have successfully used these same conditions with numerous other cell lines, including rat glial tumor C6, mouse fibroblast NIH 3T3 (3T3), B104 rat neural tumor, and human neuroblastoma SK-N-SH (SK). In other cell lines, it may be necessary to vary conditions for efficient transfection.

Twenty-four hours prior to transfection, cell monolayers are trypsinized [0.25% trypsin in 1 mM ethylenediaminetetraacetic acid (EDTA); Gibco], cells are counted using a hematocrit, and replated at 1×10^5 cells per well in six-well culture plates with 1 ml of growth media [Dulbecco's modified Eagle medium (DMEM) containing 10% fetal bovine serum (FBS), 100 units penicillin, 100 units streptomycin, and 292 μg L-glutamine (Invitrogen Corporation)] per well. After culturing for 24 h at 37°, with 5% CO_2, and 98% humidity, cells are transiently transfected. For each well, 100 μl of Opti-MEM® I reduced serum media (Gibco™ Invitrogen Corporation) is mixed with 3 μl FuGENE® 6 Transfection Reagent (Roche Diagnostics) and incubated for 5 min at room temperature. Then 0.5 μg of reporter plasmid is added to the Opti-MEM/FuGENE 6 mix, along with 0.2 μg of the pCMV-β plasmid (Clontech), which is included as a measure of transfection efficiency. The mixture is incubated at room temperature for 15 min and applied to a cell monolayer, which has had the growth media removed, covering the monolayer as evenly as possible with the solution. One milliliter of fresh growth media is then applied and the cells cultured for an additional 24 h, after which the growth media are removed and the cells are washed with approximately 0.25 ml of a phosphate-buffered saline solution. Cells are lysed using 0.25 ml of 1× Passive Lysis buffer (Promega) for 15 min on a rocking platform, followed by manual scraping with a rubber policeman. Lysates are then transferred to 1.5-ml Eppendorf tubes.

Photinus luciferase assays are performed using 20 μl of cell lysate in 96-well plates using the Luciferase Assay System (Promega). In our studies, we use a MicroLumat LB P luminometer (EG&G Berthold) to measure luciferase activity. *Photinus* luciferase activities are normalized for transfection efficiency using the β-galactosidase activity of the control construct. To measure this activity, 20 μl of cell lysate is assayed in transparent 96-well plates (Falcon) using the FluoReporter lacZ kit (Invitrogen) on a Millipore Cytofluor™ 2350 measurement system. Translation efficiencies are then expressed as normalized luciferase activities in raw light units divided by normalized luciferase mRNA levels, which are determined using ribonuclease protection assays (RPAs).

2.1.2. Procedure: Ribonuclease protection assays

For RPAs, total cellular RNA is isolated from transfected cells using 1 ml of TRIzol® Reagent (Invitrogen) per well. RNA is extracted according to the manufacturer's instructions and resuspended in 30 μl of diethyl pyrocarbonate (DEPC)-treated H_2O. Due to the high sensitivity of the RPA, it is important to remove any residual reporter DNA in the RNA preparation. This is accomplished using the DNA-*free*™ Kit (Ambion) according to the manufacturer's instructions.

RPA probes are generated by PCR amplification of 1 ng of a plasmid DNA template that contains the target sequence. Plasmid templates include the pGL3c plasmid (Promega) for the *Photinus* luciferase probe and the pCMV-β plasmid for the lacZ probe. The *Photinus* luciferase RPA probe corresponds to nucleotides 1839 to 1939 of the *Photinus* luciferase coding sequence as numbered in the pGL3c plasmid and can be amplified efficiently using the following oligonucleotides:

(F)   5′-ATATATGAATTCGAAGTACCGAAAGGTCTTA-3′   and
(R) 5′-ATATATACTAGTCTAGAATTACACGGCGATCT-3′

The β-galactosidase RPA probe corresponds to nucleotides 875 to 1095 of the β-galactosidase coding sequence as numbered in the pCMV-β plasmid and can be amplified using the following oligonucleotides:

(F)   5′-ATATATAAGCTTGTCGTTTACTTTGACCAAC-3′   and
(R) 5′-ATATATCTCGAGACTGTTGGGAAGGGCGAT-3′

The PCR reactions contain 30 pmol of each oligonucleotide and 2.5 units of *pfu*Ultra™ DNA polymerase (Stratagene). PCR amplification proceeds for 35 cycles at the following cycling conditions: 94° for 1 min, 56° for 1 min, and 72° for 1 min. Amplified products are isolated from 2% agarose gels, restricted using the appropriate restriction sites as indicated below, and cloned into the pBluescript® KS plasmid (Stratagene).

The amplified products are cloned downstream of the T7 or T3 RNA polymerase promoters, which are contained within the pBluescript® KS plasmid. In our studies, we cloned *Photinus* luciferase sequences downstream of the T7 promoter (using *Eco*RI and *Spe*I restriction sites) and the *lacZ* sequences downstream of the T3 promoter (using *Hin*dIII and *Xho*I restriction sites). For transcription reactions from the T7 promoter, the plasmid is linearized with *Hin*dIII and for the T3 promoter it is linearized with *Bam*HI. Linearized plasmid DNA is precipitated with 1/20th volume 0.5 M EDTA, 1/10th volume 3 M sodium acetate, and 2 volumes ethanol at −20° for 30 min. Precipitated plasmids are resuspended in RNase-free water at a concentration of approximately 1 μg/μl. Of each linearized DNA template 1 μg is used to generate radiolabeled riboprobes using the MAXIscript® *In Vitro* Transcription Kit (Ambion) with either T7 or T3 RNA polymerases, as appropriate. The reactions contain 5 μl of Easy Tides [α-^{32}P]CTP (3000 Ci/mmol, Dupont/NEN) and are incubated at 37° for 1 h. Unincorporated nucleotides are removed by spin-column chromatography using Sephadex B-25 columns (Pharmacia).

A *Photinus* luciferase riboprobe generated as described above is 162 nucleotides long (see below) and contains 101 nucleotides that should be protected by the *Photinus* luciferase mRNA (underlined). The other 61 nucleotides are derived from the plasmid and should not be protected (in italics).

5′-*GGGCGAAUUGGAGCUCCACCGCGGUGGCGGCCGCU*
*CUAGAACUA*GUCUAGAAUUACACGGCGAUCUUUCCGCCC
UUCUUGGCCUUUAUGAGGAUCUCUCUGAUUUUUCUUGC
GUCGAGUUUUCCGGUAAGACCUUUCGGUACUUCGAAUUC
GAUAUCAAGCU-3′

The *lacZ* riboprobe generated as above is 296 nucleotides long (see below) and contains 221 nucleotides that should be protected by the *lacZ* mRNA (underlined) and 75 nucleotides derived from the plasmid (italics) that should not be protected.

5-*CUAAAGGGAACAAAAGCUGGGUACGCGGCCCCCCGAG*
*CUCA*CUGUUGGGAAGGGCGAUCGGTGCGGGCCUCUUCG
CUAUUACGCCAGCUGGCGAAAGGGGGAUGUGCUG
CAAGGCGAUUAAGUUGGGUAACGCCAGGGUUUUCCCAGU
CACGACGUUGUAAAACGACGGGAUCGCGCUUGAGCAG
CUCCUUGCUGGUGUCCAGACCAAUGCCUCCCAGACCGG
CAACGAAAAUCACGUUCUUGUUGGUCAAAGUAAACGAC
AAGCUUGAUAUCGAAUUCCUGCAGCCCGGGGGAUC-3′

RPAs are performed using the RPA III[TM] Ribonuclease Protection Assay Kit (Ambion). Briefly, *in vitro* transcribed riboprobes are isolated from 6% denaturing polyacrylamide gels using 350 μl probe elution buffer. We routinely ethanol precipitate approximately 80,000 cpm of each radiolabeled riboprobe with 1 μg of DNase-treated RNA from transfected cells at −20° for 1 h. The precipitated RNA/probe mixtures are resuspended in 10 μl of Hybridization III buffer, heated to 90° for 3 min, and allowed to hybridize overnight at 45°. Riboprobe that remains unhybridized is removed by the addition of 150 μl of RNase Digestion III buffer containing a 1:100 dilution of an RNase A/RNase T1 mixture and incubated at 37° for 30 min. RNase digestion is terminated and the protected riboprobes are precipitated by the addition of 225 μl of RNase Inactivation/Precipitation III Solution at −20° for 15 min. Precipitated probes are then resuspended in 5 μl of Gel loading buffer II, heated to 90° for 3 min, and loaded onto 6% denaturing polyacrylamide gels alongside molecular weight markers. Following electrophoresis, gels are dried and visualized on a Storm 860 PhosphorImager (Molecular Dynamics). The gel bands corresponding to RNase-protected fragments are then quantified using AlphaEaseFC Stand-Alone software (Alpha Innotech, San Leandro, CA).

2.2. Other approaches to block cap-dependent translation

Cap-dependent translation can also be assessed in rabbit reticulocyte or other cell-free lysates using *in vitro* transcribed mRNAs. In such an experiment, the *in vitro* transcribed mRNA is translated in the absence or presence of a cap analogue, e.g., m7GpppG, which blocks cap-dependent translation by binding to eIF4E and preventing it from binding to the capped mRNAs. As in transfected cells, cap-dependent translation can also be assayed in cell-free lysates by comparing the translation of mRNAs lacking or containing a 5' hairpin structure (Rogers *et al.*, 2004).

2.2.1. Procedure: cell-free translation

Capped mRNA transcripts for such experiments can be transcribed from DNA templates containing a suitable promoter sequence, e.g., T3 or T7 RNA polymerase using a system such as the mMESSAGE mMACHINE RNA transcription kit (Ambion, Austin, TX), which can yield up to 80% capped transcripts. A typical transcription reaction is performed using 1 μg of linearized template, which has been ethanol precipitated and resuspended in DEPC-treated H_2O. Following the transcription reaction, unincorporated nucleotides are removed by spin–column chromatography using Sephadex G-25 columns. The transcripts are then quantitated by ultraviolet (UV) light absorbance at A_{260}.

Translation reactions are performed in cell-free lysates using 0.5 μg of the capped mRNA templates in the presence or absence of 0.2 mM cap analogue (m^{7}G(5')ppp(5')G, Roche Diagnostics). As in the cellular experiments described above, the inclusion in these experiments of a cap-dependent positive control mRNA will provide an indication of the extent to which cap-dependent translation is blocked.

To prepare a translating cell lysate from cultured cells (we have routinely used N2a and C6 cells), 40 10-cm plates of cells are grown to approximately 90% confluency, the growth medium is removed, and each plate is trypsinized using 2 ml trypsin at 37°. The trypsinized cells are combined in 50-ml conical Falcon tubes (BD Biosciences) at 10 plates per tube on ice. Plates are washed with 10 ml of ice-cold phosphate-buffered saline per 10 plates, which is also combined with the trypsinized cells (30 ml final volume per 50-ml tube). Cells are centrifuged at 800×g at 4°, the supernatant is aspirated, and the cell pellet is washed in a further 10 ml of ice-cold phosphate-buffered saline. The cells are then pelleted again as described above. The supernatant is removed and the cells in each tube are resuspended in 1 ml of Hypotonic buffer (10 mM HEPES, pH 7.4, 1.5 mM Mg (CH$_3$OO)$_2$, 15 mM KCl), which is supplemented with 0.5 μl 1 M dithiothreitol (DTT) per 1 ml of cell volume on ice. Cells are allowed to swell for approximately 10 min and then homogenized with 110 strokes of a 10-ml Dounce homogenizer. Cell lysates are then transferred to prechilled microfuge tubes and centrifuged at 16,000×g in a refrigerated microcentrifuge for 10 min at 4°. The supernatants are then removed, combined, and can be stored at −80° until use.

In vitro translation reactions are performed in a total volume of 20 μl, which contains 10 μl of translating cell lysate with 0.5 μg of capped mRNA, 50 μM of each amino acid (Complete Amino Acids Mixture

(Promega), 10 units of SUPERase • In^{TM} (Ambion), 100 mM (CH$_3$OO)K, 1.3 mM Mg(CH$_3$OO)$_2$, and 2 μl of 10× reaction mixture (250 μM HEPES, pH 7.4, 250 μg/ml creatine kinase, 100 mM creatine phosphate, 20 mM DTT, 10 mM ATP, 20 μM GTP, 750 μg/ml tRNA, 2.4 mM spermidine, and 25 mM cAMP). Reactions are incubated for 1 h at 30°.

In vitro translations using Rabbit Reticulocyte Lysate (Promega) are performed in a total volume of 20 μl using 0.5 μg of capped mRNAs with 14 μl of lysate, 20 μM of each amino acid (Amino Acids Mixture's Minus Leucine, and Minus Methionine [Promega]), and 4 units SUPERase • In for 15 min at 30°. *In vitro* translations to assess cap-dependent translation are performed in the presence or absence of 0.2 mM cap analogue (m7G(5′) ppp(5′)G, Roche Diagnostics).

Parallel translation reactions are set up to assess mRNA integrity over the incubation period to rule out that the observed results may be due to mRNA degradation. mRNA is purified from reaction mixtures at the start and end of the incubation period and is analyzed by Northern blot analysis using a *Photinus* luciferase probe. Such a riboprobe can be generated from the *Photinus* luciferase reporter constructs, which contain a T7 RNA promoter at the 3′ end of the coding sequence, by linearizing with *Nco*I and transcribing using the MAXIscript In Vitro Transcription Kit.

Based on tests as described above, it should be possible to determine whether a 5′ leader facilitates translation initiation by a cap-dependent mechanism. If this is not the case (i.e., if translation is only partly cap-dependent or completely cap independent), it may indicate that an IRES is contributing to the observed translation.

2.3. Assessment of cap-independent translation

Putative IRESs can be evaluated using established criteria, which include testing the putative IRES in the intercistronic region of a dicistronic mRNA to determine whether it can drive expression of the second cistron in a manner that is independent of the translation of the first cistron, e.g., by inserting a hairpin structure into the 5′ leader to block translation of the first cistron and measuring whether expression of the second cistron persists. Another important test involves showing that the activity of the putative IRES depends on the production of the dicistronic mRNA, rather than on the production of monocistronic mRNAs corresponding to the second cistron. The latter may arise by cryptic promoter activity, or by splicing or cleavage of the dicistronic mRNA. Convincing tests involve showing that expression of the second cistron does not occur when the promoter driving transcription of the dicistronic mRNA is deleted (Chappell and Mauro, 2003; Dobson *et al.*, 2005). Another approach involves showing that expression of the second cistron is proportional to the amount of dicistronic mRNA produced, for example, when the dicistronic mRNA is transcribed using promoters of different strengths (Fig. 15.2), or by using a promoter the activity of which can be regulated (Martineau *et al.*, 2004; Mauro *et al.*, 2004). Inasmuch as the levels of monocistronic mRNAs transcribed from a cryptic promoter should not increase in response to the increased transcription of the dicistronic mRNA, this approach has an additional advantage in that it makes it possible to measure IRES activity even from sequences that may have some cryptic promoter activity. The possibility that monocistronic mRNAs are produced by splicing of the dicistronic mRNA can be determined by performing reverse transcriptase polymerase chain reaction (RT-PCR) using oligonucleotide primers located at the 5′ and 3′ ends of the mRNA. These reactions will yield a product corresponding to the full-length mRNA. Monocistronic mRNAs arising from cryptic splicing of the mRNA will yield a shorter PCR product or shorter products. A final

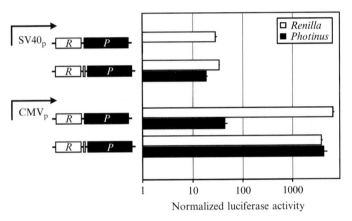

Figure 15.2 Assessing IRES activity using promoters of different strengths to drive various levels of dicistronic mRNAs. The dicistronic mRNA constructs indicated encode *Renilla* luciferase as the first cistron and *Photinus* luciferase as the second cistron. The gray bar in the intercistronic region of two of the constructs represents a 22-nt sequence element from the 5′ leader of the *Rbm3* mRNA (Chappell *et al.*, 2004). The top two constructs are transcribed by the SV40 promoter (SV40$_p$); the bottom two constructs are transcribed by the more active CMV promoter (CMV$_p$). The constructs were tested in mouse Neuro 2a (N2a) cells. The results show that the expression of the *Renilla* luciferase enzyme is more than 100-fold higher in cells transfected with the CMV construct compared to the SV40 construct. The results (shown on a log scale) also show that the *Photinus* luciferase enzyme is expressed at a low level relative to *Renilla* luciferase expression in the control constructs, but is expressed at a higher level in constructs containing the *Rbm3* sequence, and is proportional to the *Renilla* luciferase levels. This result indicates that the activity of the *Rbm3* sequence depends on production of the dicistronic mRNA, and that the translation of the second cistron occurs from the dicistronic template. This result is consistent with IRES activity and is inconsistent with the possibility of cryptic promoter activity. The luciferase activities represent raw light units normalized for transfection efficiencies and are set at 1.0 for the *Photinus* luciferase activity of the SV40 *RP* construct. Horizontal lines indicate standard error of the mean (SEM).

possibility regarding apparent IRES activity is that cleavage may occur between the two cistrons, leading to some 5′ end-dependent translation of the second cistron. This possibility can be addressed by performing RNA analyses to detect the cleaved RNA fragments. Appropriate assays may include RNase protection assays of the intercistronic region or Northern blots.

In all of the above experiments, it is important to note that the detection of a shorter mRNA does not necessarily rule out IRES activity. The latter requires knowing that the shorter mRNA contains the second cistron and is present at levels sufficient to account for the observed activity. In the cases of cryptic promoter activity or splicing, mRNA levels can be estimated by performing RNA analyses in a quantitative manner using monocistronic mRNAs corresponding to the second cistron as a reference (e.g., see Cornelis *et al.*, 2000).

To determine the location of an IRES within the 5′ leader, its boundaries can be mapped. The 5′ boundary can be defined by deletion analysis (i.e., by progressively deleting nucleotides from the 5′ end to determine which nucleotides are required for IRES activity). Likewise, progressive deletions from the 3′ end can define the 3′ boundary. Deletion analyses of this type can be most efficiently performed in several steps (i.e., starting with relatively large deletions to roughly map the boundaries, followed by a second and perhaps additional smaller deletions to more finely map these boundaries). The size of the deletions and the number of steps required to define boundaries will be determined in large part by the length of the 5′ leader.

Note that it may be difficult or impossible to define distinct IRES boundaries if the IRES has a modular composition (i.e., if it is composed of shorter IRES elements that can function independently). For example, in our analysis of translational elements in the 5′ leader of the *Rbm3* mRNA (Chappell and Mauro, 2003), we began with deletions of approximately 100 nucleotides. Based on the results of these deletions, we identified fragments with discrete IRES elements and defined the boundaries of four *cis*-acting sequences within one 100-nt fragment. These boundaries were defined using 20-nt deletions, followed by 5-nt deletions, and finally individual nucleotide deletions where required.

3. EXPERIMENTAL APPROACHES TO DETERMINE WHICH SEGMENTS OF AN MRNA ARE SHUNTED

Stable hairpin structures and upstream AUG codons can block translation initiation when inserted between the site of ribosomal recruitment and the initiation codon. The inability of either obstacle to block translation suggests that they are bypassed by ribosomal subunits. It should be noted, however, that although these elements may block translation, they do not necessarily provide evidence for the mechanism of ribosomal movement. For example, a hairpin structure may inhibit translation because it blocks scanning ribosomes. Alternatively, a hairpin structure may inhibit translation because it sterically blocks a ribosomal subunit, which is tethered to the cap, from interacting with the initiation codon (Chappell et al., 2006b). Likewise, an upstream AUG codon may block translation because it diverts scanning ribosomal subunits, or because it competes for ribosomal subunits that are either tethered to mRNA or clustered in the vicinity. This notion of ribosomal tethering and clustering was described in an earlier publication (Chappell et al., 2006b) to explain how ribosomal subunits reach an AUG codon. Tethering suggests that ribosomal complexes that are bound to the mRNA, e.g., via the eIF4F complex at the cap structure, may reach an AUG codon by bypassing intervening sequences. Alternatively, clustering suggests that movement to an AUG codon may involve the dynamic binding and release of ribosomal subunits at internal sites.

3.1. Use of hairpin structures as obstacles

An RNA hairpin structure can be inserted at various locations within the 5′ leader and its effects on translation measured to determine whether particular sites in the mRNA are bypassed. The sequence of a hairpin structure used in our studies is shown below. The restriction sites used for this construction should be unique as far as possible and will depend on those present in the mRNA of interest:

(restriction site) CCAGCGUAAUCGGGAACGUCGUAGGGGUAA GCCAUUGUACGACCACCGGCUCGAGGGGCCC (restriction site) GGGCCCCUCGAGCCGGUGGUCGUACAAUGGCUUACCCCU ACGACGUUCCCGAUUACGCUGG (restriction site)

A hairpin structure can be introduced at various sites within the mRNA using restriction sites as described earlier for the evaluation of cap-dependent translation. Alternatively, DNA fragments corresponding to nucleotide sequences located upstream and downstream of the hairpin structure can be generated by PCR amplification using appropriate oligonucleotide primers. In this strategy, three fragments (upstream, hairpin, and downstream) are ligated together in a reporter plasmid. The use of different restriction sites at each location will ensure that the fragments will assemble in the right order and correct orientation.

An additional obstacle that we have embedded in the stem of our hairpin structure (Chappell et al., 2006a) is an AUG codon (underlined) in excellent nucleotide context (A at −3 and G at +4, relative to the A of the AUG codon, which is designated as +1; Kozak, 1986). When out of frame with the initial condon, this second obstacle provides a failsafe mechanism (i.e., it should be utilized if the hairpin structure is melted and it is exposed) and should result in decreased reporter activity. Translation initiating at the embedded AUG codon also provides a means to monitor the double-stranded status of the hairpin structure in that utilization of the embedded AUG codon can be measured. If the embedded AUG codon is in frame with the initiation codon, translation initiating at this site will produce an extended reporter protein the expression of which can be monitored by Western blot analysis of the reporter protein. Alternatively, introduction of an epitope tag (e.g., an hydroxyapatite [HA] tag into the open reading frame [ORF] derived from the embedded AUG codon) may enable translation initiating at the embedded AUG codon to be monitored, even if the derived ORF is out of frame with the initiation codon. The hairpin structure described above contains such a tag. Detection of the epitope-tagged protein can be detected (e.g., with an anti-HA tag antibody, provided that that this protein is of sufficient length for Western blot analysis). The stability of short epitope-tagged proteins may also be an issue in such studies. However, this can be assessed with a control construct that expresses the epitope-tagged protein.

The caveats discussed earlier for using hairpin structures to assess cap-dependent translation also apply when using these structures to assess shunting; specifically, hairpin structures may affect translation by altering other conformations in the 5' leader, which may affect the structure or activity of an IRES or the accessibility of the AUG codon. Also, our observation of efficient translation of mRNAs in which the RNA hairpin structure is cleaved in transfected cells provides strong evidence for shunting between noncovalently linked RNAs, but underscores the need to perform RNA analyses to look for such cleavage. Such cleavage may in some cases lead to degradation of the mRNA, or to differential degradation of one of the fragments.

3.1.1. Procedure: Assessing the integrity of an RNA hairpin structure

The integrity of hairpin structures contained within mRNAs can be assessed by performing RPAs on DNase-treated mRNA using two riboprobes: one complementary to the 3' stem of the hairpin (hp) structure (hp probe) and the other to the *Photinus* luciferase coding sequence (discussed above). The latter is included as a control to monitor the presence of the mRNA targets. The RPA probe for the hairpin structure can be generated using 1 ng of a plasmid containing the 5' half of the hairpin structure as a template for PCR amplification using reaction conditions as described above. The 5' half of the hairpin sequence is amplified using a forward primer containing the *Spe*I restriction site at its 5' end, and a reverse primer containing the *Xho*I restriction site at its 3' end. After isolation from a 2% agarose gel, this fragment is cloned upstream of the T7 RNA polymerase promoter contained within the pBluescript KS plasmid using *Spe*I and *Xho*I restriction sites. The hairpin-encoding plasmid is then linearized with *Xho*I, precipitated, and transcribed as described above. The resulting 128-nucleotide hairpin riboprobe (see below) contains 67 nucleotides of sequence that should be protected by the 3′ hairpin sequence (underlined) and 61 nucleotides derived from the pBluscript plasmid that should not be protected (italics).

> 5′-*GGGCGAAUUGGAGCUCCACCGCGGUGGCGGCCGCU*
> *CUAGAACUAGCAGCUGGAAUUC*CCAGCGUAAUCGGGAAC
> GUCGUAGGGGUAAGCCAUUGUACGACCACCGGCUC
> GAGGGGCCCGACGUC*UCGA*-3′

In our studies, we have also included two *in vitro* transcribed control RNAs in the RPA (Chappell *et al.*, 2006a). One of the RNAs contained the full hairpin sequence and luciferase coding sequences (full hp RNA); the other RNA is similar, but lacks the 5′ half of the hairpin structure (half hp RNA). When these control RNAs were tested in the RPA with the hp and *Photinus* luciferase probes, we found that the half hp RNA was protected by both the hp probe and the *Photinus* luciferase probe. In contrast, the full hairpin RNA was protected only by the *Photinus* luciferase probe, presumably because base pairing of the hairpin structure precluded its hybridization. Consequently, the extent to which the hp probe is protected and the length of the protected product provide indications as to the integrity of the hairpin structure and can be used as references for hairpin integrity. For example, as in our previous studies (Chappell *et al.*, 2006a), if the RPA of RNA from transfected cells shows that the hp and *Photinus* luciferase probes protect fragments of relative intensity similar to the RPA of the control half hp RNA (i.e., the ratios of the two protected bands are the same), it would suggest that the hairpin structure was completely clipped in the transfected cells.

Both the half hp and full hp RNAs are transcribed *in vitro* from plasmids linearized using a *Bam*HI restriction site that is located downstream of the *Photinus* luciferase coding sequence. As described above, transcriptions are performed using the MAXIscript® T7 *In Vitro* Transcription Kit.

3.2. Use of upstream AUG codons as obstacles

An indication that an mRNA may use a shunting mechanism for translation initiation is the presence of one or more AUG codons in its 5′ leader, particularly if the mRNA is translated efficiently despite the presence of upstream AUG codons that reside in a good nucleotide context. This conclusion is based on the suggestion that scanning ribosomal subunits will initiate translation almost exclusively at an AUG codon that resides in a good context (Kozak, 1986, 2002).

The first step to evaluate shunting in an mRNA with upstream AUG codons would be to mutate these upstream AUG codons, individually and in combination and measure the effects of these mutations on translation. These experiments can be performed in transiently transfected cells using reporter constructs as described above. If segments of the 5′ leader containing an upstream AUG codon are shunted, mutation of the upstream AUG codon should not affect translation of the reporter protein. In contrast, if an upstream AUG codon is utilized, its mutation is expected to increase translation of the reporter protein.

We have used this approach of mutating upstream AUG codons in our analysis of the 5′ leader of the β-secretase (*BACE1*) mRNA. The results showed that mutation of any or all of four upstream AUG codons had no effect on translation in rat B104 cells (Rogers *et al.*, 2004), even though the contexts of all four AUG codons were capable of facilitating relatively efficient levels of translation initiation when tested individually in the 5′ leader of the β-globin mRNA. These results provided strong evidence for ribosomal shunting in the *BACE1* 5′ leader.

Upstream AUG codons can also be introduced into mRNAs as obstacles, much like hairpin structures. For such experiments, it is important to use an AUG codon that resides in a good context. Important nucleotides include a purine at −3 (A is best) with a G at +4, as long as it is not followed by a U at +5 (Kozak, 1997).

Ideally, an upstream AUG codon is introduced such that the resulting ORF overlaps the reporter cistron. Such upstream ORFs will either be in frame with the initiation codon or out of frame with it. In either case, translation initiating at an upstream AUG codon is expected to divert ribosomes from the initiation codon and reduce translation of the authentic reporter protein. A potential problem with the use of an in-frame AUG codon with a reporter protein that is monitored by measuring its enzymatic activity is that the fusion protein, e.g., an N-terminal extended *Photinus* luciferase protein, may have enzyme activity that is different from that of the authentic protein. Consequentially, it is necessary to physically monitor the expression of the extended protein by Western blot analysis using an antibody to the reporter protein. We have used these complementary approaches in our earlier studies (Chappell *et al.*, 2006a). Note that the molecular mass of the *Photinus* luciferase protein is approximately 60 kDa and that the size of the extended products will vary depending on the location of the upstream AUG codon. In our earlier studies, we were able to detect an extension of 17 amino acids, which increased the molecular mass by approximately 2 kDa (Chappell *et al.*, 2006a).

Depending on the sequence of the 5′ leader, it may not be possible to generate an ORF that overlaps the initiation codon from all locations within the 5′ leader. Stop codons that occur between an introduced upstream AUG codon and the authentic initiation codon will generate upstream ORFs that terminate before reaching the initiation codon. In some cases, a stop codon can be avoided by shifting the upstream AUG codon by one or two nucleotides. If this cannot be done, it may still be possible to perform a shunting analysis by introducing upstream AUG codons that generate upstream ORFs that do not overlap the initiation codon. Ribosomes initiating translation at such an upstream ORF will be diverted from the main cistron and should still result in its decreased translation. However, if the upstream ORF is short, a fraction of ribosomes may remain associated with the mRNA after its translation and reinitiate translation at the main cistron (Kozak, 2002). Another possibility for upstream AUG codons that yield only short ORFs, none of which overlaps the initiation codon, is to mutate one or more of the upstream stop codons. These mutations of the stop codons should also be tested in the absence of the upstream AUG codon to determine whether these mutations themselves affect translation.

As with the introduction of a hairpin structure into the 5′ leader of an mRNA, a consideration in these studies is that the introduction of some mutations that generate upstream AUG codons may have unintended effects on translation. For example, some mutations may alter the conformation of the 5′ leader. However, we expect that the introduction of upstream AUG codons will be much less disruptive than the introduction of RNA hairpin structures. Note that it may be possible to minimize potential problems associated with mutating the 5′ leader by introducing point mutations at sites where some of the nucleotides already fit the consensus.

3.2.1. Procedure: *Photinus* luciferase Western blotting

To monitor *Photinus* luciferase levels by Western blot analysis, 70 μl of cell lysate, in Passive Lysis buffer (Promega), is preheated to 70° for 10 min in the presence of 5 μl 1 M DTT and 25 μl 4× sample loading buffer (Invitrogen). Then 30 μl of denatured protein is loaded onto a 7% Tris–acetate gel in Tris–acetate buffer (Invitrogen). Proteins are transferred to poly(vinylidene difluoride) membranes (Bio-Rad) and probed with goat anti-*Photinus* luciferase polyclonal IgG first antibody (1:1000 dilution) and donkey antigoat IgG second antibody conjugated to alkaline phosphatase (1:5000 dilution) (both from Promega). Western blots are developed using the Western Breeze chemiluminescent Western blot immunodetection kit (Invitrogen).

4. IDENTIFICATION OF RIBOSOMAL SHUNT SITES

The experimental approaches of positioning barriers to translation within an mRNA 5′ leader can identify candidate shunt sites. A fragment containing such candidate shunt sites can be tested in isolation to determine if it can facilitate shunting across a hairpin structure or an upstream AUG codon when flanking the shunt site, or when tested on one side of the obstacle, with a bona fide shunt site (e.g., the *Gtx* element; Chappell *et al.*, 2006a) on the other side of it. This approach is based on synthetic shunting constructs developed in our analysis of the *Gtx* element. The *Gtx* element was initially identified as an IRES element (Chappell *et al.*, 2000), and subsequently shown to facilitate ribosomal shunting independent of its ability to recruit ribosomes (Chappell *et al.*, 2006a). In the shunting analysis, we tested multiple copies of the *Gtx* element upstream of the obstacle, with a single element downstream. In this particular example, multiple copies of the *Gtx* element were used to increase the signal-to-noise ratio by increasing ribosomal recruitment upstream of the obstacle.

This same shunting assay can be used to define the boundaries of shunt sites contained within fragments that show positive results in this assay. These boundaries are defined by deletion analysis, as was described above for IRES modules. A consideration in these studies is that some shunt sites may not function when tested in this assay because their activities require other sequences not contained within the fragment being tested, or because specific RNA conformations are required for activity and may be altered in this context. These possibilities can be addressed to some extent by testing larger segments. A candidate shunt site can be further analyzed in the context of the authentic 5′ leader by mutating this sequence and determining whether the mutation affects shunting, for example, across an upstream AUG codon or hairpin structure.

 ## 5. DETERMINING WHETHER PUTATIVE SHUNT SITES BIND TO RIBOSOMAL SUBUNITS

Shunt sites presumably facilitate the nonlinear movement of 40S ribosomal subunits by interacting with these subunits either directly or indirectly. To determine whether binding is direct, binding assays can be performed using a radiolabeled RNA probe corresponding to the defined shunt site or larger segments of the 5′ leader containing shunt sites. These probes can be incubated with purified 40S ribosomal subunits, and binding assessed using nitrocellulose filter binding assays.

5.1. Procedures

5.1.1. Isolation of 40S ribosomal subunits
40S ribosomal subunits can be isolated from lysates of cultured cells prepared as described above for the preparation of cell-free lysates. Cell lysates are allowed to thaw on ice and each milliliter of lysate is supplemented with 1 μl

of 200 mM ethyleneglycoltetraacetic acid (EGTA), 10 μl of 1 mg/ml aprotinin, and 10 μl of 10 mM phenylmethylsulfonyl fluoride. The lysates are
then transferred to 3.2-ml-thick welled polycarbonate tubes (#362305,
Beckman) and centrifuged at 176,000$\times g$ for 90 min at 4°. Unless otherwise
indicated, we perform such centrifugation steps using an Optima^{TM} TLX
Ultracentrifuge with a TLA 100.4 fixed angle rotor. Pellets are gently
resuspended in 200 μl Resuspension buffer (0.25 M sucrose, 20 mM
HEPES, pH 7.4, 50 mM KCl, 2 mM MgCl$_2$, 0.1 mM EDTA) with a Teflon
pestle on ice. The A_{260} of the resuspended polysomes is measured and the
solution diluted to 50 A_{260} units/ml in a solution with a final concentration
of 500 mM KCl, 50 mM HEPES, pH 7.4, 2 mM MgCl$_2$, and 2 mM DTT;
1 mM GTP and 1 mM puromycin are then added and the resuspended
polysomes are incubated on ice for 15 min followed by 37° for 15 min.

To prepare isolated ribosomal subunits, the resuspended puromycin-
treated ribosomes (see above) are loaded onto 15% to 35% linear sucrose
gradients (sucrose in 500 mM KCl, 1.5 mM MgCl$_2$, 20 mM HEPES, pH
7.4, 2 mM DTT) and centrifuged at 74,000$\times g$ in an SW28 rotor for 13.5 h at
4° with slow deceleration in a Beckman Optima^{TM} ultracentrifuge.
The gradient is fractionated using an ISCO TRIS^{TM} fractionator at an
absorbance of A_{260}. Fractions corresponding to 40S ribosomal subunits are
pooled and diluted with a buffer containing 20 mM HEPES, pH 7.4,
1.5 mM MgCl$_2$, and 2 mM DTT. Diluted fractions are centrifuged at
338,000$\times g$ for 20 h at 4°. The supernatant is removed and isolated ribosomal
subunits are resuspended in 50 μl Resuspension buffer. The A_{260} of the
suspension is measured and stored in 5-μl aliquots at −80°. This protocol
can yield relatively pure populations of 40S ribosomal subunits that can be
used for binding assays (see below).

5.1.2. Nitrocellulose filter binding assays

Nitrocellulose filter binding assays can be performed using either DNA or
RNA probes of various lengths. 5′ end-labeled oligonucleotide probes are
generated by incubating 20 pmol of oligonucleotide with 20 units of polynucleotide kinase (PNK) in 1× PNK buffer (New England Biolabs) in a reaction
containing 5 μl [γ-^{32}P]ATP (20 μCi, Dupont/NEN) for 30 min at 37°.
Unincorporated nucleotides are removed by passing the reaction through
two Sephadex Microspin G-25 columns. Longer radiolabeled RNA probes
are transcribed *in vitro* from DNA templates, e.g., using T7 RNA polymerase,
as described earlier. Radiolabeled probes (40,000 cpm) are incubated with 40S
ribosomal subunits in Binding buffer 1 (20 mM Tris–HCl, pH 7.4, at 4°,
6 mM NaCl, 5 mM 2-mercaptoethanol, 1 mM Na$_3$EDTA, pH 8.8, 10%
glycerol, and 1.8 mM MgCl$_2$) for 30 min at 37° in a final volume of 10 μl.
Samples are then diluted to 100 μl in the same buffer and filtered through
nitrocellulose filters (0.45 μm) using either a dot blot or slot blot apparatus.
Prior to filtering, the nitrocellulose filters are prepared by presoaking the

membrane in 0.4 M KOH for 10 min, rinsing with H_2O until the pH is neutral, and equilibrating in Binding buffer 1 for 1 h before use. After filtering the binding reactions, the wells are washed several times with Binding buffer 1 prewarmed to 37°. The filter will bind to protein but not nucleic acids and thus will retain the ribosomal subunits, along with any bound probes, while any unbound probe will pass through. The radiolabeled filter-bound probe can then be visualized and quantified. For our studies we use a Storm 860 PhosphorImager and AlphaEaseFC Stand-Alone software. Filter-bound ^{32}P can also be measured by liquid scintillation counting. For these experiments, background is the counts retained by the filter in the absence of ribosomal subunits. Binding specificity can be determined by competing with various unlabeled RNAs, including the probe itself as a specific competitor. Nonspecific competitors may include a scrambled or mutated probe sequence or tRNA (see Hall and Kranz, 1999; Woodbury and von Hippel, 1983; Zang and Romaniuk, 1995). In addition, by incubating the radiolabeled RNA probes with increasing amounts of isolated ribosomal subunits, the binding strengths of any interactions can be estimated.

If the results of the binding assays described above indicate that particular shunt sites do not bind directly to isolated 40S ribosomal subunits, this may suggest that a cellular *trans*-acting factor or factors are mediating interactions between the mRNA and the ribosomal subunit. The identification of such factors is not the focus of this chapter, but an example of such an analysis can be found in references (Xi *et al.*, 2004, 2005). If the results of these studies indicate direct binding, the nature of this binding can be pursued as follows:

5.1.3. UV crosslinking and localization of crosslinked probes

Probes determined to bind specifically to ribosomal subunits can be further investigated by UV crosslinking analyses (see Hu *et al.*, 1999; Tranque *et al.*, 1998) to distinguish between binding that occurs to rRNA or to ribosomal proteins. Crosslinking is performed using RNA probes containing the crosslinking reagent 4-thiouridine (s^4U, USB). Such probes can be transcribed *in vitro*, e.g., using T7 RNA polymerase. A typical reaction would include 30 mM DTT, 400 μM each of GTP and ATP, a 280 μM:120 μM mixture of s^4UTP:UTP, 40 μM CTP (Ambion), 50 μCi of [α-^{32}P]CTP (3000 Ci/mmol, Dupont NEN), and 25 units of T7 RNA polymerase (Stratagene). Nonradioactive competitor RNAs are transcribed using 500 μM each of GTP, CTP, ATP, and UTP at 37° for 3 h. Transcription reactions are terminated by the addition of 20 units of DNase I (Ambion) to digest the template. RNA is extracted with phenol/chloroform (1:1) and further purified by passage through Sephadex Microspin G-25 columns. In addition to various specific competitor RNAs, nonspecific competitors, e.g., poly(C) RNA (50 nt), which was used in our previous studies (Tranque *et al.*, 1998), can be generated similarly by *in vitro* transcription.

Crosslinking reactions contain 20 pmol of s⁴U labeled probe and 2 pmol of ribosomal subunits in Binding buffer 2 (10 mM Tris–HCl, pH 7.4, 50 mM KCl, 10 mM MgCl$_2$) in the presence of 150 pmol of tRNA. Samples are incubated at 37° for 10 min, cooled on ice, and crosslinked by exposure to 365 nm UV light for 10 min. The ribosomal proteins are extracted from the rRNA using TRIzol reagent. If the radioactivity is associated with the aqueous fraction (RNA) and not with the extracted protein, the RNA is then precipitated and electrophoresed on 1.2% agarose/formaldehyde gels. RNA in the gel is visualized with ethidium bromide and the gel is then transferred to HybondN⁺ nylon membrane (Amersham) and exposed to film to confirm association with rRNA.

For probes that appear to crosslink to a ribosomal protein or proteins, these proteins can be identified using various approaches, including two-dimensional gel electrophoresis and mass spectrometry. Numerous published protocols describe these approaches and they are not covered in this chapter.

For probes that crosslink to rRNA, the site of interaction can be localized using oligonucleotide-directed RNase H digestion of the rRNA (Hu *et al.*, 1999; Tranque *et al.*, 1998). In this method, the rRNA is specifically cleaved at various sites, and for each case, the fragment bound to the radiolabeled probe is identified. To cleave the rRNA, short complementary DNA oligonucleotides are annealed to the rRNA in RNase H buffer (40 mM Tris–HCl, pH 7.9, 10 mM MgCl$_2$, 60 mM KCl, and 1 mM DTT). The mixture is heated to 50° for 3 min, and then incubated at 30° for an additional 30 min. Then 0.5 μg of RNase H is added and the incubation is continued at 30° for another 30 min. The RNA is purified and electrophoresed as described above.

A consideration in the crosslinking studies is that the extent of UV crosslinking is affected by the number and position of the s⁴U residues within the probes (Dubreuil *et al.*, 1991). For individual sequences that lack U residues, an s⁴U residue can be introduced at either end of the sequence corresponding to the shunt site.

Inhibition of reverse transcriptase (toeprinting) can be used to further define the location of a crosslinked probe in the 18S rRNA that has been roughly mapped by oligonucleotide-directed RNase H digestion. This procedure can be performed using RNA extracted from ribosomal subunits that have been crosslinked to a nonradioactive s⁴U-containing probe. The RNA is extracted with phenol/chloroform (1:1) and toeprinted using oligonucleotide primers located downstream of the crosslinked site. See Ringquist and Gold (1998) for detailed protocols and considerations when using this method.

Putative base pairing interactions can then be tested using a hybrid yeast system that was used in our earlier studies (Chappell *et al.*, 2006a; Dresios *et al.*, 2006) and is described below.

6. Assessing mRNA–rRNA Base Pairing in Yeast

Binding that occurs between an mRNA shunt site and 18S rRNA can be further evaluated by altering both mRNA and rRNA sequences and determining whether an intact complementary match is required for shunting. If so, mutations that disrupt putative mRNA–rRNA base pairing should abolish shunting mediated by these sites. Similarly, mutations that restore complementarity should restore shunting. In an earlier study, we used this approach to determine whether a specific mRNA–rRNA base pairing interaction could facilitate ribosomal shunting in model mRNAs (Chappell *et al.*, 2006a). To perform these studies, we applied a yeast experimental system that we developed earlier to evaluate mRNA–rRNA base pairing (Dresios *et al.*, 2006). This system employs ribosomes harboring hybrid mouse–yeast 18S rRNA sequences and enables analysis of a putative shunt site with complementarity to mouse but not yeast 18S rRNA sequences.

In our analysis of shunting mediated by the *Gtx* element, we showed that efficient shunting required an intact element on both sides of the obstacle. Inasmuch as there is only one functional binding site for this element in the 40S ribosomal subunit (Dresios *et al.*, 2006), we were able to conclude that shunting mediated by this element involved dynamic binding and dissociation of ribosomal subunits upstream of the obstacle, and rebinding at a downstream element (Chappell *et al.*, 2006a).

It is also possible to analyze putative shunt sites with complementarity to the yeast wild-type 18S rRNA that may occur in yeast or mammalian mRNAs; however, this analysis requires a different approach that is beyond the scope of this chapter. Ideally, shunt sites from mammalian mRNAs would be analyzed in a mammalian system; however, a suitable mammalian system that enables alterations in the 18S rRNA has not yet been developed.

An advantage of analyzing mRNA–rRNA base pairing in the yeast hybrid system is that alternative explanations for reporter activity, such as cryptic promoter activity in the DNA constructs, are highly unlikely because any cryptic promoter activity should occur to a similar extent in yeast cells that express either wild-type or recombinant 18S rRNAs.

In earlier studies, we assessed mRNA–rRNA base pairing in *Saccharomyces cerevisiae* strain NOY908, which lacks all chromosomal copies of the 35S rDNA gene. The survival of these cells requires the expression of the 35S rRNA from a high-copy number episomal vector (pNOY373, Fig. 15.3A; Wai *et al.*, 2000). The expression of mouse–yeast hybrid or mutated yeast 18S rRNAs is achieved by transforming these cells with a second plasmid, again in the context of the 35S rRNA. In plasmid pVM1, which is derived from pNOY353, the yeast 18S rRNA sequences from nucleotides 31 to 1625 were replaced with the corresponding mouse sequences using

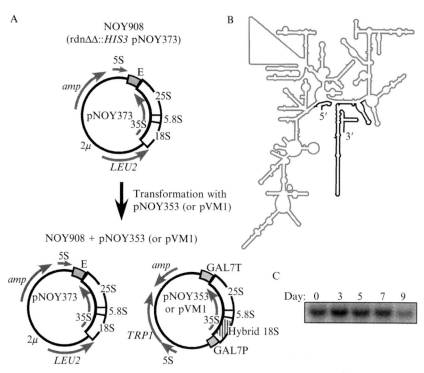

Figure 15.3 Yeast system for expressing recombinant 18S rRNAs. (A) Mouse–yeast hybrid 18S rRNAs are expressed in yeast strain NOY908 (rdnDDHIS3 pNOY373), which grows by transcription via RNA polymerase I of a single rDNA repeat on plasmid pNOY373 (Wai *et al.*, 2000). This strain was transformed with a second plasmid, pNOY353, carrying wild-type yeast sequences in the 35S rDNA gene or pVM1, carrying mouse–yeast hybrid 18S rDNA sequences. In the schematic, the mouse rDNA sequences contained within the hybrid 18S rDNA in pVM1 are indicated by the hatched gray bar. In both pNOY353 and pVM1, transcription of the 35S rRNA is driven by the *GAL7* promoter in the presence of galactose. Transformants are plated on galactose media and cells carrying pNOY353 or pVM1 are selected for their ability to grow in the absence of tryptophan. Yeast strain NOY908 and plasmids pNOY353 and pNOY393, which are used for this analysis, were obtained from Dr. M. Nomura at the University of California, Irvine. Schematic representation of pNOY373 and pNOY353 from Fig. 2 of Wai *et al.* (2000), by permission of Oxford University Press. (B) Schematic representation of a mouse–yeast hybrid 18S rRNA. The mouse sequences are indicated by the gray line; the yeast sequences are indicated by the black lines. The 5′ and 3′ ends of the rRNA are indicated. The secondary structures were adapted from those obtained on the Comparative RNA Web Site (http://www.rna.icmb.utexas.edu/) of Robert Gutell, Ph.D., University of Texas. (C) Northern blot analysis of total RNA prepared from yeast cells expressing the hybrid 18S rRNA at different time points. Northern blots were hybridized with an oligonucleotide probe that recognizes the mouse 18S rRNA at nts 770 to 795, but does not recognize the yeast molecule.

conserved *Nde*I and *Bsr*GI restriction sites. This extensive exchange of approximately 90% of the rRNA sequences was designed to maximize the number of nucleotide differences between the yeast and hybrid 18S rRNAs and thus the number of putative mRNA elements that could be analyzed using this system. In addition, this extensive exchange of rRNA sequences was designed to maintain the higher order interactions of the rRNA as much as possible (Thompson *et al.*, 2001; Fig. 15.3B). The hybrid 18S rRNA contains the 5' domain, central domain, and 3' major domain of the mouse 18S rRNA, and the 3' minor domain (helices 44 and 45) of the yeast 18S rRNA.

Ribosomes generated in this manner were active as judged by various criteria including functional assays and their presence in actively translating polysomes (Dresios *et al.*, 2006). However, it should be noted that we have not been able to establish cell lines expressing only the mouse–yeast hybrid 18S rRNA using standard plasmid shuffling techniques, suggesting that the hybrid ribosomes alone cannot support viability. In addition, we noted that levels of the hybrid rRNA are substantially reduced over time (Fig. 15.3C), making it necessary to perform these studies with cells that are freshly transformed.

6.1. Procedure

6.1.1. Evaluation of base pairing interactions between mRNA and 18S rRNA

To assess a particular base pairing interaction, NOY908 is transformed with pVM1 or with pNOY353 (wild type). Cultures of NOY908 cells started from a single colony are grown overnight in 2% glucose synthetic media lacking histidine (to select for maintenance of the deleted chromosomal rDNA locus) and leucine (to select for pNOY373) to an optical density of 1.0 at 600 nm. Cells (1 ml per transformation) are harvested by centrifugation at $6000 \times g$ for 5 min at room temperature, washed twice with sterile water (10 ml), and resuspended in 0.2 ml of Solution A per transformation (10 m*M* Tris–HCl, pH 7.5, 1 m*M* EDTA, 200 m*M* lithium acetate [pH 7.5, adjusted with acetic acid]). For higher transformation efficiencies, cultures that have reached an optical density of 1.0 at 600 nm are diluted to 0.5 in fresh synthetic media and grown for another generation (approximately 5 h) before harvesting. Then 50 μg of high-molecular-weight carrier DNA (herring testis or salmon sperm) that has been denatured by boiling for 5 min prior to use is added to the cell suspension followed by 1 μg of pVM1, pNOY353, or pNOY353 mutated plasmids. The solution is mixed, supplemented with 1.2 ml of Solution B (Solution A plus 40% PEG (3350 or 4000)], and incubated at 30° with shaking. Cells are heat shocked at 42° for 15 to 30 min and pelleted by centrifugation at $12,000 \times g$ for 15 sec. Cells are resuspended in 0.2 ml of TE buffer (10 m*M* Tris–HCl, pH 7.5, 1 m*M* EDTA), plated on 2% galactose synthetic media lacking tryptophan (to select for the presence of pVM1) and histidine, and incubated at 30° for 3 to 5 days until transformants appear.

To determine whether particular transformants express the mouse–yeast hybrid 18S rRNA (from plasmid pVM1), or the mutated yeast 18S rRNA (from a plasmid derived from pNOY353), single colonies are selected and grown in 5 ml galactose synthetic media lacking tryptophan and histidine. Cells are harvested at an optical density of 0.8 to 1.0 at 600 nm and the RNA is extracted following the hot phenol method. Briefly, cells are spun down at $6000 \times g$ for 5 min at 4°, washed once with ice-cold water, and resuspended in 0.4 ml TES solution (10 mM Tris–HCl, pH 7.5, 5 mM EDTA, 1% sodium dodecyl sulfate [SDS]). An equal volume of acidic phenol (pH 4.3) is then added and the mixture is vortexed for 10 sec, followed by incubation at 65° for 1 h with gentle agitation. After this incubation, the mixture is placed on ice for 5 min and centrifuged at $12,000 \times g$ for 5 min at 4°. Acidic phenol (0.4 ml) is added to the top (aqueous) phase, the mixture is vortexed for 30 sec, cooled on ice for 5 min, and centrifuged at $12,000 \times g$ for 5 min at 4°. The aqueous phase is transferred to a new tube, 0.4 ml of chloroform is added, the mixture is vortexed for 30 sec, and centrifuged at $12,000 \times g$ for 5 min at 4°. Sodium acetate (pH 5.3) is added to the aqueous phase to a final concentration of 0.3 M, followed by 2 volumes of ethanol, and the sample is centrifuged at $12,000 \times g$ at 4° for 10 min. The pellet (total RNA) is washed twice with 70% ethanol and resuspended in sterile water. Hybrid 18S rRNA can be detected by Northern blot analysis of total RNA, e.g., using a 1% agarose-formaldehyde gel transferred to a Nylon membrane, and probed with a 5′ end-labeled (^{32}P) oligonucleotide probe that is specific to the mouse–yeast hybrid 18S rRNA or to the mutated yeast 18S rRNA. We have successfully used a probe complementary to nucleotides 775 to 791 of the mouse 18S rRNA to detect the hybrid 18S rRNA (TGAGTGTCCCGCGGGGC).

Quantitative Northern blots can be used to assess the amount of hybrid or mutated 18S rRNA relative to the yeast wild-type 18S rRNA. To this end, the recombinant yeast RNA is resolved alongside various dilutions of total mouse or wild-type yeast RNA. Wild-type yeast RNA can be obtained as described above and total mouse RNA can be obtained from cultured mouse cell lines, e.g., mouse N2a cells. In our studies, we extracted mouse RNA using TRIzol reagent. Northern blots of the recombinant yeast RNA and wild-type yeast or mouse RNAs are probed with rRNA-specific oligonucleotide probes complementary to mouse 18S rRNA at nucleotides 775 to 791 (see above) and the hybridization intensities are compared to determine how much mouse RNA gives a signal comparable to the recombinant yeast RNA. This comparison with the mouse rRNA is valid because total yeast RNA contains approximately the same amount of 18S rRNA as total mouse RNA based on ethidium bromide staining (Dresios et al., 2006). In our earlier studies, we showed that up to approximately 10% of the 18S rRNA in NOY908-pVM1 cells was the mouse–yeast hybrid 18S rRNA, and that approximately 5% of the 18S rRNA in cells with pVM7, which contains a point mutation of the yeast 18S rRNA (Dresios *et al.*, 2006). Although these levels of hybrid ribosomes were sufficient for our earlier analyses (Chappell *et al.*, 2006a; Dresios *et al.*, 2006), they may be a limitation for some analyses of mRNA–rRNA base pairing interactions, which have smaller effects on translation. An advantage of this system is that it can be used to assess the effects of mutations in the 18S rRNA that are not compatible with cell viability because cell survival does not depend on the recombinant ribosomal subunits.

To confirm that the recombinant 18S rRNA is present in ribosomal subunits, ribosomes are prepared from yeast cells and tested for its presence. To prepare ribosomes from transformed NOY908 cells, 50 ml cultures are grown in galactose synthetic media lacking tryptophan and histidine. Cells are pelleted and resuspended in Homogenization buffer (20 mM Tris–HCl, pH 7.5, 100 mM KCl, 10 mM MgCl$_2$, 2 mM DTT) using 2 to 3 ml of buffer per gram of cells, and lysed at 4° using glass beads (0.425 to 0.6 μm diameter, Sigma). Three to four grams of glass beads are used per milliliter of cell suspension. Each such lysis is performed in a capped SS34 centrifuge tubes (35 ml). The tube is shaken by hand at 4° using a vertical motion of approximately 50 cm; it is shaken two times per second for 1 min, then chilled on ice for 1 min. This shaking procedure is repeated a total of five times. The lysate is centrifuged at 10,000×g for 15 min at 4°. The supernatant (postnuclear fraction) is centrifuged at 100,000×g for 3 h at 4° to obtain a P100 pellet containing ribosomes and an S100 supernatant. RNA is extracted with TRIzol from these fractions and analyzed on Northern blots using an oligonucleotide probe specific for the recombinant 18S rRNA. The presence of the recombinant rRNA in the P100 fraction suggests that it is incorporated into ribosomes.

To obtain further evidence that the recombinant 18S rRNAs are present in active ribosomes, polysome profiles of cells transformed with these rRNAs are performed, followed by RNA isolation and blotting using rRNA-specific probes. Yeast cultures (50 ml) grown in galactose selective media lacking tryptophan and histidine and treated with cycloheximide (100 μg/ml) at mid-log phase are immediately harvested by centrifugation at 6000×g for 5 min at 4°. Cells are lysed in SS34 tubes using glass beads and a postnuclear fraction is prepared as described above. The supernatant is layered onto a 10% to 50% (w/w) linear sucrose gradient in Solution C (50 mM Tris acetate, pH 7.5, 50 mM NH$_4$Cl, 12 mM MgCl$_2$) and centrifuged at 4° for 2.5 h at 260,000×g in an SW41 rotor. Gradients are collected, e.g., by the use of an ISCO fractionator. RNA in each fraction is precipitated with 1/10th volume 5 M ammonium acetate and 2 volumes ethanol, extracted with TRIzol, and reprecipitated by the addition of

1/10th volume 3 M sodium acetate (pH 5.3) and 2 volumes of ethanol. Finally, RNA is resuspended in water, and analyzed on Northern blots using an oligonucleotide probe specific for the recombinant 18S rRNA.

It should be noted that the doubling time of the NOY908 cells (untransformed or transformed with another plasmid expressing either wild-type or recombinant rRNA) is approximately 5 h (Wai *et al.*, 2000, and unpublished observations) compared to wild-type strains containing chromosomal rDNA genes, which divide approximately every 90 min (Chuang *et al.*, 1997; Dong *et al.*, 2004; Yu and Warner, 2001). This delayed growth is reflected in polysome profiles with fewer and smaller polysome peaks than those obtained with wild-type strains.

An alternative approach to identify colonies with active ribosomal subunits containing mouse–yeast hybrid 18S rRNA involves selecting yeast cells for their ability to grow on media lacking a specific auxotrophic marker, the translation of which requires ribosomes harboring mouse 18S rRNA sequences. Our preliminary studies using this approach indicated that cells transformed with a construct expressing a *TRP1* mRNA, which contains a 5' hairpin structure and five *Gtx* elements, grew only in cells containing the mouse–yeast hybrid rRNA. Northern blot analysis of these cells revealed the presence of the hybrid rRNA in ribosomes (data not shown).

7. Assessing Ribosomal Shunting Mediated by mRNA–rRNA Base Pairing Interactions

A typical shunting experiment designed to determine whether a putative shunt site functions by an mRNA–rRNA base pairing mechanism can be performed using constructs containing either a hairpin structure or upstream AUG codon as a shunting obstacle, as described above (also see Chappell *et al.*, 2006a). The key to these studies is to determine whether an intact complementary match between the mRNA element and the 18S rRNA is required for reporter gene activity, which would indicate that ribosomal subunits shunted across the obstacle. Such an analysis involves showing that shunting occurs when the match is intact, is disrupted by point mutations of the 18S rRNA that disrupt complementarity, and is restored when the mRNA element is mutated to restore complementarity.

Depending on the specific mRNA or sequence element under investigation, experiments can be performed using constructs that either contain or lack a hairpin structure at their 5' ends. In earlier studies, we used constructs with 5' hairpin structures to increase the signal-to-noise ratio by reducing the contribution of the cap structure (Chappell *et al.*, 2006a; Dresios *et al.*, 2006). Figure 15.4 shows an analysis of ribosomal recruitment via mRNA–rRNA base pairing using reporter mRNAs that lack a hairpin structure at their 5' ends. The results obtained with these constructs were

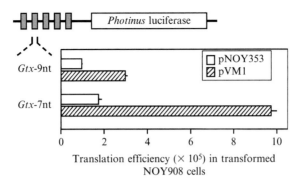

Figure 15.4 Assessing ribosomal recruitment via mRNA–rRNA base pairing using reporter mRNAs that lack a hairpin structure at their 5′ ends. Monocistronic constructs are indicated schematically. Gray bars represent *Gtx* elements containing either 9-nt or 7-nt elements, which are complementary to mouse 18S rRNA (Dresios *et al.*, 2006). Constructs were transformed into strain NOY908 expressing wild-type 18S rRNA (pNOY353; white bar) or hybrid 18S rRNA (pVM1; hatched bar). The data are represented as translation efficiency (luciferase activity per unit mRNA). Horizontal lines represent standard deviations.

comparable to those obtained with reporter mRNAs containing a hairpin structure at their 5′ ends (Dresios *et al.*, 2006) (i.e., cells expressing the hybrid 18S rRNA translated mRNAs containing five copies of the *Gtx* translational enhancer element with a higher relative efficiency than cells expressing yeast wild-type 18S rRNA). The only difference is that the effect on translation in this experiment was approximately 3-fold for the 9-nt element while in the published study it was approximately 160-fold because the contribution of the cap was effectively eliminated by the hairpin structures. It is also interesting that the 3-fold increase in translation efficiency is similar to what we had observed in transfected mammalian cells with these constructs (Chappell *et al.*, 2004).

For such experiments, NOY908 cells harboring either pVM1 or pNOY353 are transformed with hairpin or upstream AUG-containing shunting constructs. In our studies, we have performed these studies using *Photinus* luciferase as the reporter gene, in an expression vector that contains the *Gal1* promoter in the pYES2 vector (further information regarding the construction of these vectors can be found in Dresios *et al.*, 2006). Depending on the specifics of the 5′ leader and mRNA elements under investigation, such studies can be performed using constructs that either contain a stable hairpin structure at their 5′ ends to minimize the contribution of the 5′ cap structure or that lack such a structure, e.g., for a cap-dependent mRNA. Yeast transformed with these constructs are cultured at 30° on synthetic galactose media lacking tryptophan, histidine, and uracil (to select for the pYES2 reporter plasmid). Single colonies are selected and grown in 5 ml of the same media until they reach an OD_{600} reading of 1.0. Cells are pelleted by centrifugation at $12,000 \times g$ for 1 min and resuspended in sterile water. Each suspension is divided into two equal parts: one to test for luciferase activity and the other for RNA analysis. To measure luciferase activity, cells are centrifuged at $12,000 \times g$ for 1 min and resuspended in 300 μl of Reporter Lysis buffer (Promega). Suspensions are then transferred to microfuge tubes containing 0.3 g glass beads. Cells are lysed by vortexing three times for 30 sec at room temperature with 1 min intervals on ice. The lysed cells are centrifuged at $12,000 \times g$ for 5 min and 10 to 20 μl of the supernatant is assayed for luciferase activity as described above. Reporter mRNA levels are determined by probing Northern blots with an *in vitro* transcribed *Photinus* luciferase probe. RNA is extracted from transformed yeast using the hot phenol method described earlier and probed as described above using a full-length *Photinus* luciferase probe. Hybridization signals can be visualized and quantified using various approaches. We have used a Storm 860 PhosphorImager for visualization and ImageQuant software (Molecular Dynamics). It is also possible to quantify mRNA from such studies using other methods, including RPAs, which was discussed above, or quantitative RT-PCR (Chappell *et al.*, 2004).

To test whether the activity of a candidate shunt site requires the presence of an intact complementary match within the hybrid 18S rRNA, both the candidate shunt site and the 18S rRNA can be mutated and the effects of these point mutations tested in cells. The large size of plasmid pNOY353 makes site-directed mutagenesis difficult. For this reason, we found it necessary to subclone a smaller fragment of the 18S rDNA into the pBluescript II KS(+) vector and perform site-directed mutagenesis in this smaller construct using the Quick-Change II XL kit (Stratagene). In our studies, we used a 1045 nucleotide *Bss*HII restriction fragment of the 18S rDNA (nucleotides 485 to 1530), which contained our putative mRNA-binding site. However, depending on the location of the complementary match within the 18S rRNA, other restriction fragments may have to be subcloned for mutagenesis. Other unique restriction sites that can be used to subclone various fragments of the 18S rRNA are *Nde*I and *Nco*I, which reside near the 5′ end of the 18S rDNA, and *Eco*RI and *Bsr*GI, which reside near the 3′ end of the 18S rDNA. The mutagenized DNA fragment is then cloned back into the pNOY353 backbone using the appropriate restriction sites.

8. CONSIDERATIONS IN USING THE MOUSE–YEAST HYBRID RRNA SYSTEM

The mouse–yeast hybrid rRNA system described in this chapter differs from the yeast wild-type 18S rRNA at 418 out of 1865 nucleotides, and can be used only to test putative base pairing interactions between mRNA sequences with complementarity to mammalian but not to yeast 18S rRNAs.

To evaluate the levels of hybrid 18S rRNA over time, NOY908-pVM1 transformants are selected on galactose synthetic media lacking tryptophan and histidine as described earlier. A single colony is inoculated in liquid media lacking the aforementioned biosynthetic markers and the culture is grown at 30° to an optical density of 1.0 at 600 nm. Of this culture 5 ml is used to extract total yeast RNA, while 50 μl of the same culture is re-inoculated in 5 ml of the same liquid media and grown at 30° to an optical density of 1.0 at 600 nm. This process is repeated several times and the various RNAs are analyzed by Northern blot analysis using a mouse-specific 18S rRNA probe. As shown in Fig. 15.3C, the amount of hybrid 18S rRNA in NOY908-pVM1 cells declined with time. In addition, recombination between plasmids encoding wild-type and mouse–yeast hybrid 18S rDNA sequences in the NOY908-pVM1 yeast cells may lead to a loss of DNA fragments encoding 18S rRNA hybrid sequences. Such recombination events may also yield a heterogeneous population of rRNA transcripts bearing yeast and mouse sequences at different ratios. Recombination events of this type cannot yield false-positive results in that the sequence elements under investigation must be inactive with the wild-type yeast 18S rRNA; however, they may lower the sensitivity of the system.

ACKNOWLEDGMENTS

Funding was provided by the National Institutes of Health (GM61725) and the G. Harold and Leila Y. Mathers Charitable Foundation to V.P.M., and from the Skaggs Institute for Chemical Biology (J.D., S.A.C.).

REFERENCES

Chappell, S. A., and Mauro, V. P. (2003). The internal ribosome entry site (IRES) contained within the RNA-binding motif protein 3 (Rbm3) mRNA is composed of functionally distinct elements. *J. Biol. Chem.* **278,** 33793–33800.

Chappell, S. A., Edelman, G. M., and Mauro, V. P. (2000). A 9-nt segment of a cellular mRNA can function as an internal ribosome entry site (IRES) and when present in linked multiple copies greatly enhances IRES activity. *Proc. Natl. Acad. Sci. USA* **97,** 1536–1541.

Chappell, S. A., Owens, G. C., and Mauro, V. P. (2001). A 5' leader of Rbm3, a cold-stress induced mRNA, mediates internal initiation of translation with increased efficiency under conditions of mild hypothermia. *J. Biol. Chem.* **276,** 36917–36922.

Chappell, S. A., Edelman, G. M., and Mauro, V. P. (2004). Biochemical and functional analysis of a 9-nucleotide RNA sequence that affects translation efficiency in eukaryotic cells. *Proc. Natl. Acad. Sci. USA* **101,** 9590–9594.

Chappell, S. A., Dresios, J., Edelman, G. M., and Mauro, V. P. (2006a). Ribosomal shunting mediated by a translational enhancer element that base pairs to 18S rRNA. *Proc. Natl. Acad. Sci. USA* **103**, 9488–9493.

Chappell, S. A., Edelman, G. M., and Mauro, V. P. (2006b). Ribosomal tethering and clustering as mechanisms for translation initiation. *Proc. Natl. Acad. Sci. USA* **16**, 16.

Chuang, R. Y., Weaver, P. L., Liu, Z., and Chang, T. H. (1997). Requirement of the - DEAD-Box protein ded1p for messenger RNA translation. *Science* **275**, 1468–1471.

Cornelis, S., Bruynooghe, Y., Denecker, G., Van Huffel, S., Tinton, S., and Beyaert, R. (2000). Identification and characterization of a novel cell cycle-regulated internal ribosome entry site. *Mol. Cell* **5**, 597–605.

Dobson, T., Minic, A., Nielsen, K., Amiott, E., and Krushel, L. (2005). Internal initiation of translation of the TrkB mRNA is mediated by multiple regions within the 5′ leader. *Nucleic Acids Res.* **33**, 2929–2941.

Dong, J., Lai, R., Nielsen, K., Fekete, C. A., Qiu, H., and Hinnebusch, A. G. (2004). The essential ATP-binding cassette protein RLI1 functions in translation by promoting preinitiation complex assembly. *J. Biol. Chem.* **279**, 42157–42168.

Dresios, J., Chappell, S. A., Zhou, W., and Mauro, V. P. (2006). An mRNA–rRNA base-pairing mechanism for translation initiation in eukaryotes. *Nat. Struct. Mol. Biol.* **13**, 30–34.

Dubreuil, Y. L., Expert-Bezançon, A., and Favre, A. (1991). Conformation and structural fluctuations of a 218 nucleotides long rRNA fragment: 4-Thiouridine as an intrinsic photolabelling probe. *Nucleic Acids Res.* **19**, 3653–3660.

Gingras, A. C., Gygi, S. P., Raught, B., Polakiewicz, R. D., Abraham, R. T., Hoekstra, M. F., Aebersold, R., and Sonenberg, N. (1999). Regulation of 4E-BP1 phosphorylation: A novel two-step mechanism. *Genes Dev.* **13**, 1422–1437.

Hall, K. B., and Kranz, J. K. (1999). Nitrocellulose filter binding for determination of dissociation constants. *Methods Mol. Biol.* **118**, 105–114.

Hellen, C. U., and Sarnow, P. (2001). Internal ribosome entry sites in eukaryotic mRNA molecules. *Genes Dev.* **15**, 1593–1612.

Hu, M. C.-Y., Tranque, P., Edelman, G. M., and Mauro, V. P. (1999). rRNA-complementarity in the 5′ UTR of mRNA specifying the Gtx homeodomain protein: Evidence that base-pairing to 18S rRNA affects translational efficiency. *Proc. Natl. Acad. Sci. USA* **96**, 1339–1344.

Kapp, L. D., and Lorsch, J. R. (2004). The molecular mechanics of eukaryotic translation. *Annu. Rev. Biochem.* **73**, 657–704.

Keiper, B. D., and Rhoads, R. E. (1997). Cap-independent translation initiation in *Xenopus* oocytes. *Nucleic Acids Res.* **25**, 395–402.

Kozak, M. (1986). Point mutations define a sequence flanking the AUG initiator codon that modulates translation by eukaryotic ribosomes. *Cell* **44**, 283–292.

Kozak, M. (1997). Recognition of AUG and alternative initiator codons is augmented by G in position +4 but is not generally affected by the nucleotides in positions +5 and +6. *EMBO J.* **16**, 2482–2492.

Kozak, M. (2002). Pushing the limits of the scanning mechanism for initiation of translation. *Gene* **299**, 1–34.

Martineau, Y., Le Bec, C., Monbrun, L., Allo, V., Chiu, I. M., Danos, O., Moine, H., Prats, H., and Prats, A. C. (2004). Internal ribosome entry site structural motifs conserved among mammalian fibroblast growth factor 1 alternatively spliced mRNAs. *Mol. Cell Biol.* **24**, 7622–7635.

Mauro, V. P., Edelman, G. M., and Zhou, W. (2004). Reevaluation of the conclusion that IRES-activity reported within the 5′ leader of the TIF4631 gene is due to promoter activity. *RNA* **10**, 895–897.

Pinkstaff, J. K., Chappell, S. A., Mauro, V. P., Edelman, G. M., and Krushel, L. A. (2001). Internal initiation of translation of five dendritically-localized neuronal mRNAs. *Proc. Natl. Acad. Sci. USA* **98,** 2770–2775.

Ringquist, S., and Gold, L. (1998). Toeprinting assays. Mapping by blocks to reverse transcriptase primer extension. *Methods Mol. Biol.* **77,** 283–295.

Rogers, G. W., Jr., Edelman, G. M., and Mauro, V. P. (2004). Differential utilization of upstream AUGs in the beta-secretase mRNA suggests that a shunting mechanism regulates translation. *Proc. Natl. Acad. Sci. USA* **101,** 2794–2799.

Stoneley, M., Paulin, F. E. M., Le Quesne, J. P. C., Chappell, S. A., and Willis, A. E. (1998). C-Myc 5′ untranslated region contains an internal ribosome entry segment. *Oncogene* **16,** 423–428.

Thompson, J., Tapprich, W. E., Munger, C., and Dahlberg, A. E. (2001). Staphylococcus aureus domain V functions in *Escherichia coli* ribosomes provided a conserved interaction with domain IV is restored. *RNA* **7,** 1076–1083.

Tranque, P., Hu, M. C.-Y., Edelman, G. M., and Mauro, V. P. (1998). rRNA complementarity within mRNAs: A possible basis for mRNA-ribosome interactions and translational control. *Proc. Natl. Acad. Sci. USA* **95,** 12238–12243.

Vagner, S., Galy, B., and Pyronnet, S. (2001). Irresistible IRES: Attracting the translation machinery to internal ribosome entry sites. *EMBO Rep.* **2,** 893–898.

Wai, H. H., Vu, L., Oakes, M., and Nomura, M. (2000). Complete deletion of yeast chromosomal rDNA repeats and integration of a new rDNA repeat: Use of rDNA deletion strains for functional analysis of rDNA promoter elements *in vivo. Nucleic Acids Res.* **28,** 3524–3534.

Woodbury, C. P., Jr., and von Hippel, P. H. (1983). On the determination of deoxyribonucleic acid-protein interaction parameters using the nitrocellulose filter-binding assay. *Biochemistry* **22,** 4730–4737.

Xi, Q., Cuesta, R., and Schneider, R. J. (2004). Tethering of eIF4G to adenoviral mRNAs by viral 100k protein drives ribosome shunting. *Genes Dev.* **18,** 1997–2009.

Xi, Q., Cuesta, R., and Schneider, R. J. (2005). Regulation of translation by ribosome shunting through phosphotyrosine-dependent coupling of adenovirus protein 100k to viral mRNAs. *J. Virol.* **79,** 5676–5683.

Yu, X., and Warner, J. R. (2001). Expression of a micro-protein. *J. Biol. Chem.* **276,** 33821–33825.

Zang, W.-Q., and Romaniuk, P. J. (1995). Characterization of the 5 S RNA binding activity of *Xenopus* zinc finger protein p43. *J. Mol. Biol.* **245,** 549–558.

Zhou, W., Edelman, G. M., and Mauro, V. P. (2001). Transcript leader regions of two *Saccharomyces cerevisiae* mRNAs contain internal ribosome entry sites that function in living cells. *Proc. Natl. Acad. Sci. USA* **98,** 1531–1536.

Zuker, M., Mathews, D. H., and Turner, D. H. (1999). Algorithms and thermodynamics for RNA secondary structure prediction: A practical guide. *In* "In RNA Biochemistry and Biotechnology" (J. Barciszewski and B. F. C. Clark, eds.), pp. 11–43. Kluwer Academic Publishers, Amsterdam.

Author Index

Subject Index

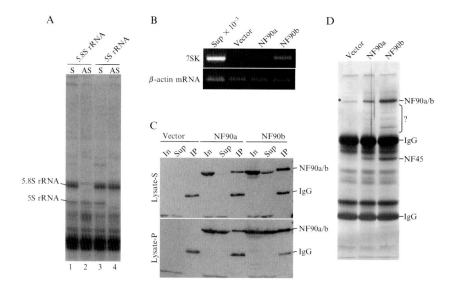

Andrew M. Parrott _et al._, Figure 12.3   Analysis of immunoprecipitated RNA and protein. (A) $3'$ End-labeled RNA immunoprecipitated from NF90b Lysate-P (G_2/M, 0.5% HCHO) was digested with RNase H in the presence of 5.8S or 5S rRNA sense (S) or antisense (AS) oligonucleotides and resolved in a urea/acrylamide gel. (B) RT-PCR of 7SK and β-actin mRNA immunoprecipitated from Lysate-P (G_2/M, 0.15% HCHO). RNA extracted from the vector cell line supernatant (Sup) served as a positive control. (C) Western blot of Lysate-S and -P (G_2/M, 0.15% HCHO) probed with Omni-probe antibody. Input (In), supernatant (Sup), and Omni-probe antibody immunoprecipitate (IP) lanes are present and heavy IgG chain is denoted. (D) Immunoprecipitate from Lysate-S (G_2/M, 0.5% HCHO) was resolved in an SDS–PAGE gel and silver stained. Asterisk denotes a nonspecific protein that comigrates with tagged NF90, and IgG heavy and light chains are denoted.

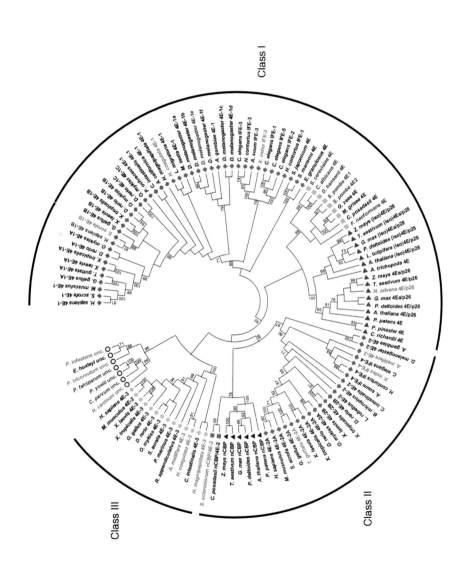

Robert E. Rhoads *et al.*, Figure 13.1 A radial cladogram describing the overall relationship of selected eIF4E family members from multiple species. The figure is taken from Joshi *et al.* (2005) and can also be viewed at http://umbicc 3–215.umbi.umd.edu/iisstart.asp. The topology of a neighbor-joining tree visualized in radial format derived from an alignment of nucleotide sequences representing the conserved core regions of the indicated eIF4E family members. eIF4E family member names in black or red indicate whether or not the complete sequence of the conserved core region of the member could be predicted from consensus cDNA sequence data, respectively. eIF4E family member names in blue indicate that genomic sequence data were used to either verify or determine the nucleotide sequence representing the core region of the member. The shape of a "leaf" indicates the taxonomic kingdom from which the species containing the eIF4E family member derives: Metazoa (diamonds), Fungi (squares), Viridiplantae (triangles), and Protista (circles), respectively. The color of a "leaf" indicates the subgroup of the eIF4E family member: metazoan eIF4E-1 and IFE-3-like (red), fungal eIF4E-like (gold), plant eIF4E and eIF(iso)4E-like (green), metazoan eIF4E-2-like (cyan), plant nCBP-like (blue), fungal nCBP/eIF4E-2-like (purple), metazoan eIF4E-3-like (pink), and atypical eIF4E family members from some protists (white). eIF4E family members within structural Class I, Class II, and Class III are indicated.